Analog and Digital Communication Systems

FIFTH EDITION

MARTIN S. RODEN
California State University, Los Angeles

Los Angeles, CA

Roden, Martin S.
Includes bibliographical references
ISBN 0-9646969-7-5

Production Supervisor: **Raymond B. Landis**
Cover Design: **David McNutt**
Editorial Consultant: **Dennis J. E. Ross**
Distribution: **Legal Books Distributing: (800) 200-7110**

Printed in the United States of America.

10 9 8 7 6 5 4 3 2 1

0-9646969-7-5

Analog and Digital Communication Systems

Table of Contents

Contents

Preface

Why would anyone want to write another introductory communications textbook? Indeed, I have written seven previous Analog and Digital Communications texts, and the market might appear to be saturated with over one dozen effective textbooks.

The answer to this question can be found in the table of contents of this text. This text takes an integrated approach toward communications, with little dichotomy between *Analog* and *Digital*.

Studies of telecommunications in undergraduate Engineering education were traditionally analog. In fact, until the late 1960s, very few schools were teaching digital communication concepts to undergraduates. As digital communications rapidly replaced analog communications during the 1970s and 1980s, some universities attempted to keep up with the times by incorporating *some* digital communications into a first course in communications. Others proposed separate courses dealing with digital communications, often with analog communications as the prerequisite.

Academicians have debated the importance of analog communications and, indeed, some programs have dropped analog in favor of a more thorough approach toward digital. Others were cautious of this approach since history has proven that recycling often occurs (I wish I had never thrown my skinny neckties away). As one example, communications has cycled between wireless and wired. A long time ago, ti started as wired (e.g., telegraph and early telephone) until we found signals could be propagated. Then the big push was wireless (e.g., radio). Then with the advent of fiber optics and broadband cable, we cycled back to wired (e.g., cable TV and cable modems). And the new millennium is bringing a big push toward wireless...until the next big change occurs. So some predict that "analog may rise again."

Throughout this dialog, textbooks have failed to keep pace. Some authors (I was one of them) wrote separate texts in digital communications hoping the trend would follow a natural progression with many courses dealing exclusively with digital. The *standard* text resembles a committee project, with about half or more on analog communication, and then separate chapters or sections on digital.

Our evolution continues with the approaches of the 1990s toward Engineering education. With programs resisting a move toward a 5-year BS degree, there was simply not enough room in the undergraduate curriculum to deal separately with analog and digital. The trend is to *collapse* the two into a single presentation. Indeed, this makes great pedagogical sense since there is really no fundamental difference between them. For example, frequency shift keying (FSK) is simply a special case of frequency modulation (FM).

The text you are currently holding thoroughly intertwines analog and digital communications. While the two forms of communications are treated simultaneously, the distinctions between them are made very clear, and faculty may still choose to separate the presentation.

In addition to the integrated approach, this text strikes a balance between theory and practice. While the undergraduate engineering student needs a firm foundation in the theory, the student must also be exposed to the real world of *Engineering design*. This serves two important purposes. The first is that an introduction to the real world serves as a strong motivating factor. Students need some "touchy-feely" to motivate them to spend hours digging through mathematical formulae. The second purpose of real world engineering is to ease the transition from academia to the profession. A graduate's first experience with real-world design should not be a great shock, but instead it should be a pleasant experience and a natural transition from the classroom environment. Consistent with this balanced approach, portions of chapters 2, 3, 4, 5, and 6 deal with contemporary applications. Discussion of these applications is clearly tied to the theory presented in the chapters. The text also contains numerous computer application examples. In particular, many of the graphs are prepared and formulas are solved using MATLAB™. In such cases, the instruction set is presented in the text. Many of the exercises are interactive (i.e., the student is asked to enter various parameters). Block diagram simulations using Tina™ software are also spread throughout the text. An enclosed CD contains copies of all the programs discussed in the text, including M-files for all the MATLAB examples.

The book is intended as an introductory text for the study of analog and/or digital communication systems, with or without noise. Although all necessary background material has been included, prerequisite courses in linear systems analysis and in probability are helpful. Comprehensive appendices explore Fourier analysis, linear systems theory, probability and random processes.

The text stresses a mathematical systems approach to all phases of the subject matter. The mathematics used throughout is as elementary as possible, but is carefully chosen so as not to contradict any more sophisticated approach that may eventually be required. An attempt is made to apply intuitive techniques prior to grinding through the mathematics. The style is informal, and the text has been thoroughly tested in the classroom with excellent success.

Chapter 1 forms an introduction to the text with basic definitions of systems and signals and a discussion of techniques to change an analog signal into a digital signal.

With the stage set, the remaining chapters consist of thorough presentations of the broad classes of signal transmission techniques, starting with baseband and working through the various forms of modulation. Within each chapter, analog and digital signals are treated, and design and applications are emphasized. The applications present contemporary uses of the theory described within the chapters, and the presentation emphasizes the thought processes used to reduce the theory to practice.

Appendices review Fourier series and transform, linear systems and filter theory, probability, and noise analysis. Additional appendices present tables of Fourier transforms, tables of error functions and Q functions, a discussion of the software on the enclosed CD, and a list of references for further study.

While this text is intended for use in an integrated approach, separating into analog and digital is possible. If it is used for an introductory course in analog communications, the following sections should be covered:

1.3, 1.4, 2.1, 2.2, 2.4.1, 2.4.2, 2.5.1, 2.5.2, 3.1, 3.2.1, 3.2.2, 3.2.3, 3.2.4, 3.3.1, 3.3.2, 3.3.3, 3.4, 3.5, 3.6.1, 4.1.1, 4.1.2, 4.1.3, 4.2.1, 4.3.1, 4.4.1, 5.1, 5.3.1, 5.4.1, and 5.5.1.

On the other hand, to use this text for a course in Digital Communications, cover the following sections:

1.3, 1.4, 1.5, 1.6, 1.7, 2.3, 2.4.3, 2.5.3, 3.2.5, 3.3.4, 3.4.1, 3.4.3, 3.5.1, 3.5.2, 3.6.2, 3.6.3, 4.1.4, 4.1.5, 4.2.2, 4.3.2, 4.4.2, 4.5.2, 5.2, 5.3.2, 5.4.2, 5.5.2, 5.6.2, 6.2, and 6.3.

It gives me a great deal of pleasure to acknowledge the assistance and support of many people without whom this text would not have been possible.

To the many classes of students who are responsive during lectures and helped indicate the clearest approach to each topic.

To the many users of previous editions of the text who took the time to contact me with suggestions and comments.

To my colleagues at Bell Telephone Labs, Hughes Aircraft Ground Systems, and California State University, Los Angeles. Special thanks go to Professors Fred Daneshgaran, George Killinger, and Lili Tabrizi for their many helpful suggestions.

To three of my students who made major contributions to the MATLAB exercises: Ouaboaman (Serge) Ouattara, Francisco Lam and Anselmo Martinez.

To Dennis J. E. Ross for continual guidance and encouragement.

To the late Professor A. Papoulis, who played a key role during the formative years of my education.

I sincerely hope this text is the answer to your prayers. If it is, please let me know. If it isn't please also communicate with me so, together, we can improve engineering education.

Martin S. Roden
California State University, Los Angeles
mroden@calstatela.edu

CHAPTER 1

INTRODUCTION

What we will cover, and why you should care:

You will not encounter your first communication system until Chapter 2 of this text. This first chapter (plus the Appendices if you need brushing up on related theory) forms the framework and defines the parameters under which we operate.

We begin with a brief history of the exciting communications revolution. It is a revolution which is accelerating at an ever-increasing rate.

We then turn our attention to an investigation of the environment within which our systems must operate. In particular, the characteristics of various communication channels are explored, and we present analysis techniques for characterizing the resulting signal distortion. It would not make much sense to start designing communication systems without knowing something about the environment in which they must operate.

The third section of this chapter defines the three types of signals which we will encounter. We begin with analog signals, and make a gradual transition to digital through the intermediate step of discrete time (sampled) signals.

The block diagram of a communication system is presented in the fourth section. The various blocks in the transmitter and receiver form the road map for our excursion through this exciting subject. We present a single comprehensive system block diagram which applies to both analog and digital communication systems.

It often proves advantageous to process analog signals using digital signal processing techniques, so we must find ways to move easily between the two domains. The conversion of an analog continuous time signal to a digital signal is a two-step process. First the waveform must be sampled, and then the samples must be digitized. This two-step process is the subject of sections 1.5 and 1.6. We examine both the theory and practice of analog-to-digital conversion and the reverse procedure, digital-to-analog conversion.

Necessary background:

There are no prerequisites to understanding Sections 1.1, 1.3 and 1.4. The other sections require the following:

Section 1.2: You need some knowledge of Fourier analysis (see Appendix A) and linear system theory (see Appendix B) to appreciate the mathematical characterization of the channel.

Section 1.5: To understand sampling, you need some knowledge of Fourier analysis (see Appendix A). A familiarity with basic electronics is helpful to understand the practical implementations of samplers.

Section 1.6: Some electronics background is helpful to understand the operation of practical converters.

Section 1.7: Although a broad understanding of channel capacity requires no special background, familiarity with random processes and noise analysis helps you gain a deeper appreciation of the mathematical results.

1.1. The Need to Communicate–History

Among the earliest forms of communication were vocal-chord sounds generated by animals and human beings, with reception via the human ear. The earliest languages consisted of grunts (which I still hear during some of my exams). Then more sophisticated languages developed with larger and larger vocabularies. When greater distances were required, the sense of sight was used to augment that of sound. In the second century B.C., Greek telegraphers used torch signals to communicate. Different combinations and positions of torches were used to represent the letters of the Greek alphabet. These early torch signals represent the first example of digital communications! Later, drum sounds were used to communicate over greater distances, again calling on the sense of sound. Increased distances were possible since the drum sounds were easily distinguished from the background noises.

In the 18th century, communication of letters was accomplished using semaphore flags. Like the torches of ancient Greece, these semaphore flags relied on the human eye to receive the signal. This reliance severely limited the transmission distances.

In 1753, Charles Morrison, a Scottish surgeon, suggested an electrical transmission system using one wire (plus ground) for each letter of the alphabet. A system of pith balls and paper with letters printed on it was used at the receiver.

In 1835, Samuel Morse began experimenting with telegraph. Two years later, in 1837, telegraph was invented by Morse in the United States and by Sir Charles Wheatstone in Great Britain. The first public telegram was sent in 1844, and electrical communication was established as a major component of life. These early forms of communication consisted of individual message components such as the letters of the alphabet.

It was not until Alexander Graham Bell invented the telephone in 1876 that analog electrical communication became common. Experimental radio broadcasts began about 1910 with Lee De Forest producing a program from the Metropolitan Opera House in New York City. Five years later, an experimental radio station opened at the University of Wisconsin in Madison. Station WWJ in Detroit and KDKA in Pittsburgh were among the first to conduct regular broadcasts in 1920.

Public television had its beginnings in England in 1927. In the United States, it started three years later in 1930. During the early period, broadcasts did not follow any regular schedule. Such regular scheduling did not begin until 1939, during the opening of the New York World's Fair.

Satellite communications was launched in the 1960s with *Telstar I* being used to relay TV programs starting in 1962. The first commercial communications satellites were launched in the mid 1960s.

The 1970s saw the beginning of the *computer communications revolution*. Data transfer became an integral part of our daily lives and has led to a merging of disciplines among Communications and Computer Engineering.

The *personal communications revolution* began in the 1980s. Before the decade of the 90s ended, the average professional had a cellular telephone, a pager, a high-speed digital connection to the internet from home for use in paying bills or accessing daily news, and a home FAX machine. Consumers used fully interactive compact disk technology, networked laptops with worldwide data services, and the global positioning satellite system (GPS) to assist in navigating cars through traffic jams.

The new millennium is bringing a unique set of applications and innovations as communications continues to have a significant impact on our lives. As wideband universal wireless access becomes more common, cellular telephone concepts will be applied in many new situations, and direct satellite transmission will become more common. At the same time, advances in fiber optics drove a move toward high data rate worldwide communication. Things could not possibly be more exciting.

1.2. The Environment

Before we can begin designing systems to communicate information, we need to know something about the channel through which the signals must be transmitted. We start by exploring the ways in which the channel can change our signals, and then we discuss some common types of channels.

Distortion

Anything that a channel does to a signal other than pure delay and constant multiplication is considered to be *distortion* (See Appendix B for a discussion of *distortionless linear systems*). Let us assume that the channels we encounter are *linear*. Therefore the output frequencies are the same as those at the input. Although we are assuming linearity, we must acknowledge that some *nonlinear* forms of distortion are significant at higher transmission frequencies. Indeed, the higher frequencies are affected by air turbulence, which causes a frequency variation (i.e., Doppler shift).

Linear distortion is characterized by *time dispersion* (spreading) due either to multipath effects (echoes) or to the characteristics of the channel. For now, we look at the effects that can be readily characterized by the system function of the channel.

If you are not familiar with linear systems analysis and Fourier transform, we suggest you skip this section. Doing so will not affect your understanding of the remainder of this text.

The channel can be characterized by a system transfer function of the form,

$$H(f) = A(f)e^{-j\theta(f)} \tag{1.1}$$

The *amplitude factor* is $A(f)$ and the *phase factor* is $\theta(f)$.

Distortion arises from these two frequency-dependent terms as follows. If $A(f)$ is not a constant, we have what is known as *amplitude distortion*. If $\theta(f)$ is not linear in f, we have *phase distortion*.

Amplitude Distortion

Let us first assume that $\theta(f)$ is linear with frequency. The transfer function is therefore of the form,

$$H(f)=A(f)e^{-j2\pi ft_o}$$ (1.2)

The phase proportionality constant has been denoted as t_o since it represents the channel delay.

Although it will require some mathematical manipulation, we shall take the time to examine the effects of amplitude distortion. The strategy we use represents one of many possible approaches toward obtaining mathematical results. The best approach depends on the particular form of the distortion,

We analyze the expression of Eq. (1.2) by expanding $A(f)$ in a series. For example, it may be possible to expand $A(f)$ in a Fourier series. This can be done if $A(f)$ is bandlimited to a certain range of frequencies (i.e., make believe it repeats periodically). In such cases, we can write,

$$H(f)=\sum_{n=0}^{\infty} H_n(f)$$ (1.3)

where the terms in the summation are of the form,

$$H_n(f)=a_n\cos\left(\frac{n\pi f}{f_m}\right)e^{-j2\pi t_o}$$ (1.4)

These terms each characterize a *cosine filter* whose amplitude characteristic follows a cosine wave in the passband. This is shown in Fig. 1.1 for $n=2$ and $t_o = 0$.

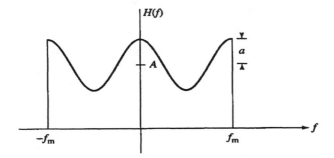

Figure 1.1 - Cosine Filter

The system function for this filter is

$$H(f) = \left(A + a\cos\frac{2\pi}{f_m}f \right)e^{-j2\pi ft_o}$$

$$= Ae^{-j2\pi ft_o} + \frac{a}{2}\exp\left[j2\pi f\left(\frac{1}{f_m}-t_o \right) \right] \tag{1.5}$$

$$+ \frac{a}{2}\exp\left[j2\pi f\left(-\frac{1}{f_m}-t_o \right) \right]$$

The system function of Eq. (1.5) indicates that the impulse response consists of three delayed and scaled impulses.

If the input to the cosine filter, $r(t)$, is bandlimited, the output is given by

$$s(t) = Ar(t-t_o) + \frac{a}{2}r\left(t+\frac{1}{f_m}-t_o \right) + \frac{a}{2}r\left(t-\frac{1}{f_m}-t_o \right) \tag{1.6}$$

Equation (1.6) indicates that the response is in the form of an undistorted version of the input added to two time-shifted versions (*echoes* or *multipath*).

Returning to the general filter case, we see that the output of a system with amplitude distortion is a sum of shifted inputs. Thus with,

$$H(f) = \sum_{n=0}^{\infty} a_n\cos\left(\frac{n\pi f}{f_m} \right)e^{-j2\pi ft_o} \tag{1.7}$$

the output due to an input, $r(t)$, is given by

$$s(t) = \sum_{n=0}^{\infty} \frac{a_n}{2}\left[r\left(t+\frac{n}{2f_m}-t_o \right) + r\left(t-\frac{n}{2f_m}-t_o \right) \right] \tag{1.8}$$

Equation (1.8) can be computationally difficult to evaluate. This approach is therefore usually restricted to cases where the Fourier series contains relatively few significant terms.

Example 1.1
Consider the triangular filter characteristic shown in Fig. 1.2.

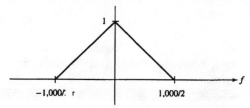

Figure 1.2 - Triangular Filter

Assume that the phase characteristic is linear, with slope $-2\pi t_o$. Find the output of this filter when the input signal is

$$r(t) = \frac{\sin 400\pi t}{t}$$

Solution: We must first expand $H(f)$ in a Fourier series to get

$$H(f) = \frac{1}{2} + \frac{4}{\pi^2}\cos\frac{\pi f}{1000} + \frac{4}{9\pi^2}\cos\frac{2\pi f}{1000} + \frac{4}{25\pi^2}\cos\frac{5\pi f}{1000} + \ldots$$

$r(t)$ is bandlimited so that all frequencies are passed by the filter. This is true since $R(f)$ is zero at frequencies above 200 Hz and the filter cuts off at $f = 1000$ Hz. If we retain the first three nonzero terms in the series, the output becomes,

$$s(t) = \frac{1}{2}r(t-t_o) + \frac{2}{\pi^2}\left[r\left(t - \frac{1}{1000} - t_o\right) + r\left(t + \frac{1}{1000} - t_o\right)\right]$$

$$+ \frac{2}{9\pi^2}\left[r\left(t - \frac{3}{1000} - t_o\right) + r\left(t + \frac{3}{1000} - t_o\right)\right]$$

The result is sketched as Fig. 1.3 for $t_o = 0.005$ seconds.

The undistorted function is plotted as a dashed line on the same set of axes. Examining the differences between the two curves, we see that the distortion causes an attenuation and a

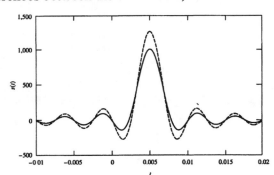

Figure 1.3 - Result of Example 1.1

slight time delay. That is, the peaks of the output are smaller and slightly to the right of the corresponding peaks of the undistorted waveform.

MATLAB EXAMPLE: In the following, we redo example 1.1 for various input parameters. We suggest that you start by choosing the input values proposed in the program. Then you can vary these and observe the effect on the result.

```
clc
disp('Plot the output of a triangular filter from time t_min to time t_max.')
disp('   Input signal to filter: s(t) = sin(400*pi*t)/t ')
disp('------------ PRESS A KEY TO CONTINUE --------------')
pause
disp(' ')
tmin = input('Please input t_min (example: -0.03): ');
tmax = input('Please input t_max (example: 0.03): ');
disp(' ')
fm = input('Please input the cutoff frequency of the filter fm (example: 1000):');
t0 = input('Please input the time delay through the filter t) (example: 0.005)
(example t0 = 0.005): ');
t = tmin:(tmax-tmin)/10000:tmax;    % time range and steps size
%-----compute arguments t's of r(t) in the expression of s(t)--------------
t1 = t-t0;          t2 = t-1/fm-t0;      t3 = t+1/fm-t0;
t4 = t-3/fm-t0;  t5 = t+3/fm-t0;
%--------compute 1st term r1 in the expression of s(t)--------------
if t1 == 0
  r1(t1) = 1;         % l'hopital's rule for lim sin(t)/t as t => 0
else
  r1 = sin(400*pi.*t1)./t1;
end
%--------compute 2nd term r2 in the expression of s(t)--------------
if t2 == 0
  r2(t2) = 1;         % l'hopital's rule
else
  r2 = sin(400*pi.*t2)./t2;
end
%--------compute 3rd term r3 in the expression of s(t)--------------
if t3 == 0
  r3(t3) = 1;
else
  r3 = sin(400*pi.*t3)./t3;
end
%--------compute 4th term r4 in the expression of s(t)--------------
if t4 == 0
```

```
  r4(t4) = 1;
else
  r4 = sin(400*pi.*t4)./t4;
end
%--------compute 5th term r5 in the expression of s(t)--------------
if t5 == 0
  r5(t5) = 1;
else
  r5 = sin(400*pi.*t5)./t5;
end
s = 0.5*r1+(2/pi^2)*(r2+r3)+(2/(9*pi^2))*(r4+r5);
r = r1;
plot(t,s,t,r));
xlabel('t, sec.'); ylabel('s(t)');
```

Phase Distortion

Deviations of the phase away from the distortionless case (linear phase) can be characterized by variations in the slope of the phase characteristic and in the slope of a line from the origin to a point on the curve. We define *phase delay* and *group delay* (also known as *envelope delay*) as follows:

$$t_{ph}(f) \triangleq \frac{\theta(f)}{2\pi f} \qquad t_{gr}(f) \triangleq \frac{d\theta(f)}{2\pi df} \tag{1.9}$$

Figure 1.4 illustrates these definitions for a representative phase characteristic. If the channel were ideal and distortionless, the phase characteristic would be linear (a straight line intercepting the origin) and the group and phase delays would both be constant for all f. In fact for this ideal case, both of these delays would be equal to the time delay from input to output, t_o.

The phase characteristic can be approximated as a piecewise linear curve. As an example, examine the phase characteristic of Fig. 1.4. If we were to operate in a relatively narrow band around f_o, the phase could be approximated as the first two terms in a Taylor series expansion.

$$\Theta(f) = \theta(f_o) + \frac{d\theta(f_o)}{df}(f-f_o)$$

$$= t_{ph}(f_o)f_o + (f-f_o)t_{gr}(f_o) \tag{1.10}$$

Equation (1.10) applies for positive frequency, and the negative of this equation applies for

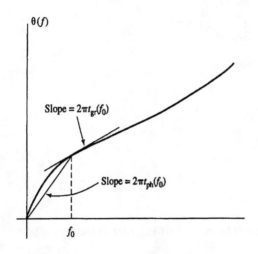

Figure 1.4 - Group
and Phase Delay

negative frequency. This is true since the phase characteristic for a real system must be an odd function of frequency.

Now suppose that the amplitude factor is constant, $A(f)=A$, and a wave of the form, $r(t)\cos 2\pi f_o t$ forms the input to the system. The Fourier transform of the input is found from the modulation property of the transform (see Appendix A).

$$\mathscr{F}\{r(t)\cos 2\pi f_o t\} = \frac{1}{2}\left[R(f-f_o) + R(f+f_o)\right] \tag{1.11}$$

The Fourier transform of the output, $S(f)$, is given by the product of the input transform with the system function. Thus,

$$S(f) = \frac{1}{2}R(f-f_o)A\exp[jt_{ph}(f_o)2\pi f_o]\exp[j2\pi(f-f_o)t_{gr}(f_o)] \tag{1.12}$$

Equation (1.12) has been written for positive f. For negative f, the value of the expression is the complex conjugate of its value at positive f.

We now find the time function corresponding to this Fourier transform. The transform can be simplified by taking note of the following three Fourier transform relationships.

$$\mathscr{F}\{r(t-t_o)\cos2\pi f_o t\} = \frac{1}{2}R(f-f_o)e^{-j2\pi(f-f_o)t_o} + \frac{1}{2}R(f+f_o)e^{-j2\pi(f+f_o)t_o}$$

$$\mathscr{F}\{r(t-t_1)\cos2\pi f_o(t-t_1)\} = \frac{1}{2}[R(f-f_o)+R(f+f_o)]e^{-j2\pi ft_1}$$

$$\mathscr{F}\{r(t-t_o)\cos2\pi f_o(t-t_1)\} = \frac{1}{2}R(f-f_o)e^{-j2\pi(f-f_o)(t_o-t_1)}$$

$$+ \frac{1}{2}R(f+f_o)e^{-j2\pi(f+f_o)(t_o-t_1)}\ e^{-j2\pi ft_1}$$

(1.13)

Using these relationships to simplify Eq. (1.12), we find

$$s(t) = Ar[t-t_{gr}(f_o)]\cos2\pi f_o[t-t_{ph}(f_o)]$$

(1.14)

Recall that we assumed that input was $r(t)\cos2\pi f_o t$. The output of Eq. (1.14) indicates that the time-varying amplitude of the input sinusoid is delayed by an amount equal to the group delay, and the oscillating portion is delayed by an amount equal to the phase delay. Both group and phase delay are evaluated at the frequency of the sinusoid. This result will prove significant later. Getting ahead of the game, we will work with two types of receiver: the coherent and incoherent. Incoherent receivers operate only on the amplitude of the received signal, so the group delay is critical to the receiver operation. On the other hand, coherent receivers use all of the waveform information, so phase delay is also important.

1.2.1 Typical Communication Channels

All communication systems contain a *channel*. It is the medium which connects the receiver to the transmitter. The channel can consist of copper wires, coaxial cable, fiber optic cable, waveguides, air (including the upper atmosphere in the case of satellite transmission), or a combination of these. All channels have a maximum frequency beyond which input signal components are almost entirely attenuated. This is due to the presence of distributed capacitance and inductance. As frequencies increase, the parallel capacitance tends to "short out" the signal and the series inductance "open circuits".

Many channels also exhibit a low-frequency cutoff due to the dual of the effects mentioned above. If there is a low-frequency cutoff, the channel can be modeled as a bandpass filter. If there is no lower-frequency cutoff (this is sometimes known as   *dc* coupled), the channel model is a lowpass filter.

Communication channels are categorized according to bandwidth. There are three traditional grades of channel: the narrowband, voiceband, and wideband (although other terms such as *broadband* are often used to characterize wireless channels)..

Bandwidths up to 300 Hz are in the *narrowband*, or *telegraph grade*. Channels with these bandwidths can be used for slow data transmission, on the order of 600 bits per second (bps). Narrowband channels cannot reliably be used for unmodified voice transmissions without compression..

Voice-grade channels have bandwidths between 300 Hz and 4 kHz. While they were originally designed for analog voice transmission, they are regularly used to transmit data. Some forms of compressed video can be sent on voice-grade channels. The public telephone (subscriber loop) circuits are voiceband.

Wideband channels have bandwidths greater than 4 kHz. These can be leased from a carrier (e.g., telephone company). They can be used for high-speed data, video, or multiple voice channels.

The remainder of this section gives a brief overview of the variety of communication channels in use today. We then focus on telephone channels since their use for both analog and digital communication predominates over the other types of channels.

Wire, Cable and Fiber

Copper wire, coaxial cable or optical fibers can be used in *point- to-point* communication. That is, if we know the fixed location of the transmitter and the location of the receiver, and if these can be conveniently connected to each other, a wire connection is possible. Copper wire pairs, twisted to reduce the effects of incident noise, can be used for low-frequency communication. The bandwidth of this system is dependent on length. The attenuation (in dB/km) follows a curve similar to that shown in Fig. 1.5.

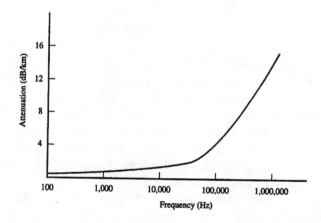

Figure 1.5 - Copper Wire Attenuation vs. Distance

An improvement over twisted copper pair is realized by moving to *coaxial cable.* The bandwidth of this channel is much higher than that of twisted wire, and multiple pairs of wires can be enclosed within a single cable sheath. The sheath which surrounds the wires shields them from incident noise, so coaxial cables can be used over longer distances than can twisted pair.

Fiber optics offers advantages over metal cable, both in bandwidth and in noise immunity. Fiber optics is particularly attractive for data communications, where the bandwidth permits much higher data rates than those achievable with metallic connectors.

Air (*terrestrial* communications) has both advantages and disadvantages when used as a transmission channel. The most important advantage is the ability to *broadcast* signals. You do not need to know the exact location of the receiver in order to set up a communication link. Mobile communication would not be practical without that capability (picture trailing a long wire behind your vehicle). Among the disadvantages are channel characteristics that are highly dependent on frequency, additive noise, limited allocation of available frequency bands, susceptibility to intentional interference (jamming), and lack of security.

Attenuation (at sea level) is a function of frequency, barometric pressure, humidity, and weather conditions. A typical curve for fair-weather conditions would resemble Fig. 1.6.

Visible light occupies the range of frequencies from about 40,000 GHz to 75,000 GHz.

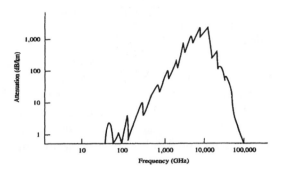

Figure 1.6 - Attenuation vs. Frequency

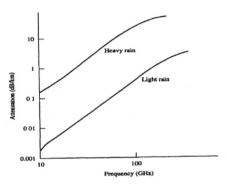

Figure 1.7 - Low Frequencies

Figure 1.7 amplifies the lower frequency portion of the attenuation curve under conditions of rain.

Additive noise and transmission characteristics also depend on frequency. The higher the frequency, the more the transmission takes on the characteristics of light. For example, at radio frequencies (*rf*) in the range of 1 MHz, transmission is *not* line-of-sight, and reception beyond the horizon is possible. However, at *ultra high frequencies (uhf)* in the range of 500 MHz and above, transmission starts acquiring some of the characteristics of light. Line-of-sight is sometimes needed, and humidity and obstructions degrade transmission. At *microwave* frequencies over several hundred MHz, transmission is line-of-sight. Antennas must be situated in a manner to avoid obstructions.

Satellites provide advantages in long-distance communication. The signal is sent to the satellite via an *uplink* and the electronics in the satellite (transponder) retransmits this signal to the *downlink*. It is just as easy for the satellite to retransmit the signal to a receiver immediately adjacent to the uplink as it is to transmit to a receiver 4000 km away from the uplink. The only requirement is that the receiver be within the *footprint* of the satellite–that is, the area of the earth's surface covered by the satellite transmitting antenna. Satellite communication has several major disadvantages. Satellites are often located in assigned orbital slots in geosynchronous orbit, about 35,000 km (22,300 miles) above the earth's surface. This results in a round-trip travel time of approximately ½ second, making rapid interactive two-way communication difficult. This delay is driving a move toward Low Earth Orbit (LEO) satellites. There have even been proposals to use airplanes in place of satellites where the planes would occupy an airspace in shifts.

All broadcast transmission systems suffer from a lack of privacy. Anyone within the reception area of the transmitting antenna can "tap in" to the conversation. This shortcoming can be partially alleviated with scrambling and encryption.

Telephone Channels

When we refer to telephone channels, we are talking about the fixed hardware (wires and fiber optics). The wireless portion (e.g., in cellular telephone) is considered separately. There are two types of phone lines in use today. The *dial-up line* is routed through voice switching offices. The alternative is the *leased line*, which is a permanent circuit that is not subject to the public type switching. Since the same line is used every time, some types of distortion can be predicted and compensated for.

In the dial-up circuit, assorted problems arise due to procedures adopted for voice channels. These problems can be broadly categorized into *distortion*, *additive interference (noise)*, and *crosstalk*.

The phone system is specifically tailored to audio signals with an upper frequency in the vicinity of 4 kHz. When such lines are used for data, the upper cutoff is often stretched to provide data rates above 10 kbps. *Loading coils* in the line improve performance in the voice band, but cause additional amplitude distortion above 4 kHz, thus making higher bit rates more difficult to achieve.

Long distance phone channels contain *echo suppressors* which are voice activated. These prevent a speaker from receiving an echo due to reflections from transitions in the channel. The time delay in activating these suppressors can make certain types of data operation difficult.

Phone lines have amplitude characteristics which are not constant with frequency, and they therefore contribute amplitude distortion. Figure 1.8 shows a typical attenuation, or loss curve. The loss is given in decibels (dB) and is relative to attenuation at about 1000 Hz, where the minimum loss occurs.

Phase distortion also occurs in the phone line. A typical phase characteristic for about 7 km of phone line is shown in Fig. 1.9. Since this is not a straight line, distortion is present.

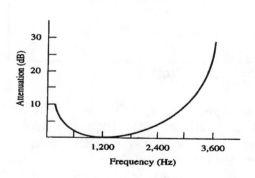 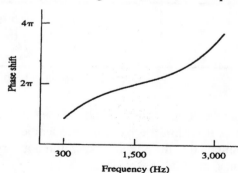

Figure 1.8 - Typical Telephone Channel Attenuation   Figure 1.9 -Typical Telephone Channel Phase

Voice grade channels are classified according to the maximum amount of attenuation distortion and maximum envelope delay distortion within a particular frequency range. Telephone companies can provide *conditioning* to reduce the effects of particular types of distortion. In leasing a telephone channel, a particular conditioning is specified and the company guarantees a certain performance level. Naturally, the better the channel, the higher the cost.

Table 1.1 lists typical channel conditioning characteristics for representative types of conditioning. We illustrate the C-Conditioning specifications. Other types include the BELLCORE Voice-Grade, and AT&T Service types.

Conditioning	Attenuation Frequency Range	Distortion Variation	Envelope Delay Frequency Range	Distortion Variation μs
Basic	500-2500 300-3000	-2 to +8 -3 to +12	800-2600	1750
C1	1000-2400 300-2700 300-3000	-1 to +3 -2 to +6 -3 to +12	1000-2400 800-2600	1000 1750
C2	500-2800 300-3000	-1 to +3 -2 to +6	1000-2600 600-2600 500-2800	500 1500 3000
C4	500-3000 300-3200	-2 to +3 -2 to +6	1000-2600 800-2800 600-3000 500-3000	300 500 1500 3000
C5	500-2800 300-3000	-0.5 to +1.5 -3 to +3	1000-2600 600-2600 500-2800	100 300 600

Table 1.1 - Typical Telephone Channel Parameters

As an example, please examine the C2 channel. If you were to purchase such a channel and use it within the band of frequencies between 500Hz and 2800 Hz, you would be guaranteed that the attenuation would not vary beyond the range of -1 to +3 dB (relative to response at 1004 Hz). If you use it in the wider band between 300 Hz and 3000 Hz, the guaranteed attenuation range increases to -2 to +6 dB. Similarly, the envelope delay variation will be less than 3000 microseconds if you operate within the band between 500 Hz and 2800 Hz.

In addition to the parameters given in Table 1.1, the various lines have specified losses, loss variations, maximum frequency error, and phase jitter. For example, the C1 channel specifies a loss of 16 dB ± 1 dB at 1000 Hz. The variation in loss is limited to 4 dB over long periods of time. The frequency error is limited to 5 Hz and the phase jitter to 10°.

1.3 Types of Signals

The signals we wish to transmit either come directly from the source or result from signal processing operations. Examples of signals directly from the source are the pressure wave emitted by human vocal chords (e.g., grunting, speech or singing) and the electrical signal resulting from a power source connected to a computer keyboard. Processed signals can result from analog to digital converters, encoding and encryption devices, and signal conditioning circuitry. The manner in which we communicate information is dependent on the form of the signal. There are two broad signal classifications–analog and digital. Within these are a number of more detailed subdivisions.

1.3.1. Analog Signals

An *analog signal* can be viewed as a waveform which can take on a continuum of values for any time within a range of times. Although your measuring device may be limited in resolution (e.g., it may not be possible to read an analog voltmeter more accurately than the nearest 1/100 of a volt), the actual signal can take on an infinity of possible values. For example, at a particular time you might read the value of a voltage waveform to be 13.45 volts. If the voltage is an analog signal, the actual value could be expressed as an extended decimal with an infinite number of digits to the right of the decimal point.

Just as the ordinate of the function contains an infinity of values, so does the time axis. Although we may conveniently resolve the time axis into points (e.g., every microsecond on an oscilloscope), the function has a defined value for any of the infinity of time points within any interval.

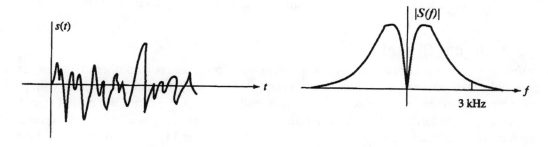

Figure 1.10 - Representative Speech Waveform

An example of an analog signal is a human speech waveform. We illustrate a representative waveform and its Fourier transform in Figure 1.10. Note that we show only the magnitude of the Fourier transform. If the speech waveform resulted from someone whistling into a microphone, the time waveform would be a sinusoid, and the Fourier transform would be an impulse at the whistling frequency (and a second impulse at the mirror image of this frequency value). If the person hummed into a microphone, the time waveform would be periodic with fundamental frequency equal to the frequency at which the person is humming. The Fourier transform would consist of impulses at the fundamental and at harmonics of this fundamental. [Wouldn't it be fun to try this in the lab!]

1.3.2. Analog Sampled Signal

Suppose that an analog time signal is defined only at discrete time points. For example, suppose you measured a voltage waveform by sending values to a voltmeter every microsecond. The resulting function is only known at these discrete points in time. This results in a *discrete time function*, or a *sampled waveform*. It is distinguished from an analog waveform by the manner by which we specify the function. In the case of the analog waveform, we must either display the function (e.g., graphically, on an oscilloscope), or give a functional relationship between the variables. In contrast to this, the discrete signal can be thought of as a list, or sequence of numbers. Thus, while an analog waveform can be expressed as a function of time, $v(t)$, the discrete waveform is a sequence of the form, v_n, where n is an integer, or index.

Discrete signals can be visualized as pulse waveforms. Figure 1.11 shows an analog time function and the resulting sampled pulse waveform (we will refer to this sampled waveform later as Pulse Amplitude Modulation–PAM).

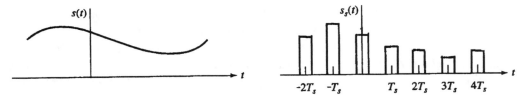

Figure 1.11 - Discrete waveform derived from analog time function

1.3.3. Digital Signal

A *digital signal* is a form of sampled or discrete waveform, but each number in the list can now only take on specific values. For example, if we were to take a sampled voltage waveform and round each value to the nearest tenth of a volt, the result is a digital signal.

We can use a thermometer as an example of all three types of signals. If the thermometer has a dial or a tube of mercury (caution–it's toxic!), the output is analog. We can read the temperature at any time, and to any desired degree of accuracy (limited, of course, by the resolution of the reader–human or mechanical).

Suppose now that the thermometer consists of a dial, but that it is only updated once every minute. The result is an analog sampled signal.

If the display now takes the form of a numerical readout, the thermometer becomes digital. The readout is the result of sampling the temperature (perhaps every minute), and then displaying the sampled temperature to a predetermined resolution (perhaps the nearest 1/10 degree).

Digital signals result from many devices. For example, dialing [1] a telephone number results in one of 12 possible signals depending on which button is pressed. Other examples include pressing keys on a bank automated teller machine (ATM), or using a computer keyboard. Digital signals also result from performing analog to digital conversion operations.

1.4. Elements of a Communication System

Acommunication system consists of a transmitter, a channel, and a receiver, as shown in Figure 1.12. The purpose of the *transmitter* is to change the raw information into a format that is matched to the characteristics of the channel. For example, if the information is a speech waveform and the channel is the air, the signal must be modified prior to insertion into the channel. This is true since the basic speech waveform will not propagate efficiently through the air–its frequency range is below that of the passband of the channel.

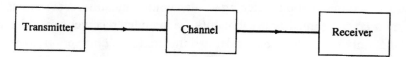

Figure 1.12 - Block Diagram of Communication System

The *channel* connects the transmitter to the receiver. It may take the form of one of the channels described in Section 1.2 of this chapter.

The *receiver* (sometimes called the *sink*) accepts the signal from the channel, and processes it to permit interfacing with the final destination (e.g., the human ear, a computer monitor).

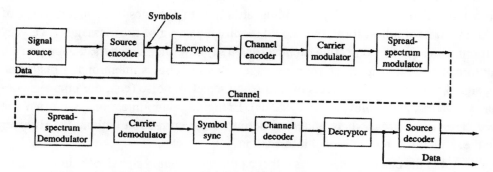

Figure 1.13 - Expanded Communication System Block Diagram

---

[1]The word "dialing" is as obsolete as the term "service station". It is a carryover from the early days of telephony when telephone sets contained a rotary dial used to enter numbers.

Figure 1.13 shows an expanded version of the simplified block diagram of a communication system. We will take a few minutes to briefly describe the function of each block. These functions will be expanded on in later chapters of the text.

The *signal source* (or *transducer*) is the starting point of our system. It may be a microphone driven by a pressure wave from a human source or a musical instrument, a measuring device that is part of a monitoring system, or a data source such as a computer keyboard or a numeric pad on an automated teller machine. Its output is a time waveform–usually electrical.

The *source encoder* operates on one or more signals to produce an output which is compatible with the communication channel. It could be as simple as a lowpass filter in an analog communication system, or it could be as complex as a converter which accepts analog signals and produces a periodic train of output *symbols*. These symbols may be binary (ones and zeros), or may be members of a set with more than two elements. When channels are used to communicate signals from more than one source at the same time, the source encoder also contains something called a *multiplexer*.

With electrical communication replacing written communication, security has become increasingly important. We must assure that only the intended receiver can understand the message, and that only the authorized sender can transmit it. *Encryption* provides such security. As unauthorized receivers and transmitters become more sophisticated and computers become larger and faster, the challenges of secure communication become greater. In analog systems, security is often provided using *scrambling* systems used in some pay-per-view television channels and privacy devices for telephones.

The *channel encoder* provides a different type of communication security than that provided by the encryptor. It decreases the effects of transmission errors. Whenever noise is introduced into a communication channel, errors occur at the receiver. These errors can take the form of changes in an analog signal. In a digital system, they can make it possible for one transmitted symbol to be interpreted as a different symbol at the receiver. We can decrease the effects of errors by providing structure to the messages in the form of redundancy. In its simplest form, this would require repeating of the messages (in a very noisy room, you might try repeating each word you say several times so the person you are talking to can correctly interpret your message). We sometimes intentionally distort an analog signal to decrease the effects of frequency sensitive noise (e.g., pre-emphasis/de-emphasis systems, and Dolby sound systems). In digital communications, we often use *forward error correction*, where encoding permits error correction without the receiver asking the transmitter for additional information.

The output of the channel encoder is either a processed analog signal, or a digital signal composed of symbols. For example, in a binary digital system the output would be a train of ones and zeros. We need to modify the channel encoder output signal in a manner which matches the characteristics of the channel. For example, you couldn't just send ones and zeros down the channel–Indeed, it's not even clear what is meant by "send ones and zeros down the channel". Do we mean that you speak these words into a microphone? Or do you write a sequences of ones and zeros on a piece of paper and hold it at the input to the channel?

The *carrier modulator* produces an analog waveform which transmits efficiently through the channel. The waveform is selected for efficiency, and also to permit use of the channel by several transmitters simultaneously. For analog signal sources, the carrier modulator modifies the range of signal frequencies in order to allow efficient transmission. For digital signal sources, the modulator produces signal segments (bursts) corresponding to the discrete symbols at its input.

Spread spectrum is a technique for providing some immunity to frequency selective effects such as interference and fading. A signal is *spread* over a wide range of frequencies so that narrowband interference affects only a small portion of the signal. Spread spectrum has other advantages related to simplified methods of sharing a channel among multiple users. Since an unauthorized listener may mistake a spread spectrum signal for wideband noise, this technique provides some (limited) additional security beyond that afforded by encryption.

We have been describing the blocks that form the upper half of Figure 1.13. The lower half of this figure comprises the receiver, which is simply a mirror image of the transmitter. It must "undo" each operation that was performed at the transmitter. The only variation from this one-to-one correspondence is that the *carrier modulator* of the transmitter has been replaced by two blocks in the receiver: the *carrier demodulator* and the *symbol synchronizer*. The symbol synchronizer is needed only in digital systems–it represents a major distinction between analog and digital communication systems. Once the waveforms are reproduced at the receiver, the overall signal must be properly partitioned into segments corresponding to each symbol and to each message. This partitioning is the function of the *synchronizer*.

1.5. Sampling

A *discrete* signal is a signal which is *not* continuous in time. It has values only at specific points of the time axis. If the discrete signal is analog, its values at any time for which the signal is defined may lie within a continuum of possible values. We wish to find a way of converting a continuous (not discrete) analog waveform into a discrete signal. To do this, the time axis must somehow be *discretized*. The conversion of the continuous time axis into a discrete axis is accomplished by time sampling.

The *sampling theorem* states the following:

> If the Fourier transform of a time function is zero above a certain frequency (call it f_m) and the values of the time function are known for $t = nT_s$ (for all integer values of n), then the time function is exactly known for ALL values of t.

This is remarkable! Knowing the value of the time function at discrete time points allows us to fill in the curve between these points *precisely and accurately*! Of course, something this remarkable must have some limitations. You couldn't be given two values separated by hours and expect to fill in the curve between these points. Indeed, for the samples to provide all of the information, they must be "close enough" to each other. The restriction is that the spacing between samples (i.e., seconds per sample), T_s, be less than $1/2f_m$.

The upper limit on T_s can be expressed in a more meaningful way by taking the reciprocal of T_s. This reciprocal is the sampling frequency, denoted $f_s = 1/T_s$ in samples per second. The restriction then becomes,

$$T_s < \frac{1}{2f_m}$$

$$\frac{1}{T_s} > 2f_m$$

$$f_s > 2f_m$$

So the sampling rate must be more than twice the highest frequency of the signal. For example, if a voice signal has 4 kHz as a maximum frequency, it must be sampled at least 8000 times per second to comply with the conditions of the sampling theorem. The minimum sampling rate, $2f_m$, is known as the *Nyquist sampling rate*.

Before going further, let's observe that the spacing between the sample points is inversely related to the maximum frequency, f_m. This is intuitively satisfying, since the higher f_m, the faster we would expect the function to vary. The faster the function varies, the closer together the sample points should be in order to permit reconstruction of the function.

We present two proofs of the sampling theorem. The first proof is physical and intuitive while the second is more mathematical.

Proof 1: Figure 1.14 shows a signal, *s(t)*, a pulse train, *p(t)*, and the product of the pulse train with the signal, *s(t)p(t)*. If the pulse train consists of narrow pulses, one would say that the product of *s(t)* with *p(t)* is a *sampled version* of the original waveform. In actuality, the output depends not only on the sample values of the input, but on a range of values around each sample point. The theory does not require these extra values, which represent added redundant information. However, practical systems often sample over a small range of time surrounding the actual sample points. As we prove the theorem, it should become obvious that the multiplying function need not consist of perfect square pulses. In fact, the function can be any periodic signal and the pulse widths can approach zero (i.e., multiply by a train of impulses).

Multiplying *s(t)* by *p(t)* of the type shown in Fig. 1.14 is a form of *time gating*. It can be viewed as the opening and closing of a gate, or switch.

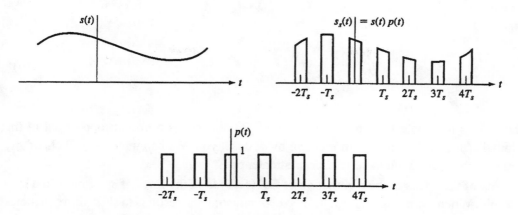

Figure 1.14 - Product of Pulse Train with *s(t)*

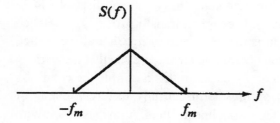

Figure 1.15 - Representative *S(f)*

Our goal is to show that the original signal can be recovered from the sampled waveform, $s_s(t)$. We do this by examining the Fourier transform of $s_s(t)$. The sampling theorem requires that we assume $s(t)$ is limited to frequencies below f_m (the subscript m stands for *maximum*). The Fourier transform of $s(t)$, $S(f)$, therefore cuts off at f_m. Figure 1.15 shows a representative shape for this Fourier transform. While we use this triangular shape throughout the text, we do not mean to restrict the actual transform to follow this curve.

Since the multiplying pulse train is assumed to be periodic, it can be expanded in a Fourier series. The $p(t)$ shown in Fig. 1.14 is an even function, so we can use a trigonometric series containing only cosine terms (though this is not necessary to prove the theorem. It just makes the equations look simpler).

$$s_s(t) = s(t)p(t)$$

$$= s(t) \left[a_0 + \sum_{n=1}^{\infty} a_n \cos 2\pi n f_s t \right] \qquad (1.15)$$

$$= a_0 s(t) + \sum_{n=1}^{\infty} a_n s(t) \cos 2\pi n f_s t$$

where

$$f_s = \frac{1}{T_s} \tag{1.16}$$

The goal is to isolate the first term in Eq. (1.15), since this term is proportional to the original $s(t)$. Don't let the multiplication by a_0 upset you. We can undo the effects of any constant multiplier with an amplifier or attenuator.

Each of the terms in the summation of Eq. (1.15) is of the form of $s(t)$ multiplied by a sinusoid. When a time signal is multiplied by a sinusoid, the result is to shift all frequencies of the signal by an amount equal to the frequency of the sinusoid. This is known as the *modulation theorem* (See Appendix A). So the frequency content of each term in Eq. (1.15) is centered around the frequency of the multiplying sinusoid (when we discuss AM in Chapter 3, we will call this the *carrier frequency*). The Fourier transform of $s_s(t)$ is sketched in Fig. 1.16. The shape centered at the origin is the transform of $a_o s(t)$, and the shifted versions represent the transforms of the various harmonic terms. We see that the various terms do not overlap in frequency provided that $f_s > 2f_m$. But this is simply the condition given in the sampling theorem. Since the various terms occupy different bands of frequency, they can be separated from each other using linear filters. A lowpass filter with a cutoff frequency of f_m can be used to recover the $a_o s(t)$ term.

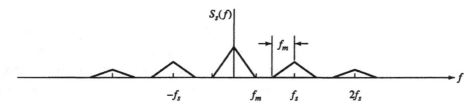

Figure 1.16 - Fourier transform of sampled wave

Proof 2: The second proof we present is less intuitive than the first. We take the time to explore this since the approach supplies insight into the mathematical principles of sampling.

Since $S(f)$ is non-zero along a finite portion of the f-axis (see Fig. 1.15), we can expand it in a Fourier series in the interval

$$-f_m < |f| < f_m$$

In expanding $S(f)$ in a Fourier series, you should be careful not to let the change in notation confuse the issue. The "t" usually used in the Fourier series is an independent functional variable, and any other letter could be substituted for it. Performing the Fourier series expansion, we find

$$S(f) = \sum_{n=-\infty}^{\infty} c_n e^{jn2\pi t_o f} \tag{1.17}$$

where

$$t_o = \frac{1}{2f_m} \tag{1.18}$$

The c_n in the Eq. (1.17) are given by

$$c_n = \frac{1}{2f_m} \int_{-f_m}^{f_m} S(f)e^{-jn2\pi t_o f}df \tag{1.19}$$

However, the Fourier inversion integral tells us that

$$s(t) = \int_{-\infty}^{\infty} S(f)e^{j2\pi t}df = \int_{-f_m}^{f_m} S(f)e^{j2\pi ft}df \tag{1.20}$$

In the final expression in Eq. (1.20), the limits of integration have been reduced since $S(f)$ is equal to zero outside of the interval between $-f_m$ and $+f_m$. Upon comparing Eq. (1.19) with Eq. (1.20), we see that

$$c_n = \frac{1}{2f_m}s(-nt_o) = \frac{1}{2f_m}s\left(-\frac{n}{2f_m}\right) \tag{1.21}$$

Equation (1.21) is the result we have been seeking! It says that the c_n are specified by the values of $s(t)$ at the points $t = n/2f_m$. Once the c_n are known, $S(f)$ is known, and once $S(f)$ is known, $s(t)$ is known. *We have thus proven the sampling theorem.* (Please reread the last two sentences and convince yourself we have proven the theorem–don't just shrug and say, "Roden is correct".).

Although this completes the proof, we will carry the mathematics a step further to actually solve for $s(t)$ in terms of the sample values. We substitute the c_n of Eq. (1.21) into Eq. (1.17) to get

$$S(f) = \frac{1}{2f_m} \sum_{n=-\infty}^{\infty} s\left(-\frac{n}{2f_m}\right) e^{jn\pi f/f_m} \tag{1.22}$$

We now find the inverse Fourier transform of this $S(f)$.

$$s(t) = \sum_{n=-\infty}^{\infty} \frac{1}{2f_m} \int_{-f_m}^{f_m} s\left(-\frac{n}{2f_m}\right) e^{jn\pi f/f_m} e^{j2\pi ft} df$$

$$= \sum_{n=-\infty}^{\infty} s\left(-\frac{n}{2f_m}\right) \left[\frac{\sin(2\pi f_m t + n\pi)}{2\pi f_m t + n\pi}\right] \tag{1.23}$$

Equation (1.23) is the final statement of the sampling theorem. We can use it to find the value of $s(t)$ at any point in time simply by knowing the sample values of $s(t)$. That is, the only unknowns on the right side of the equation are the sample values.

You can get some feeling for the sampling theorem by sketching a few terms in Eq. (1.23). We do this in Figure 1.17 for a representative $s(t)$ and 3 sample points.

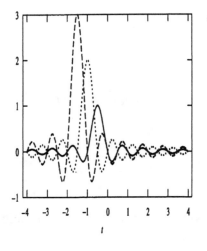

Figure 1.17 - Terms From the Sampling Reconstruction

The actual waveform is the sum of the various component time functions. But if you look at each sample point, *every* waveform is identically zero except the one centered at that sampling point. Between sampling points, we must calculate the sum of the various terms from adjacent points.

MATLAB EXAMPLE:

This MATLAB program allows you to develop a plot similar to Fig. 1.17. You get to input five sample values and the program plots the five waveforms and their sum.

```
clc;
disp('Reconstruct a signal in time domain from five sampled values');
disp('The program displays the five terms in Eq. (1.23) and the sum of these terms');
k5=input('Please input the first sample value:');
k4=input('Please input the second sample value:');
k3=input('Please input the third sample value:');
k2=input('Please input the fourth sample value:');
k1=input('Please input the fifth sample value:');

fm=1;
t=-4:.01:+4;

t1=2*pi*fm*t+pi;
t2=2*pi*fm*t+2*pi;
t3=2*pi*fm*t+3*pi;
t4=2*pi*fm*t+4*pi;
t5=2*pi*fm*t+5*pi;

s1=k1*(sin(t1))./t1;
s2=k2*(sin(t2))./t2;
s3=k3*(sin(t3))./t3;
s4=k4*(sin(t4))./t4;
s5=k5*(sin(t5))./t5;
ssum=s1+s2+s3+s4+s5;
plot(t,s1,t,s2,t,s3,t,s4,t,s5,t,ssum,'*m');
```

Example 1.2

A bandlimited signal occupies the frequency range between 1000 Hz and 1010 Hz. A typical Fourier transform is shown in Figure 1.18. Although the sampling theorem indicates that the sampling rate must be higher than 2020 samples per second, investigate the possibilities of sampling at rates as low as 20 samples per second.

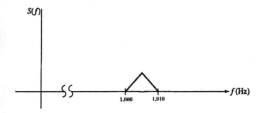

Figure 1.18 - Bandlimited Transform for Example 1.2

Solution: Figure 1.19 shows the Fourier transform that results from sampling at twice the highest frequency (2020 samples per second) and also at twice the bandwidth (20 samples per second). We are assuming that all harmonics have equal amplitude (i.e., assume sampling with an ideal impulse train).

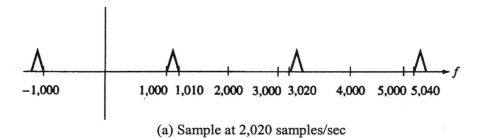

(a) Sample at 2,020 samples/sec

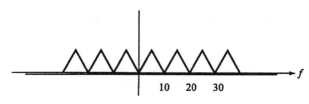

(b) Sample at 20 samples/sec

Figure 1.19 - Fourier transform of sampled waveforms

Note that it is possible to recover the original signal when sampling at only 20 samples per second, *provided* the exact band of frequencies occupied is known. At this lower sampling rate, it is not sufficient to know only the maximum frequency. Also note that if the limits of the band were not multiples of the bandwidth, we would need to sample at a higher rate to avoid overlap.

This example shows that it is possible to recover a bandlimited signal by sampling at a rate as low as twice the *bandwidth* of the signal.

1.5.1 Errors in Sampling

The sampling theorem indicates that $s(t)$ can be perfectly recovered from its samples. This is fine if we are talking only about theory. However, if sampling is attempted in the real world, errors result. This happens since our real-life systems are not ideal.

Round-off errors occur when the various sample values are rounded off in the communication system. This rounding takes place in digital communication where we only send discrete level values. We will later call this error *quantization noise*, and it is examined in detail in Chapter 2.

Truncation errors occur if the sampling is done over a finite time. The sampling theorem requires that samples be taken for all time in the infinite interval, and every sample is used to reconstruct the value of the original function at any particular time. The theorem *does not say*, "give me a few sample values and I'll tell you exactly how to draw a line connecting these points". In a real system, the signal is observed over a limited time interval. Figure 1.17 shows the effect of a particular sample decreases as the samples get farther away.

We define an error as the difference between the reconstructed time function and the original function. Upper bounds can be placed upon the magnitude of this error function. Such bounds involve sums of the rejected time sample values, and some examples are included in the exercises at the end of this chapter.

A third error results if the sampling rate is not high enough. This situation can be intentional or accidental. For example, if the original time signal has a Fourier transform which asymptotically approaches zero with increasing frequency, a conscious decision is often made to define a maximum frequency beyond which signal energy is negligible. In order to minimize the resulting error, the input signal is often lowpass filtered prior to sampling. These filters do not completely cut signals outside the passband (due to finite roll off). But even if we design a system with a high enough sampling rate, an unanticipated high frequency signal (or noise) may appear at the input. In either case, the error caused by sampling too slowly is known as *aliasing*, a name which is derived from the fact that the higher frequencies disguise themselves in the form of lower frequencies. This is the same phenomenon that occurs if a rotating device is viewed as a sequence of individual frames, as in a television picture or an object illuminated by a strobe light. As the device rotation speed increases, a point is reached where the perceived angular velocity starts decreasing. Eventually, a speed is reached (matched to the frame rate) where the device appears to be standing still. Further increases make the rotation appear to reverse direction.

Analysis of *aliasing* is most easily performed in the frequency domain. Before doing that, we illustrate a simple example of aliasing in the time domain. Figure 1.20 shows a sinusoid at a frequency of three Hz. Suppose we sample this sinusoid at four samples per second. The sampling theorem tells us that the minimum sampling rate for unique recovery is six samples per second, so four samples per second is not fast enough. The samples at the slower rate are indicated by *x's* on the figure. But alas, these are the same samples that would result from a sinusoid at 1 Hz as shown by the dashed curve. The 3 Hz signal is disguising itself (aliasing) as a 1 Hz signal.

The Fourier transform of the sampled wave is found by periodically shifting and repeating the Fourier transform of the original signal. If the original signal has frequency components above one-half of the sampling rate, these components *fold back* into the frequency band of interest. Thus, in Fig. 1.20, the 3 Hz signal folded back to fall at 1 Hz.

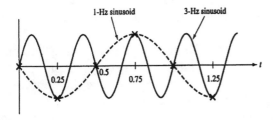

Figure 1.20 - Example of Aliasing

Figure 1.21 illustrates the case where a representative signal is sampled by an ideal train of impulses $s_\delta(t)$ (we use this as the ideal theoretical limit of narrow pulses) at less than the Nyquist rate. Note that the transform at the output of the lowpass filter is no longer the same as the transform of the original signal. If we denote the filter output as $s_o(t)$, the error is defined as

$$e(t) = s_o(t) - s(t) \tag{1.24}$$

Taking the Fourier transform of both sides of Eq. (1.24) yields

$$\begin{aligned} E(f) &= S_o(f) - S(f) \\ &= S(f - f_s) + S(f + f_s) \quad for \quad f < f_m \end{aligned} \tag{1.25}$$

Note that if $S(f)$ were limited to frequencies below $f_s/2$, the error transform would be zero. Without assuming a specific form for $S(f)$, we cannot carry this example further. In general, various bounds can be placed upon the magnitude of the error function based on properties of $S(f)$ for $f > f_s/2$.

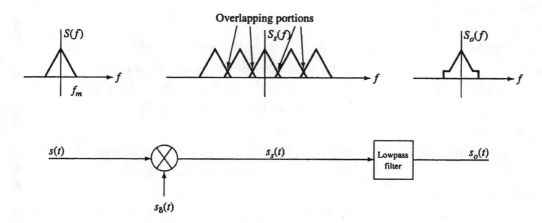

Figure 1.21 - Impulse Sampling at less than Nyquist

MATLAB EXAMPLE: This example calculates the actual aliasing error when a sinusoidal signal is sampled too slowly. You get an opportunity to input the sampling frequency for sampling a sinusoid. The program plots the error. Since the sinusoid continues indefinitely as time extends in both directions, we expect the error to increase as we move toward the endpoints of the sampling interval. This can be seen in the result, particularly when you sample at exactly the Nyquist rate.

```
function aliaserr()

%Reconstruct a sinusoid s(t) = cos(6t) from its sampled values and check
%the effect of sampling below the Nyquist frequency.
%Note frequency is 3/pi. This was chosen to avoid having all the samples fall
%at zero crossings of the function.

clc;                % clear command window
close all;              % close all (previously) figures - plots windows
disp('------------ PRESS A KEY TO CONTINUE --------------')
pause
disp(' ')
k = input('Please input the ratio fs/fm. For example, a value of 2 would indicate sampling at Nyquist');
fm=3/pi;
fs=k*fm;
```

```
tmin = -4/fm;
tmax = 4/fm;
N = 16*fs/fm;                                    % # samples for s(t) = cos(6*t);
so = 0;                          % initialize reconstruction signal so(t)
%sof1 = 0;                        % initialize reconstruction signal so(f)
%sf1 = 0;                         % initialize reconstruction signal s(f)
for n = -N/2:N/2
   t = tmin:(tmax-tmin)/1000:tmax;     % continuous time range and steps size for s(t
  tn = -n/fs;                % discrete sampling time tn at time n
  sn=cos(6*fm*n/fs);
  so = so + (2*fm/fs)*sn.*sin(2*pi*fm.*(t+n/fs)+1E-20)./(2*pi*fm.*(t+n/fs)+1E-20);
    % reconstruct signal s(t)
 %NOTE:1e-20 added to avoid sin(x)/x problem at x=0 (i.e., divide by zero error)
end
end


t = tmin:(tmax-tmin)/1000:tmax;     % continuous time range and steps size for s(t)
s=cos(6*fm.*t);         %original signal
e = s - so;                         % aliasing error e(t)
f = -fm-fs:fm+fs;                   % continuous frequency range
subplot(2,2,1)
plot(t,s)
title('original signal'); xlabel('t, sec'); ylabel('s(t)');
subplot(2,2,2)
plot(t,so)
title('   reconstructed signal'); xlabel('t, sec'); ylabel('so(t)');
subplot(2,2,3)
plot(t,e)
title('error signal'); xlabel('t, sec'); ylabel('e(t)');
```

Example 1.3:

A 100 Hz pulse train (assume amplitude levels of 0 and 1) forms the input to the RC filter shown in Figure 1.22. The output of the filter is sampled at 700 samples per second. Find the aliasing error.

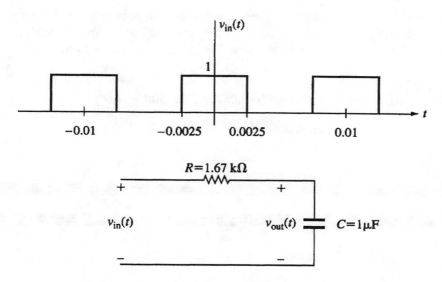

Figure 1.22 - Square Wave and Filter for Example 1.3

Solution: The square wave can be expanded in a Fourier series to yield,

$$v_{in}(t) = \frac{1}{2} + \frac{2}{\pi}\cos 2\pi \times 100t - \frac{2}{3\pi}\cos 2\pi \times 300t + \frac{2}{5\pi}\cos 2\pi \times 500t \ \ldots$$

$$= \frac{1}{2} + \sum_{n=1, n \ odd}^{\infty} (-1)^{\frac{n+3}{2}} \frac{2}{n\pi}\cos 2\pi n \times 100t$$

The filter transfer function is given by

$$H(f) = \frac{1}{1+j2\pi fRC} = \frac{1}{1+j2\pi f(0.00167)}$$

The output of the filter is found by modifying each term in the input Fourier series. The amplitude is multiplied by the transfer function magnitude and the phase is shifted by the transfer function phase. The result is

$$v_o(t)=\frac{1}{2} + 0.45\cos(2\pi\times100t-45^o) - 0.067\cos(2\pi\times300t-71.6^o)$$
$$+ 0.025\cos(2\pi\times500t-78.7^o) - 0.013\cos(2\pi\times700t-81.9^o)$$

Let us assume ideal impulse sampling. The result is that the component at 500 Hz appears at 200 Hz in the reconstructed waveform, and the component at 700 Hz appears at *dc* (zero frequency). We shall ignore the higher harmonics. The reconstructed waveform is therefore given by

$$\frac{1}{2}+0.45\cos(2\pi\times100t-45^o)-0.067\cos(2\pi\times300t-71.6^o)$$
$$+0.025\cos(2\pi\times200t-78.7^o)-0.013\cos(-81.9^o)$$

The last two terms represent the aliasing error.

Example 1.4

Assume that the bandlimited function,

$$s(t) = \frac{\sin20\pi t}{\pi t}$$

is sampled at 19 samples/second. The sampling function is a unit height pulse train with pulse widths of 1 msec. The sampled waveform forms the input to a lowpass filter with cutoff frequency of 10 Hz, as illustrated in Fig. 1.23. Find the output of the lowpass filter and compare this with the original *s(t)*.

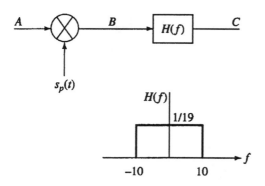

Figure 1.23 - Sampling for
Example 1.4

Solution: We only need to know the first two coefficients in the Fourier series expansion of the pulse train. These are given by

$$a_0 = \frac{0.001}{1/19} = 0.019$$

$$a_1 = 38 \int_{-5\times10^{-4}}^{5\times10^{-4}} \cos 2\pi\times 19 t\,dt = \frac{2}{\pi}\sin(19\times10^{-3}\pi t) = 0.038$$

The Fourier transforms of the signals at points A, B, and C in Fig. 1.23 are shown in Fig. 1.24.

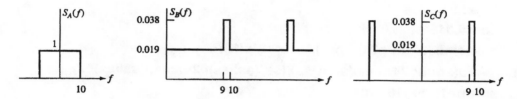

Figure 1.24 - Fourier transform of sampling signals

The output time function is the inverse Fourier transform of $S_o(f)$, and is given by

$$s_o(t) = \frac{0.019\sin 20\pi t}{\pi t} + 0.038\frac{\sin\pi t}{\pi t}\cos 19\pi t$$

The second term in the result represents the aliasing error.

MATLAB EXAMPLE:

Suppose we wish to find the maximum amplitude of the output time function in Example 1.4. It should be obvious that this occurs at $t = 0$, but if you wished to use a simple MATLAB program, the instructions would be as follows:

```
t=-5:.01:5;
s=.038*sin(pi*t)./(pi*t).*cos(19*pi*t);
MAX=max(s)
```

The maximum amplitude is 0.038 at $t = 0$, at which point the signal portion [the first term of $s_o(t)$] has amplitude of 0.019. Do not be tempted to calculate a percentage error by taking the ratio of the error amplitude to the desired signal amplitude. Since the first term goes to zero at periodic points, and the second term is not necessarily zero at these same points, the percentage error would approach infinity.

Errors are often analyzed by looking at the energy of the time function representing the error. Energy is the area under the square of the function. We could therefore find the energy of the second term in the equation. Once this energy is found, it is divided by the total energy of the desired signal (the first term) to get a percentage of error. As an example, suppose we wanted to perform this operation over two sidelobes of the main signal. That is, we compare the signal to aliasing error over the range of time between -0.2 and +0.2 seconds. A simple MATLAB program calculates the signal to noise ratio as the ratio of mean square values.

```
t=-.2:.01:.2;
sout=.019*sin(20*pi*t)./(pi*t)
salias=.038*sin*pi*t)./(pi*t).*cos*19*pi*t);
snr=(std(sout)^2+mean(sout)^2)/(std(salias)^2+mean(salias)^2)
snrdb=10*log10(snr)
```

MATLAB returns a signal to noise ratio of 14.804, or 11.704 dB. Note that we are using the fact that the mean square value is the variance plus the square of the mean.

The restriction upon $S(f)$ imposed by the sampling theorem is not very severe in practice. All signals of interest in real life do possess Fourier transforms which are approximately zero above some frequency. No physical device can transmit infinitely high frequencies since all channels contain series inductance and parallel (parasitic) capacitance. The inductance opens and the capacitance shorts as frequencies increase.

1.6 A/D and D/A Conversion

D igital communication possesses certain important advantages over analog communication. Among these advantages are improved noise immunity and processing simplicity. For these reasons, we often send analog information signals using digital techniques. But how do we convert an analog signal into a digital signal?

The first step in changing an analog continuous time signal into a digital form is to convert the signal into a list of numbers. This conversion is accomplished by *sampling* the time function. The resulting list of numbers each represents a continuum of values. That is, although a particular sample may be stated as a rounded-off number (e.g., 5.758 volts), in actuality it should be continued as an infinite decimal. To turn this into a *digital* signal, the list of analog numbers must be *coded* into discrete *code words*. The first approach toward accomplishing this might be to round off each number in the list. Thus, for example, if the samples range from zero to 10 volts, each sample could be rounded to the nearest integer. This would result in code words drawn from the 11 integers between zero and 10.

In many communication systems, the form chosen for code words is a *binary number*–that is, ones and zeros. The reasons for this choice will become clear when we discuss specific transmission techniques. With this binary restriction, we can envision one simple form of analog to digital converter for our "zero to 10 volt" example. The converter would operate on the zero- to 10-volt samples by first rounding each sample value to the nearest volt. It would then convert the resulting integer into a 4-bit binary number (the *BCD code*).

While practical A/D converters perform an operation similar to that outlined above, an appreciation of this operation is best gained by starting from scratch.

Analog to digital conversion (ADC) is also known as *quantizing*. The goal is to change a continuous variable into one that has discrete values. In *uniform quantization*, the continuum of functional values is divided into uniform-width regions, and an integer code is assigned to each region.

Figure 1.25 illustrates the concept of 3-bit quantization in two different ways. Figure 1.25(a) shows the range of functional values divided into eight regions. Each of these regions is assigned a 3-bit binary number. We have chosen eight regions because this is a power of two. All 3-bit binary combinations are used, leading to greater efficiency. Also note that the range of values is given as that between zero and unity. While this may seem restrictive, any function can be *normalized* to fall within this region through addition of a constant and scaling. We examine this in a few moments when we start designing A/D converters.

Before continuing, let's use this figure to make a general observation about binary counting. As you examine the binary numbers along the ordinate, note that the first bit (the *most significant bit*, or *msb*) is equal to one for the top half of the range, and zero for the bottom half. It oscillates with a period equal to the total range. The next bit oscillates with a period equal to half of the range, and is equal to one for the top half *of each half*, and zero for the bottom half *of each half*. This pattern continues with each successive bit subdividing the region by two, and indicating which half of the new subregion the sample value is in. Thus with each additional bit, we are doubling the *resolution*.

Figure 1.25(b) illustrates quantization by use of an input-output relationship. While the input is continuous, the output can take on only discrete values. The width of each step is constant since the quantization is uniform.

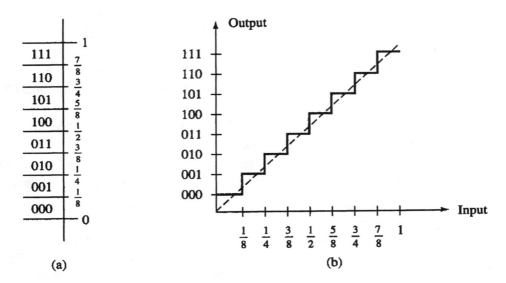

Figure 1.25 - Concept of Quantization

Figure 1.26 shows a representative $s(t)$ and the resulting digital form of the signal for both 2-bit and 3-bit ADC.

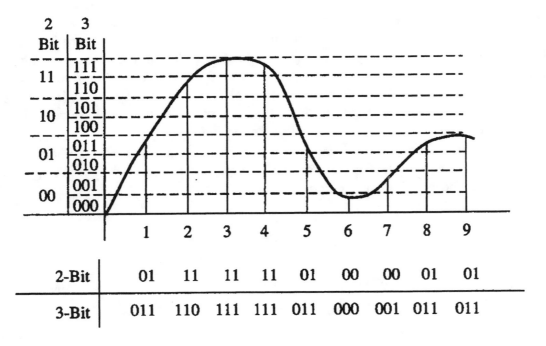

Figure 1.26 - Result of ADC for Representative $s(t)$

1.6.1 Quantizers

There are three generic types of quantizer, or A/D converter. These are:

1. *Counting quantizers*, which sequentially count through each quantizing level.

2. *Serial quantizers*, which generate a code word, bit by bit. They start with the most significant bit and work their way to the least significant bit.

3. *Parallel quantizers*, which generate all bits of a complete code word simultaneously.

Counting Quantizers

This class of A/D converter converts the sample amplitude to a proportional time (i.e., the larger the sample, the longer the period of time). It then converts the time to an equivalent binary number. Figure 1.27 illustrates one type of counting quantizer.

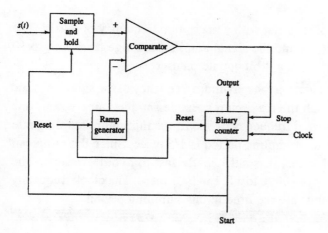

Figure 1.27 - Counting Quantizer

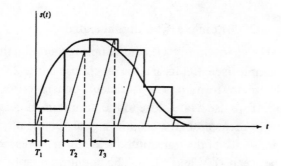

The ramp generator starts at each sampling point, and a binary counter is simultaneously started. The output of the *sample and hold* system is a staircase approximation to the original function, with stair steps that stay at the previous sample value throughout each sampling interval. A representative waveform is shown in the figure. We now start a ramp at the sampling instant and stop it when it reaches the staircase. A counter also starts at the sampling instant and continues during the entire time the ramp is increasing. The time duration of the ramp, and therefore the duration of the count, is proportional to the sample value. This is true because the ramp slope is kept constant. The clock frequency is such that the counter has enough time to count to its highest count (all ones) for a ramp duration corresponding to the maximum possible sample. The ending count on the counter corresponds to the quantizing level. Note that for this quantizer to operate properly, none of the sample values can be negative. If the original signal contains negative samples, we can simply shift the waveform up by adding a constant (e.g., *clamp* the waveform to zero).

Example 1.5

Design a counting quantizer for a voice signal with a maximum frequency of 3 kHz. The ramp slope is specified as 10^6 volts/sec, and the signal amplitudes range from zero to 10 volts. Find the required clock frequency if a 4-bit counter is used.

Solution: The only reason for considering the maximum frequency of the signal is to see if the ramp slope is sufficient to reach the maximum possible sample value within one sampling period. With a maximum signal frequency of 3 kHz, the minimum sampling rate for recovery is 6 kHz, so the maximum sampling period is 1/6 msec. Since the ramp can reach the 10-volt maximum in 0.01 msec, it is sufficiently fast to avoid overload. The counter must be capable of counting from 0000 to 1111 in 0.01 msec. The clock frequency must be 1.6 MHz, since up to 16 counts are required in this sampling period.

Example 1.6

Design a counting ADC to convert $s(t) = \sin 2000\pi t$ into a 4-bit digital signal.

Solution: The frequency of $s(t)$ is 1000 Hz, so to comply with the sampling theorem, the sampling rate must be greater than 2000 samples/sec. Let us arbitrarily choose a rate 25% above this value, or 2500 samples/sec. There are many practical tradeoff considerations which go into this choice of sampling rate. If we use a rate close to the minimum, precise lowpass filtering is required at the receiver to reconstruct the original signal. On the other hand, if we use a much higher rate, the bandwidth of the transmitted waveform increases.

The individual sample values range between -1 V and +1 V. The counting quantizer discussed in this section operates on positive samples. We therefore shift the signal by 1 volt to assure that samples never go negative. The shifted samples range from 0 volt to 2 volt. The ramp must be capable of reaching the maximum sample value within the sampling period, 0.4 msec. The slope must therefore be at least $2/0.4 \times 10^{-3} = 5000$ volts/sec. In practice, we would choose a value larger than this to account for slight jitter in the timing of the system and also to give the ramp time to return to zero prior to the next sampling point. We might choose a much larger slope value if we wanted to convert the sample in a small fraction of the period. This would apply if a converter is being shared among a number of signals. At the minimum slope, it takes the ramp function 0.4 msec to reach the maximum sample value. The counter should therefore count from 0000 to 1111 in 0.4 msec. This requires a counting rate of 40,000 counts/sec.

Serial Quantizers

The serial quantizer (also known as the *successive approximation quantizer*) successively divides the ordinate into two regions. It first divides the axis in half and observes whether the sample is in the upper or lower half. The result of this observation generates the most significant bit in the code word.

The half-region in which the sample lies is then subdivided into two regions, and a comparison is again performed. This generates the next bit. The process continues a number of times equal to the number of bits of A/D. Thus each bit increases the resolution by a factor of two.

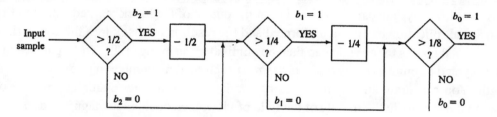

Figure 1.28 - The Serial Quantizer

Figure 1.28 shows a flow diagram of the serial quantizer for three bits of encoding and for inputs in the range of zero to one. The diamond shaped boxes are *comparators*. They compare the input to some fixed value and give one output if the input exceeds the fixed value and another output if the reverse is true. The flow diagram indicates these two possibilities as two possible output paths labeled YES and NO. Note that b_2 is the first bit of the coded sample value, known as the *most significant bit* or *msb*. The third and final bit of the coded sample, b_0, is known as the *least significant bit*, or *lsb*. The reason for this terminology is that the weight associated with b_2 is 2^2, or four, while the weight associated with b_0 is 2^0, or one. If more (or fewer) bits are required, the appropriate comparison blocks can be added (or removed) from Fig. 1.28.

If the input range of signal sample values were not zero to one, the signal could be normalized (shifted and then amplified or attenuated) to achieve values within this range. You should convince yourself that shifting and scaling do not change the digital version of the signal. If the signal ranges from V_{min} to V_{max}, normalization is accomplished as follows:

$$s_{norm}(t) = \frac{s(t) - V_{min}}{V_{max} - V_{min}}$$

Note that when $s(t)=V_{max}$, $s_{norm}(t)=1$ and when $s(t)=V_{min}$, $s_{norm}(t)=0$.

Example 1.7

Illustrate the operation of the system of Fig. 1.28 for the following two input sample values: 0.2 volt and 0.8 volt.

Solution: For an input of 0.2 volt, the first comparison with ½ yields a NO answer. Therefore, $b_2=0$. The second comparison with 1/4 yields a NO answer, so $b_1=0$. The third comparison with 1/8 yields a YES answer, so $b_0=1$. The binary code for 0.2 volt is therefore 001.

For the 0.8 volt input, the first comparison with ½ yields a YES answer, so $b_2=1$. We then subtract ½ leaving 0.3. The second comparison with 1/4 results in a YES answer, so $b_1=1$ and we subtract 1/4, leaving 0.05. The third comparison with 1/8 yields NO, so $b_0=0$. The code for 0.8 volt is 110.

A simplified system can be realized if, at the output of the block marked "-1/2" in Fig. 1.28, a multiplication by two is performed and the result is fed back into the comparison with 1/2. All blocks to the right can then be eliminated, as in Fig. 1.29. The signal sample can be cycled through as many times as desired to achieve any number of bits of code word length. You can draw an analogy to using a microscope to view a small sample. You position the sample in the middle of the field of view, then double the magnification. This process is repeated as many times as desired.

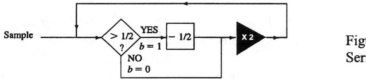

Figure 1.29 - Simplified Serial Quantizer

Successive approximation A/D converters are used in medium- to high-speed applications. They require very little circuitry. A typical 8-bit converter operates in the range of about 3000 to 100,000 samples per second.

Parallel Quantizers

The parallel quantizer (or *flash coder*) is the fastest in operation since it develops all bits of the code word simultaneously. This type of A/D converter typically operates at a rate of tens of millions of samples per second. It is also the most complex, requiring a number of

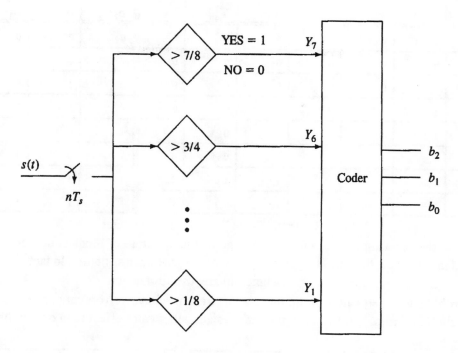

Figure 1.30 - The Parallel Quantizer

comparators that is only one less than the number of levels of quantization.[2] We illustrate this with the 3-bit encoder of Fig. 1.30. The block labeled "coder" observes the output of the seven comparators. It is a simple combinational logic circuit. If all seven outputs are one (YES), the coder output is 111 since the sample value must be greater than 7/8. If comparator outputs one through six are one and output seven is zero, the coder output is 110 since the sample must be between 6/8 and 7/8. We continue through all levels and finally, if all comparator outputs are low, the sample is less than 1/8, so the coder output is 000. The truth table for the coder is shown in Table 1.2. Only eight out of the 128 (2^8) possible coder inputs are legal, and the other 120 inputs represent illegal possibilities (e.g., there is no way a sample could be greater than 7/8 and less than 5/8). Thus, the combinational logic circuit contains a high percentage of *don't care* conditions, and the design is simplified accordingly. Alternatively, the logic can be configured to give an error signal for any illegal input combination.

---

[2]Some practical implementations operate with a two-step process to reduce the amount of hardware. For more information, refer to manufacturer data sheets such as the TLC5510 from Texas Instruments.

Y_1	Y_2	Y_3	Y_4	Y_5	Y_6	Y_7	b_2	b_1	b_0
0	0	0	0	0	0	0	0	0	0
1	0	0	0	0	0	0	0	0	1
1	1	0	0	0	0	0	0	1	0
1	1	1	0	0	0	0	0	1	1
1	1	1	1	0	0	0	1	0	0
1	1	1	1	1	0	0	1	0	1
1	1	1	1	1	1	0	1	1	0
1	1	1	1	1	1	1	1	1	1

Table 1.2 - Truth Table for Parallel Quantizer

While the serial quantizer takes advantage of the structure of binary numbers when counted in sequence, the parallel quantizer does not require such structure. In fact, the code for the quantization regions can be assigned in any useful manner.

A problem with sequential assignment is that transmission bit errors cause non-uniform reconstruction errors. A bit error in the *msb* has a much greater effect than one in the *lsb*.

Digit	Binary	Gray	Digit	Binary	Gray
0	0000	0000	8	1000	1100
1	0001	0001	9	1001	1101
2	0010	0011	10	1010	1111
3	0011	0010	11	1011	1110
4	0100	0110	12	1100	1010
5	0101	0111	13	1101	1011
6	0110	0101	14	1110	1001
7	0111	0100	15	1111	1000

Table 1.3 - The Gray Code

In 1947, F. Gray, who was working with electronic coding devices, invented a *reflected binary code* (also known as a *Gray code*) in which adjacent numbers differ in only one bit position. We illustrate a 4-bit version in Table 1.3.

An error in one bit in the code word changes the corresponding digit by only one. This code is easily implemented in the flash coder logic by modifying the truth table of Table 1.2. It can also be used in the other types of quantizers. In the counting quantizer, we simply vary the count sequence. In the serial quantizer, we follow the decision operations with a simple combinational logic circuit to convert the sequential code to the Gray code.

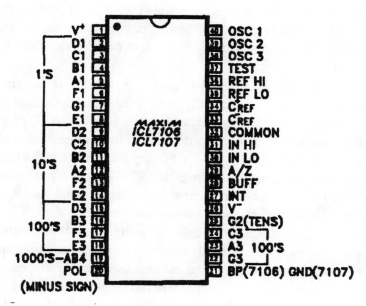

Figure 1.31 - ICL7106 A/D Converter
Courtesy of Maxim Integrated Products

Practical PCM Encoders

Practical PCM encoders are constructed to follow the block diagrams we just studied. Most forms are packaged as single integrated circuits. We discuss one practical quantizer for each class of conversion.

Counting quantizers are often implemented as *dual slope* A/D converters. The input sample is first applied to an integrator for a fixed length of time, thus yielding an integrated output which is proportional to that sample value. The input to the integrator is then switched to a reference voltage (which is opposite in sign from the signal sample), the counter is started, and the output of the integrator is compared to zero. The counter is stopped when the integrator output ramp reaches zero.

The ICL7106 A/D Converter simulates a *counting quantizer*. The IC package has 40 pins, and is illustrated in Fig. 1.31. Pins two through 25 are used for the output display. The IC is configured to directly drive LCDs as it includes seven-segment decoders and LCD drivers. The display is 3-1/2 digit, which means it can indicate numbers with magnitudes as high as 1999. The seven segment outputs for the *units* display are indicated as A1 to G1. Those for the *tens* display use a suffix of two, and the *hundreds* display uses three. The *thousands* display is indicated as AB4. Only one lead is needed for this display since this digit is either zero or one (for a 3 ½ digit display).

The analog input is applied to pins 30 and 31. The actual operation of the IC proceeds in three phases. The first is *auto-zero*, where the analog inputs are disconnected and internally shorted to a common, pin 32. The output of the comparator is shorted to the inverting input of the integrator.

The second phase occurs when the input signal value is integrated for a time corresponding to 1000 clock pulses. Finally, in the third phase, the reference voltage stored on a capacitor (which is externally connected between pins 34 and 35) is used to create the second ramp. The range of input values determines the required value of the reference, which is input to the REF HI pin, #36. If this input is one volt, the chip is capable of converting voltages with magnitudes as high as 1.999. The clock can be derived from pins 38, 39 and 40. You can use either an external oscillator, a crystal connected between pins 39 and 40, or an RC circuit configured across these pins.

A complete A/D conversion of a single sample requires 4000 counts. The signal is integrated for 1/4 of this period, or 1000 counts. The second integration and the auto-zero require the remaining 3000 counts. The internal clock is developed by dividing the oscillator input by four. Thus, for example, if you wish to perform 10 conversions per second, the oscillator input must be 160 kHz. This device is not capable of high speed conversion, and should be used only for slowly varying signals (low sampling rates), or *dc* inputs. Further details can be found at www.maxim-ic.com.

The *ADC 0804* from *National Semiconductor* is an example of an integrated circuit that performs *serial A/D conversion*. A portion of the data sheet is illustrated in Fig. 1.32. This is an 8-bit device. Its internal construction consists of a number of flip flops, shift registers, a decoder and a comparator. The full conversion takes eight internal clock pulses. The internal clock is provided by dividing the clock signal at pins 4 and 19 by eight. Thus, for example, with a 64 kHz signal on these pins, the IC can perform one conversion in 1 msec. The ADC 0804 is capable of converting a sample in about 100 microseconds, so we can convert about 10,000 samples per second.

The digital outputs, labeled DB_0 through DB_7, appear on pins 11 through 18. Other pin assignments are as follows:

Pin	Label	Function
1	CS (Chip Select)	Set LOW to initialize, HIGH to start conversion
2	RD (read)	Goes LOW to indicate microprocessor is ready to receive data
3	WR (write)	LOW to initialize, HIGH to start conversion
4	CLK	Input external oscillator or connect resistor between 4 and 19 to set oscillation frequency
5	INTR (interrupt)	Goes LOW to tell microprocessor that data is available

| 6 | V_{IN} | Part of differential input (with pin 7 for negative input) |
| 9 | $V_{REF}/2$ | Reference voltage (one half of full-scale voltage) |

ADC0801/ADC0802/ADC0803/ADC0804/ADC0805
8-Bit µP Compatible A/D Converters

General Description

The ADC0801, ADC0802, ADC0803, ADC0804 and ADC0805 are CMOS 8-bit successive approximation A/D converters that use a differential potentiometric ladder—similar to the 256R products. These converters are designed to allow operation with the NSC800 and INS8080A derivative control bus with TRI-STATE • output latches directly driving the data bus. These A/Ds appear like memory locations or I/O ports to the microprocessor and no interfacing logic is needed.

Differential analog voltage inputs allow increasing the common-mode rejection and offsetting the analog zero input voltage value. In addition, the voltage reference input can be adjusted to allow encoding any smaller analog voltage span to the full 8 bits of resolution.

Features

- Compatible with 8080 µP derivatives—no interfacing logic needed - access time - 135 ns
- Easy interface to all microprocessors, or operates "stand alone"

- Differential analog voltage inputs
- Logic inputs and outputs meet both MOS and TTL voltage level specifications
- Works with 2.5V (LM336) voltage reference
- On-chip clock generator
- 0V to 5V analog input voltage range with single 5V supply
- No zero adjust required
- 0.3" standard width 20-pin DIP package
- 20-pin molded chip carrier or small outline package
- Operates ratiometrically or with 5 V $_{DC}$, 2.5 V$_{DC}$, or analog span adjusted voltage reference

Key Specifications

Resolution	8 bits
Total error	±¼ LSB, ±½ LSB and ±1 LSB
Conversion time	100 µs

Connection Diagram

ADC080X
Dual-In-Line and Small Outline (SO) Packages

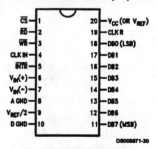

See Ordering Information

Figure 1.32 - ADC0804 A/D Converter
Courtesy of National Semiconductor Corporation

Further details and application notes can be found on the National Semiconductor web pages at www.national.com.

The *TLC5510* from *Texas Instruments* is an integrated circuit which accomplishes *parallel (or flash)* analog to digital conversion. The layout of this 24 pin IC is shown in Fig. 1.33.The circuit is capable of converting a sample in 50 nsec, yielding a maximum conversion rate of about 20 million samples per second. It contains a bank of comparators. The analog input is on pins 19 and 21, and the digital output is read from pins three to 10. The chip uses a two-step process that reduces power consumption and hardware requirements. Details can be obtained from the Texas Instruments web pages at www.ti.com.

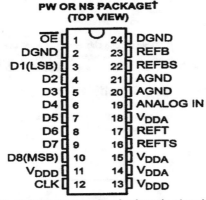

Figure 1.33 -TLC5510 Flash A/D
*Courtesy of Texas Instruments
Incorporated*

Digital to Analog Conversion

We now shift our attention to the conversion of a digital to an analog signal. This is performed by a *digital to analog converter* (D/A converter, or DAC). To perform this conversion, we need simply associate a level with each binary code word. Since each code word represents a range of sample values, the actual value chosen for the conversion is usually the center point of the region. If the A/D conversion regions are numbered sequentially (i.e., *not* a Gray code) the D/A conversion is equivalent to assigning a weight to each bit position.

Let us illustrate the procedure for a 4-bit binary word. We assume that the analog sample is normalized (i.e., it falls in the range between zero and one volt) and that sequential coding (as opposed to Gray coding) is used. Conversion to the analog sample value is accomplished by converting the binary number to decimal, dividing by 16, and adding 1/32. Thus, for example, the code 1101 represents the decimal number 13, so we convert this to $13/16 + 1/32 = 27/32$. The addition of the 1/32 takes us from the bottom of the 1/16 wide region to the middle.

Figure 1.34 illustrates the conversion. If a one appears in the msb position, a 1/2 volt battery is switched into the circuit. The second bit controls a 1/4 volt battery, and so on.

The ideal decoder of Fig. 1.34 is analogous to the serial quantizer since each bit is associated with a particular component of the sample value.

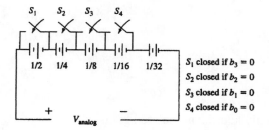

Figure 1.34 - D/A Converter

A more complex decoder results when an analogy to the counting operation is attempted. Figure 1.35 shows the counter decoder. A clock feeds a staircase generator and a binary counter simultaneously. The output of the binary counter is compared to the binary digitized input. When a match occurs, the staircase generator is stopped. The output of the generator is sampled and held until the next sample value is achieved. The final staircase approximation result is smoothed by a lowpass filter to recover an approximation to the original signal.

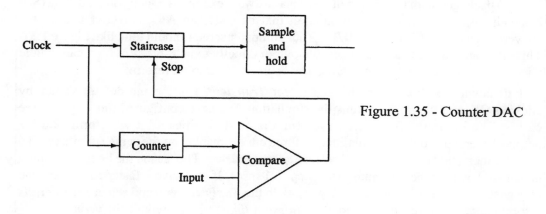

Figure 1.35 - Counter DAC

Practical Digital/Analog Converters

Figure 1.36 shows a practical version of the ideal decoder of Fig. 1.34. The gain of the operational amplifier shown in the figure is $R/5R_{in}$, where R_{in} is the parallel combination of the resistors included in the input circuit (i.e., those with associated switches closed). In terms of input conductance, the gain is $RG_{in}/5$, where G_{in} is the sum of the conductances associated with closed switches. Therefore, closing S_1 contributes 8/5 to the gain. Closing S_2 contributes 4/56 to the gain; S_3 contributes 2/5; S_4, 1/5. In order to calculate the gain due

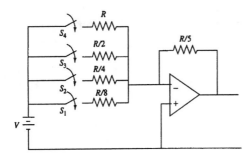

Figure 1.36 - Practical D/A Converter

to more than one switch being closed, we simply add the associated gains.

As an example, if the *dc* voltage source, *V*, is equal to five volts, we see that S_1 contributes eight volts to the output; S_2 contributes four volts, S_3 contributes two volts and s_4 contributes one volt. The output voltage is an integer voltage between zero and 15 corresponding to the decimal equivalent of the binary number controlling the switches. To convert this system into a DAC analogous to that of Fig. 1.34, we need simply divide the output by 16 and add 1/32 volt. The division can be performed by scaling *V* or by scaling the resistor values. The addition of 1/32 volt can be accomplished by adding a fifth resistor (unswitched) to the input. The value of this resistor would be 2*R*.

Computer Simulation Example

The CD packaged with this text includes Tina software examples. Once you load Tina (See Appendix F), you can run the simulation that cascades an A/D converter and a D/A converter. Open the file called ***ADDA***. Your computer screen should look like Figure 1.37. This takes an 8-bit A/D and connects the output directly to an 8-bit D/A. You can double click on each item to get more details. The source is a ramp waveform.

Pull down the ***Analysis*** menu and select ***Transient***. Choose the default values by clicking ***OK***, and the program runs the simulation. We have configured this to plot three curves: The least significant bit waveform, the most significant bit waveform, and the output. When you run the simulation, all three curves display on the same set of axes. To separate them, pull down ***View*** and select ***Separate curves***. The reason the lines are sloping is that we have set the minimum time step to 1 msec. If you have a faster computer, you may wish to reduce this number. Just go back to the ***Analysis*** menu and select ***Set Analysis Parameters...*** Then scroll down to ***TR minimum time step*** and reduce the value.

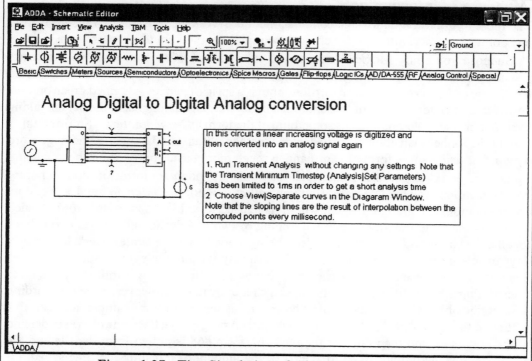

Figure 1.37 - Tina Simulation of A/D and D/A converter

Now let's have some fun. We will eliminate the least significant bit so we only have seven bits of quantization. Just click on the top wire (the one connecting the pins labeled "0"). Then delete that wire. If you rerun the simulation, you will not see any difference. Note that we have reduced the number of round off levels from 256 to 128. Your eye cannot detect the quantization noise. Keep eliminating wires until you start seeing a staircase approximation to the ramp. (If you have not reduced the time step below 1 msec, you will have to eliminate about half of the wires before you see a change). Once you have illustrated the quantization noise, close the circuit without saving the changes, and then reopen it. Now we will eliminate the most significant bit wire before running the simulation. Can you explain why you now get two ramps?

1.7. Shannon's Channel Capacity Theorem

As communication system designers, we expend considerable effort in structuring signals and in designing systems that efficiency communicate these structured signals. We need a way of evaluating the results of our efforts. Could we have done much better, or is our design pushing the theoretical limits of achievement? The Shannon channel capacity theorem provides the theoretical limits against which our designs can be judged. Before we can present the theorem, we need to establish basic concepts normally considered part of the study of *information theory*.

Measure of information

The more *information* a message contains, the more work we must do to transmit it from one point to another. Some non-technical examples should help to illustrate this.

Suppose you enjoy Mexican cuisine (as your author does). You enter your favorite restaurant to order the following: One cheese enchilada, one beef taco, one order of refried beans, one order of rice, two corn tortillas, butter, a sprig of parsley, chopped green onions, and a glass of water. Instead of this entire list being written out longhand, the person taking the order probably communicates something of the form "*#2 combination*". Stop and think about this. Rather than attempting to decide whether to transmit the specific message word by word or using some form of shorthand, a much more basic approach has been taken.

Let us look at another historical example. A popular form of contemporary communication was the telegram. Had you called   *Western Union* to send a wedding congratulations telegram, the agent would have tried to sell you a standard wording such as "Heartiest congratulations upon the joining of two wonderful people. May your lives together be happy and productive. A gift is being mailed under separate cover". In the early days of telegraphy, the company noticed that many wedding telegrams sounded almost identical to that presented above (i.e., wording was predictable). A primitive approach to transmitting this would be to send the message letter by letter. However, since the wording is so predictable, why not have a short code for this message? For example, let's call this particular telegram, "#68". The operator would then simply type the name of the sender and of the addressee and then #68. At the receiver, form #68 is pulled from the file and the address and signature are filled in. Think of the amount of transmission time saved! Of course, if a simple error were made and telegram #69 were sent by mistake, it might convey a totally different message. Alternatively, a single error in a letter-by-letter transmission would probably not be serious.

As yet a third example, consider the (atypical, I hope) university lecture in which a professor stands before the class and proceeds to read directly from the textbook for two hours. A great deal of time and effort could be saved if the professor instead read the message, "pages 103 to 120". In fact, if the class expected this and knew they were past page 100, the professor could shorten the message to "0320." The class could then read these 18 pages on their own and get the same information. The professor could dismiss the class after 20 seconds and return to more scholarly endeavors.

By now the point should be clear. A basic examination of the *information content* of a message has the potential of saving considerable effort in transmitting that message from one point to another.

Information and Entropy

The modest examples given above indicate that some relatively long messages do not contain a great deal of information. Every detail of the restaurant order or the telegram is so predictable that the person at the receiving end can almost guess what comes next.

The concept of information content of a particular message must now be formalized. Once we do this, we shall find that the less information in a particular message, the quicker we can communicate the message.

The concept of information content is related to *predictability*. That is, the more *probable* a particular message, the less information is given by transmitting that message.

If today were Tuesday and you phoned someone to tell him or her that tomorrow would be Wednesday, you would certainly not be communicating any information. This is true since the probability of that particular message is "1" (See Appendix C for a discussion of basic probability concepts).

The definition of information content of a message should therefore be such that it *monotonically decreases* with increasing probability and goes to zero for a probability of unity. Another property we would like this measure of information to have is that of *additivity*. If one were to communicate two (independent) messages in sequence, the total information content should be equal to the sum of the individual information contents of the two messages. Now if the two messages are independent, we know that the total probability of the composite message is the product of the two individual probabilities. Therefore, the definition of information must be such that when probabilities are multiplied together, the corresponding information contents are added.

The logarithm satisfies these requirements. We thus *define* the information content of a message, x, as I_x, in Eq. (1.26).

$$I_x = \log\frac{1}{P_x} \tag{1.26}$$

P_x is the probability of occurrence of the message x. The reason it's in the denominator is that we wish I_x to decrease as P_x increases. This definition satisfies the additivity requirement and the monotonicity requirement. For the *certain event*, $P_x = 1$ and $I_x = 0$. Note that this is true regardless of the base chosen for the logarithm.

We usually use base two for the logarithm. To understand the reason for this choice, let us return to the restaurant example. Suppose that there were only two selections on the menu. Further assume that past observation has shown that these two are equally probable (i.e., each is ordered half of the time). The probability of each message is therefore 1/2, and if base two logarithms are used in Eq. (1.26), the information content of each message is

$$I_x = \log\frac{1}{P_x} = \log 2 = 1$$

Thus, one unit of information is transmitted each time an order is placed.

Now let us decide on an efficient way to transmit this order digitally (assume binary). Since there are only two possibilities, one binary digit could be used to send the order to the kitchen. A zero could represent the first dinner, and a one, the second dinner on the menu.

Suppose that we now increase the number of items on the menu to four, with each having a probability of 1/4. The information content of each message is now $\log_2 4$, or two units. If binary digits are used to transmit the order, two bits would be required for each message. The various dinners could be coded as 00, 01, 10, and 11. We therefore conclude that if the various messages are equally probable, the information content of each message is exactly equal to the minimum number of bits required to send the message (provided this is an integer). This is the reason for using base two logarithms, and the unit of information is called the *bit of information*. Thus, one would say that in the four-menu selection example, each menu order contains two bits of information.

When all messages are equally likely, the information content of any single message is the same as that of any other message. In cases where the probabilities are not equal, the information content depends on which particular message is being transmitted. *Entropy* is defined as the <u>*average*</u> *information per message*. To calculate the entropy, we take the various information contents associated with the messages and weight each by the fraction of time we can expect that particular message to occur. This fraction is the probability of the message. Thus, given n messages, x_1 through x_n, the entropy is defined by Eq. (1.27).

$$H = \sum_{i=1}^{n} P_{x_i} I_{x_i} = \sum_{i=1}^{n} P_{x_i} \log\left(\frac{1}{P_{x_i}}\right) \tag{1.27}$$

The letter H is used to denote entropy.

Example 1.8

A communication system consists of six messages with probabilities 1/4, 1/4, 1/8, 1/8, 1/8 and 1/8, respectively. Find the entropy.

Solution: The information content of the six messages is 2 bits, 2 bits, 3 bits, 3 bits, 3 bits and 3 bits, respectively. The entropy is therefore given by

$$H = \frac{1}{4}\times2+\frac{1}{4}\times2+\frac{1}{8}\times3+\frac{1}{8}\times3+\frac{1}{8}\times3+\frac{1}{8}\times3$$

$$= 2.5 \; bits/message$$

Let's take a moment to examine Eq. (1.27) for the binary message case. That is, we consider a communication scheme made up of two possible messages, x_1 and x_2. For this case,

$$P_{x_2} = 1 - P_{x_1} \tag{1.28}$$

and the entropy is given by

$$H_{binary} = P_{x_1}\log\left(\frac{1}{P_{x_1}}\right) + (1-P_{x_1})\log\left(\frac{1}{1-P_{x_1}}\right) \tag{1.29}$$

This result is sketched as a function of P_{x1} in Fig. 1.37. (You should take the time to produce your own plot using MATLAB). Note that as either of the two messages becomes more likely, the entropy decreases. When either message has probability one, the entropy goes to zero. This is reasonable since, at these points, the outcome of the transmission is *certain*–we don't have to go through the trouble of sending any message. Thus, if $P_{x1} = 1$, we know message x_1 will be sent all the time (with probability one). Alternatively, if $P_{x1} = 0$, we know that x_2 will be sent all the time. In these two cases, no information is transmitted by sending the message.

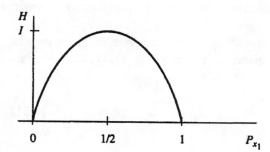

Figure 1.37 - Entropy for
Two Messages

The entropy function is symmetrical with a maximum for the case of equally likely messages. This is an important property that is used in later coding studies. If we were able to show a three-dimensional plot for the three-message case, we would find a similar result with a peak in entropy for message probabilities of 1/3.

Channel Capacity

We now investigate the rate at which information can be sent through a channel and the relationship of this rate to errors. Before we can do this, we must define the rate of information flow.

Assume that a source can send any one of a number of messages at a rate of r messages per second. For example, if the source were a terminal, the rate would be the number of symbols per second. Once we know the probability of each individual message, we can compute the entropy, H, in bits per message.

If we now take the product of the entropy, H, with the message rate, r, we get the information rate in bits per second, denoted R. Thus,

$$R = rH \ bps \qquad (1.30)$$

For example, in Example 1.8 we found the entropy to be 2.5 bits/msg. If messages were sent at two messages per second, the information rate would be 5 bps.

For a given communication system, as the information rate increases, the number of errors per second will also increase. Hopefully this agrees with your intuition. It certainly applies when you have a conversation with somebody.

C. E. Shannon[3] proved that a given communication channel has a maximum rate of information, C, known as the *channel capacity*. If the information rate, R, is less than C, one can approach arbitrarily small error probabilities by using channel encoding techniques. This is true even in the presence of noise, a fact that probably contradicts your intuition.

The negation of Shannon's theorem is also true. That is, if the information rate, R, is greater than the channel capacity, C, errors cannot be avoided regardless of the coding technique employed.

We shall not prove this fundamental theorem here. We discuss its application to several commonly encountered cases.

We consider the bandlimited channel operating in the presence of additive white Gaussian noise. In this case, the channel capacity is given by Eq. (1.31) (the *Shannon-Hartley theorem*)

$$C = B \log_2 \left(1 + \frac{S}{N} \right) \tag{1.31}$$

C is the capacity in bits per second, B is the bandwidth of the channel in Hz, and S/N is the signal to noise ratio. Equation (1.31) makes intuitive sense. As the bandwidth of the channel increases, it should be possible to make faster changes in the information signal, thereby increasing the information rate. As S/N increases, one would expect to be able to increase the information rate while still preventing errors due to the noise. Note that with no noise at all (i.e., $N=0$), the signal to noise ratio is infinity and an infinite information rate is possible regardless of the bandwidth.

Equation (1.31) might lead one to conclude that if the bandwidth approaches infinity, the capacity also approaches infinity. This last observation is not correct. Since the noise is assumed to be white, the wider the bandwidth, the more noise is admitted to the system (see Appendix C). Thus, as B increases, S/N decreases.

Suppose that the total signal and noise power per Hz were fixed and we were trying to design the best possible channel. There is an expense associated with increasing the system bandwidth, so we should attempt to find the maximum channel capacity. Suppose that the noise is white with power spectral density, $N_o/2$. Further assume that the signal power is fixed at a value S. The channel capacity is given from Eq. (1.31) as

$$C = B \log_2 \left(1 + \frac{S}{N_o B} \right) \tag{1.32}$$

Note that B is in Hz and that N_o is the noise power in watts per Hz. We now find the value that the channel capacity approaches as B goes to infinity.

---

[3]Shannon, C.E.,"A Mathematical Theory of Communication", <u>Bell System Technical Journal</u>, Vol 27, 1948.

$$C = \lim_{B \to \infty} B \log_2\left(1 + \frac{S}{N_o B}\right)$$

$$= \lim_{B \to \infty} \frac{S}{N_o} \log_2\left(1 + \frac{S}{N_o}B\right)^{N_o B/S} \qquad (1.33)$$

$$= \frac{S}{N_o} \log_2 e = 1.44\frac{S}{N_o}$$

Equation (1.33) shows the maximum possible channel capacity as a function of signal power and noise spectral density. In an actual system design, the channel capacity is compared to this figure and a decision is made whether further increase in bandwidth is worth the expense.

Suppose now that you were asked to design a binary communication system and you cranked the bandwidth of your communication channel, the maximum signal power, and the noise spectral density into Eq. (1.32). You come up with a maximum information rate, C. You plug in the sampling rate and the number of bits transmitted for each sample, and you arrive at a certain information transmission rate, R bps. Aha! You observe that R is less than C, so Shannon tells you that if you do enough work encoding this binary data train, you can achieve arbitrarily low probability of error. But you have already coded the original signal into a train of binary digits. What do you do next?

Although we will not give a detailed analysis of error control, we will hint at the approach. As applied to the example above, most procedures would entail grouping combinations of bits and calling the result a new word. For example, we can group by twos to yield four possible messages, 00, 01, 10 and 11. Each of these four possible messages can be transmitted using some code word. By so doing, the probability of error can be reduced.

As an example, if you wished to transmit this textbook to a class of students, you could read each letter aloud as, for example,

r-e-a-d-e-a-c-h-l-e-t-t-e-r-a-l-o-u-d

Alternatively, you could group letters together and send the groups as a code word from an acceptable dictionary. For the example above, you send

read-each-letter-aloud

Suppose that in reading the individual letters, you slur the "d" sound and one student receives it as "v". That student receives the erroneous message,

r-e-a-v-e-a-c-h-l-e-t-t-e-r-a-l-o-u-d

In the individual letter transmission case, the student would not know that an error occurred. However, in the word coding, the student would receive

reav-each-letter-aloud

and since "reav" is not an acceptable code word, the student could correct the error to interpret the received message as

read-each-letter-aloud

and thus make no error. The essence of coding to achieve arbitrarily small errors requires grouping into longer and longer code words. The longer the code words, the more different the dictionary entries will be from each other, and the less likely that individual errors cannot be corrected.

The idea of increasing the difference between dictionary entries will become a central consideration in designing digital communication systems. We will use this concept over the over again in each of the remaining chapter of this text.

Problems

1.1 Show that the function of time corresponding to the transform of Eq. (1.12) is

$$s(t) = Ar[t - t_{gr}(f_o)]\cos2\pi f_o[t - t_{ph}(f_o)]$$

Hint: You may find it useful to prove the following relationships first:

$$r(t-t+_o)\cos2\pi f_o t \longleftrightarrow -\frac{1}{2}[R(t-t_o)e^{-j2\pi f-f_o}t_o + R(f+f_o)e^{-j2\pi(f+f_o)t_o}$$

$$r(t-t_1)\cos2\pi f_o(t-t_1) \longleftrightarrow -\frac{1}{2}[R(f-f_o) + R(f+f_o)]e^{-j2\pi ft_1}$$

$$r(t-t_o)\cos2\pi f_o(t-t_1) \longleftrightarrow -\frac{1}{2}[f(t-t_o)e^{-j2\pi(f+f_o)(t_o-t_1)} + R(f+f_o)e^{-j2\pi(f+f_o)(t_o-t_1)}]e^{-j2\pi ft_1}$$

1.2 Find the output of the filter of Fig. 1.2 when the input is

$$r(t) = \frac{\sin200\pi t}{t} + \frac{5\sin600\pi t}{t}$$

1.3 Find the output of a typical telephone line when the input is
(a) $\cos2\pi\times500t + \cos2\pi\times1000t$
(b) A periodic triangle of frequency 1 kHz.

1.4 You are given the function

$$s(t) = \frac{\sin t}{t}$$

This function is sampled by a train of impulses, $s_\delta(t)$ as shown in Fig. P1.4.
(a) What is the Fourier transform of, $S_s(f)$, the sampled function?
(b) Find $s_s(t)$, the inverse Fourier transform of your answer to part (a).
(c) Should your answer to part (b) have been a train of impulses? Did it turn out that way? Explain any discrepancies.
(d) Design a system to recover the original $s(t)$ from $s_\delta(t)$, and demonstrate that your system works correctly for this example.

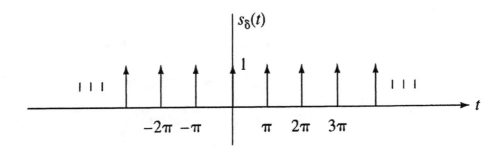

Figure P1.4

1.5 A signal $s(t)$ with $S(f)$ as shown in Fig. P1.5 is sampled by two different sampling functions, $s_{\delta 1}(t)$ and $s_{\delta 2}(t)$, where

$$s_{\delta 2}(t) = s_{\delta 1}\left(t - \frac{T}{2} \right)$$

$$T = \frac{1}{f_m}$$

Therefore, each of the two sampling waveforms is at one-half of the Nyquist minimum sampling frequency. Find the Fourier transforms of the sampled waveforms, $s_{s1}(t)$ and $s_{s2}(t)$.

Now consider $s(t)$ to be sampled by $s_{s3}(t)$, a train of impulses spaced $T/2$ apart (i.e., at the Nyquist rate). This new sampling function is the sum, $s_{s1}(t) + s_{s2}(t)$. Show that the Fourier

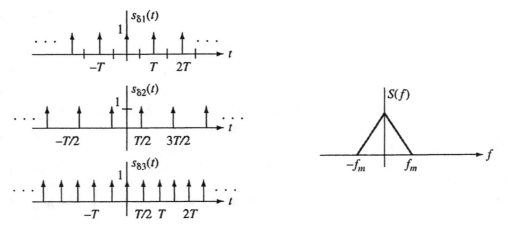

Figure P1.5

transform of $s_{s3}(t)$ is equal to the sum of the transforms of $s_{s1}(t)$ and $s_{s2}(t)$. That is, show that, although the transforms of the original two sampled functions contain aliasing error, this error is not present in the sum. [*Hint:* You will have to keep track of phases).

1.6 The function $s(t) = \cos 2\pi t$ is sampled every 3/4 second. Evaluate the aliasing error.

1.7 The signal

$$s(t) = \frac{\sin 2\pi t}{\pi t}$$

is sampled at 1.1 times the Nyquist rate. The sampling is performed for t between -1 and +1 second. The signal is reconstructed using a lowpass filter. Find the truncation error.

1.8 The function $s(t) = \cos 2\pi t$ is sampled at a rate of 2.5 samples/sec for t between 0 and 10 seconds. The signal is reconstructed using a lowpass filter. Find the difference between the original and reconstructed waveforms at the points $t = 4.9$, $t = 5$, and $t = 5.1$ sec.

1.9 You are given a low-frequency bandlimited signal, $s(t)$. This signal is multiplied by the pulse train, $s_c(t)$, as shown in Fig. P1.9 (please see next page). Find the Fourier transform of the product, $s(t)s_c(t)$. What restrictions must be imposed so that $s(t)$ can be uniquely recovered from the product waveform?

1.10 A signal is given by

$$s(t) = \frac{\sin 5\pi t}{\pi t} + \frac{\sin 10\pi t}{\pi t}$$

This signal is to be sampled with a periodic pulse train consisting of narrow pulses.
(a) Find the Nyquist sampling rate.
(b) Assuming that the sampling is done at the Nyquist rate, sketch the Fourier transform of the sampled waveform.

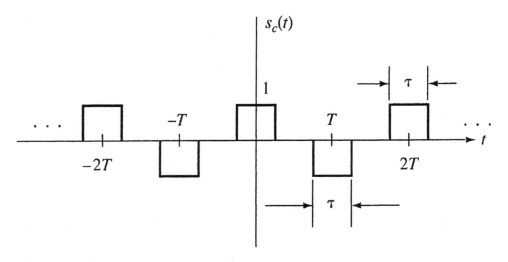

Figure P1.9

1.11 The signal $s(t) = \sin 100\pi t$ is sampled at the points $t=n/100$. This represents sampling at the Nyquist rate of 100 Hz. However, all of the sample values will be zero, and the original wave cannot be reconstructed. Explain the reason for this situation.

1.12 Derive the dual of the time-sampling theorem–that is, a Fourier transform of a time-limited signal $s(t)$ is completely known from its sample values. Find the minimum frequency spacing between samples to permit reconstruction of the Fourier transform.

1.13 Design a four-bit serial quantizer that could be used for a voice signal with a maximum frequency of 5 kHz and sample values between -3 V and +4 V. Illustrate the operation of the quantizer for sample values of —2 V and +3.4 V.

1.14 (a) Design a counting quantizer for five-bit PCM and a voice signal as the input. Assume that the voice signal has a maximum frequency of 5 kHz. Choose all parameter values and justify your choices.
(b) Repeat the design in part (a) for six-bit PCM.

1.15 Design a counting analog-to-digital converter to operate on the signal $s(t)=2\sin\pi t$. The design requires four-bit analog-to-digital conversion.

1.16 Design a flash 4-bit analog-to-digital converter to operate on the signal

$$s(t) = \sin 2\pi t + \frac{\sin 2\pi t}{t}$$

1.17 A communication system consists of three possible messages. The probability of message A is p, and the probability of message B is also p. Plot the entropy as a function of p.

1.18 A communication system consists of four possible messages. $Pr(A) = Pr(B)$ and $Pr(C) = Pr(D)$. Plot the entropy as a function of $Pr(A)$.

1.19 The probability of rain on any particular day in Las Vegas, Nevada, is 0.01. Suppose that a weather forecaster in that city decides to save effort by predicting no rain every day of the year. What is the average information transmitted in each forecast (in bits per day)? Make the (simplistic) assumption that the information content of an incorrect prediction is zero. Suppose the forecaster now decides to predict rain every day of the year. Repeat your calculations for this case.

1.20 A TV picture contains 211,000 picture elements. Suppose that this can be considered a digital signal with eight brightness levels for each element. Assume that each brightness level is equally probable.
(a) What is the information content of a picture?
(b) What is the information transmission rate (bits per second) if 30 separate pictures (frames) are transmitted per second?

1.21 The following list shows the probability of occurrence of each letter of the alphabet in a standard English text. (Note that these would be different in a technical text.)

A 0.081	B 0.016	C 0.032	D 0.037	E 0.124	F 0.023
G 0.016	H 0.051	I 0.072	J 0.001	K 0.005	L 0.040
M 0.022	N 0.072	O 0.079	P 0.023	Q 0.002	R 0.060
S 0.066	T 0.096	U 0.031	V 0.009	W 0.020	X 0.002
Y 0.019	Z 0.001				

(a) Find the entropy associated with sending a single letter.
(b) What is the information content of the three-letter word THE? Assume that the letters are independent of each other (a highly unrealistic assumption).
(c) You are told that the probability of occurrence of the word THE is 0.027. That is, out of every 1,000 words, there will be 27 occurrences of the word THE, on average. Use this result to find the information content of the word THE. What does this imply about the independence assumption of part (b)?

CHAPTER 2

BASEBAND TRANSMISSION

What we will cover, and why you should care:

Let's assume you have taken the time to absorb the most critical material from Chapter one. In particular, it is essential that you be aware of the types of signals which we will study in this text, and the concept of analog to digital conversion.

Chapter two is the first of four chapters dealing with specific methods of transmitting information through a channel. We first consider transmission of this information using relatively low frequencies—this forms the benchmark or starting point for the modifications we examine in the later chapters. Many of the functions discussed in this chapter are part of the *source encoder* block of a communication system (look back to the block diagram of a communication system in Fig. 1.13).

Section 2.1 explores baseband analog transmission. Although voice waveforms are certainly not the only analog waveforms of interest, we use them as a representative example to help solidify the concepts. Section 2.2 explores pulse modulation techniques used for transmitting discrete signals. Within this section are introduced the important topics of multiplexing, intersymbol interference and crosstalk. The third section is fairly comprehensive (i.e., meaty) and explores various formats for digital transmission.

We turn our attention to the receiver in Section 2.4. This is followed with an evaluation of the performance of various baseband communication systems.

The final section shows the application of the theory to several important functions, namely telephone, television and digital recording. We hope you will see how the theory drives practical decisions in system design.

Necessary background:

Throughout this chapter, we assume a working knowledge of Fourier analysis (see Appendix A). Some familiarity with basic linear systems (see Appendix B) is also needed to understand filtering operations. You will need a knowledge of probability theory and Gaussian densities (see Appendix C) to evaluate the performance of the digital receivers presented later in this chapter.

2.1. ANALOG BASEBAND

When we use the term *analog baseband*, we refer to analog signals with Fourier transforms occupying frequencies extending to (or close to) zero (*dc*). The reason for using the term "baseband" is that the frequencies are *not* shifted to some non-zero point on the frequency axis. In later chapters, we explore non-baseband signals where techniques are used to shift frequencies to a range centered around some non-zero frequency.

Since baseband signals occupy relatively low frequencies, they are not suited for transmission through bandpass channels. Baseband signals are typically transmitted through wires or cables.

Telephones form the backbone of traditional communication systems. Therefore transmission of voice signals represents the most prevalent application of baseband analog communications. The human ear is capable of hearing signals with frequencies in the range of about 20 Hz to 20 kHz. In fact, most people cannot hear the upper portion of this range, and a particular person's upper frequency cutoff decreases with age. It is probably no accident of evolution that signals generated by human beings and their vocal chords fall within this audible range. Just think what life would be like if our hearing range were different from our speaking range!

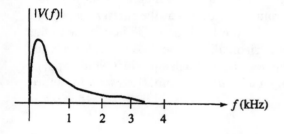

Figure 2.1 - Fourier Transform of
Typical Voice Waveform

The magnitude of the Fourier transform of a typical speech waveform is shown in Fig. 2.1. The location of the peak of this waveform depends on the physiology of the speaker (i.e., vocal cavity resonant frequency). It also depends on what the person is saying, and what language is being used. If the speaker whistles, the audio waveform is a pure sinusoid and its Fourier transform consists of a single impulse at the frequency of the whistling. If the person hums, the Fourier transform contains a fundamental frequency plus harmonics at multiples of that frequency. If you have a laboratory available with oscilloscopes and spectrum analyzers, you should try this. Don't be ashamed to whistle, hum and sing into a microphone. Those listening will respect you for this scientific approach.

In the early days of telephony, experimentation showed that the portion of the speech waveform between about 300 Hz and 3.3 kHz was sufficient both for intelligibility of speech, and for recognition. That is, although this range is not considered high fidelity, it permits both understanding of what is being said, and identification of the person saying it. Quality music requires the presence of frequencies higher than 3.3 kHz, perhaps as high as 15 kHz or more (AM radio transmits frequencies up to 5 kHz while FM radio transmits frequencies up to 15 kHz; The typical quality home entertainment system responds to frequencies above 20 kHz.).

Suppose you spoke into a microphone and amplified the resulting waveform using electronic circuitry. Believe it or not, you would have configured a simple analog baseband transmitter! If the output of the transmitter were connected to a wire channel, and the other end of the channel were connected to a loudspeaker (perhaps through some amplification if the channel contains loss or you want to annoy your neighbors), a complete baseband communication system would result. Such systems are used in wire intercoms.

2.2. DISCRETE BASEBAND

When a signal is discrete, it can be thought of as a list of numbers representing the sample values of an analog waveform. One way to send such a list through a channel is to send a pulse waveform–one pulse is placed at each sampling point. Each pulse carries information about the corresponding sample value. The sample value can be conveyed as the amplitude, width or position of the pulse.

2.2.1 Pulse Amplitude Modulation

If we vary the amplitude of the pulse, the result is known as *Pulse Amplitude Modulation* (PAM). Figure 2.2 illustrates a periodic pulse train [$s_c(t)$, known as the *carrier*], a portion of a typical analog signal [$s(t)$], and the result of controlling the pulse heights with the sample values. Note that since the pulse tops are horizontal, the modulated waveform is *not* simply the product of the pulse train with the analog signal. Such a product would appear as in Figure 2.3. It results when $s(t)$ forms the input to a *gating* circuit. Both waveforms are

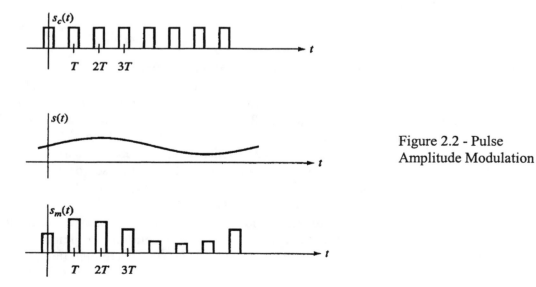

Figure 2.2 - Pulse Amplitude Modulation

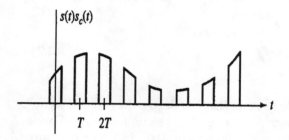

Figure 2.3 - Product of pulse train and *s(t)*

known as PAM –the $s_m(t)$ waveform of Fig. 2.2 is called *flat-top* or *instantaneous-sampled* PAM, while that of Fig. 2.3 is known as *natural-sampled* PAM. The first is known as instantaneous-sampled since the pulse height depends only on the value of *s(t)* at the exact sampling point–not on the signal values across the range of the pulse width. Flat-top PAM is generated with a *sample-and-hold* circuit. An idealized sample-and-hold (S/H) circuit is shown in Figure 2.4.

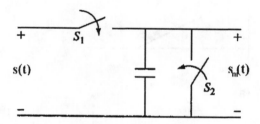

Figure 2.4 - Idealized Sample and Hold Circuit

Switch S_1 closes instantaneously at the sample point, and the capacitor charges to the sample value. The switch is then opened, and the capacitor remains at that value until the closing of Switch S_2 provides a discharge path. Practical sample and hold circuits need additional electronics to provide the energy to rapidly charge the capacitor (i.e., the series resistance is never zero) and to prevent slow discharge (leakage) prior to Switch S_2 closing.

To determine the channel requirements to transmit PAM, we must calculate the Fourier transform of a PAM waveform. We begin by evaluating the Fourier transform of the natural-sampled waveform [i.e., the product of the pulse train with *s(t)*]. If you were asked to find the Fourier transform of $s(t)s_c(t)$, you would probably first claim you could not do so without knowing the exact shape of *s(t)*. But as with most new situations, we start by expressing the terms in different ways and hope for inspiration. The unmodulated pulse train, $s_c(t)$, can be expanded in a Fourier series to obtain

$$s_c(t) = a_0 + \sum_1^\infty a_n \cos 2\pi n f_s t \qquad (2.1)$$

In Eq. (2.1), f_s is the sampling frequency and it equals $1/T_s$. When this multiplies $s(t)$, the result is a summation of products of $s(t)$ with sinusoids,

$$s(t)s_c(t) = a_0 s(t) + \sum_1^\infty a_n s(t)\cos 2\pi n f_s t \tag{2.2}$$

Figure 2.5(a) shows a representative $S(f)$. We use this shape to denote the original signal's Fourier transform, but we do not mean to restrict it to be of this exact form. The Fourier transform of each term in the summation is the signal transform, $S(f)$, shifted up and down by the frequency of the sinusoid (see the *modulation theorem* in Appendix A). The transform of $s(t)s_c(t)$ is sketched in Fig. 2.5(b).

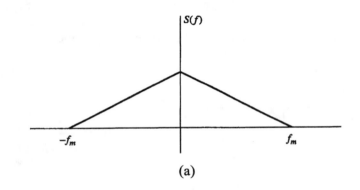

(a)

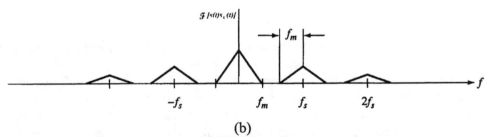

(b)

Figure 2.5 - Fourier Transform of Natural Sampled PAM

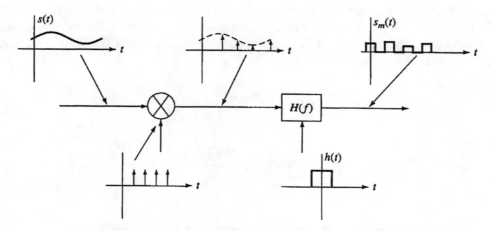

Figure 2.6 - Generation of Instantaneous Sampled PAM

The Fourier transform of instantaneous-sampled PAM is more difficult to evaluate. The evaluation is simplified by considering the *hypothetical* system of Fig. 2.6 (it's *hypothetical* since, thankfully, impulses don't exist in the real world. If a voltage could go to infinity with infinite slope, we would all have been destroyed long ago!). We begin by sampling $s(t)$ with an ideal train of impulses. We then shape each impulse into the desired pulse shape–in this case, a square pulse with a flat top.

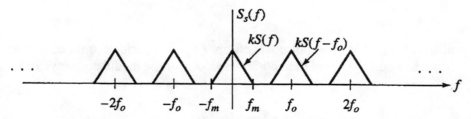

Figure 2.7 - Fourier Transform of Impulse Sampled Waveform

The Fourier transform of the sampled signal at the input to the filter is found from the sampling theorem. The Fourier series of the impulse train has equal a_n values for all n. The Fourier transform of the impulse-sampled waveform is therefore as shown in Fig. 2.7.

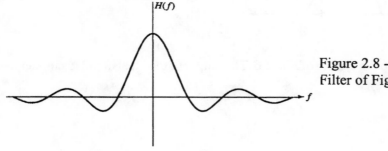

Figure 2.8 - Transfer Function of Filter of Figure 2.6

The Fourier transform of the filter output is simply the product of the transform of Fig. 2.7 with the transfer function of the filter. The transfer function of the filter is shown in Fig. 2.8, and is found from a table of Fourier transforms (see Appendix D). Finally the output transform is as shown in Fig. 2.9.

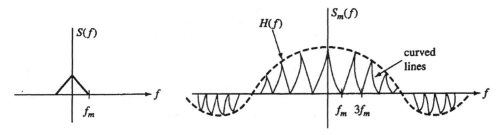

Figure 2.9 - Fourier Transform of Flat Top PAM

While the low-frequency portion of this transform may resemble $S(f)$, it is *not* an undistorted version of $S(f)$.

Example 2.1

A signal is of the form

$$s(t) = \frac{\sin \pi t}{\pi t}$$

It is transmitted using PAM. The pulse waveform, $s_c(t)$, is a periodic train of triangular pulses at shown in Fig. 2.10. Find the Fourier transform of the modulated waveform.

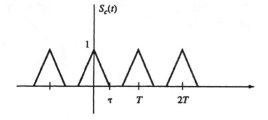

Figure 2.10 - Pulse Waveform

Solution: We refer to the system of Fig. 2.6. The output of the ideal impulse sampler has transform given by

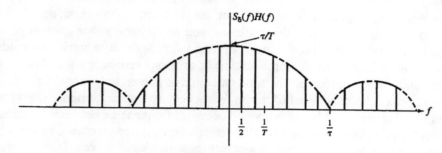

Figure 2.11 - Result for Example 2.1

$$S_\delta(f) = \frac{1}{T} \sum_{n=-\infty}^{\infty} S(f-nf_s)$$

where $S(f)$ is the transform of $\sin\pi t/\pi t$. $S(f)$ is a pulse, as shown in Fig. 2.11.

The filter must change each impulse into a triangular pulse. Its impulse response is therefore a single triangular pulse which has as its transform

$$H(f) = \frac{\sin^2 \pi f \tau}{\tau^2 \pi^2 f^2}$$

Finally, the transform of the PAM waveform is given by the product, $S_\delta(f)H(f)$ as shown in Fig. 2.11.

Please note that although the transform of a PAM wave dies down with increasing frequency, it occupies *all* frequencies from zero to infinity. This will prove to be significant later.

2.2.2. Time Division Multiplexing

It would be impractical to have a separate channel for each signal to be communicated. In telephone, this would mean a wire connection for every conversation–if up to 500,000 simultaneous calls between the U.S. and Europe were anticipated, 500,000 channels would be needed (i.e., pairs of wires or satellite channels). An even more absurd example is provided by terrestrial television. If 100 television stations wanted to broadcast simultaneously, we would need 100 parallel atmosphere systems—even the science fiction writers would have trouble with this one. So we need a way to share a channel among multiple users.

Signals can be *easily* separated from each other if they are non-overlapping in either time or frequency. An example of time separation occurs in your classroom. Hopefully, the students and the professor effectively share the communication channel. When a student talks, the professor stops, and vice versa. If you are a student in the class, and for some reason you only wanted to listen to the professor, you could hang a *time gate* on your ear and configure the gate to close (block the signal from your ear drum) when another student is talking and open when the professor is talking. If someone else only wanted to hear the other student and not the professor, the time gate on that person's ears would close when the professor speaks. An exact dual of this is frequency separation. Suppose one speaker has unusual vocal chords, and his or her voice occupies frequencies between 1 kHz and 2 kHz. A second speaker is a baritone, and all essential signal components are below 1 kHz. Filters (lowpass and bandpass) can easily separate these two signals even if the two people speak simultaneously.

We will deal extensively with frequency separation beginning with the next chapter. For now, we concentrate on time separation. Fortunately, the PAM waveform is characterized by portions of the time axis where the waveform is zero. Time separation is therefore possible.

Time division multiplexing (TDM) is the process of adding signals together such that they do not overlap in time. We first talk about TDM for signals with identical sampling rates. Then we introduce techniques to multiplex signals with unequal sampling rates.

Multiplexing of Similar Channels

Time division multiplexing of signals with identical sampling rates can be viewed as interspersing pulses, like shuffling cards. Figure 2.12 illustrates this process for two signals.

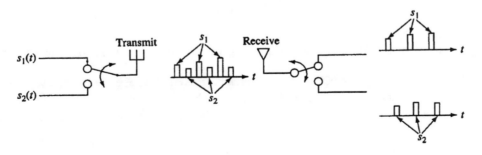

Figure 2.12 - Multiplexing of Two Channels

Note that the switches alternate between each of the two positions making sure to take no longer than one sampling period to complete the entire cycle. Two pulses are sent in each sampling period, so the pulse rate on the channel is twice the sampling rate.

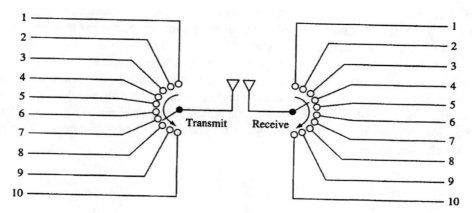

Figure 2.13 - Multiplexing of 10 Channels

Suppose that we now enlarge the system from two to 10 channels. The switch becomes a *commutator* as shown in Fig. 2.13. The switch must make one complete rotation fast enough so that it arrives at Channel #1 in time for the second sample. The receiver switch must rotate in synchronism with that at the transmitter. In practice, this synchronization (known as *frame synchronization*) requires effort. If we knew exactly what was being sent on one of the channels, we could identify its samples at the receiver. Indeed, one method of synchronization is to sacrifice one of the channels and send a known synchronizing signal in its place.

The only thing that limits how fast the switch can rotate, and therefore how many channels can be multiplexed, is the fraction of time required for each PAM signal. That fraction is the ratio of the width of each pulse to the spacing between adjacent samples of a single channel (i.e., the sampling period). The trade-off design consideration is that the more narrow each pulse, the wider the bandwidth of the resulting signal.

Multiplexing of Dissimilar Channels

The commutator approach toward multiplexing requires that the sampling rate of the various channels be identical. If signals with different sampling rates must be multiplexed, there are two general approaches that can be taken. One uses a buffer to store sample values and then intersperse these and spit them out at a fixed rate. The buffer approach is also effective if sampling rates contain variation (timing jitter). This is known as *asynchronous multiplexing*. The system must be designed so that the buffer always has samples to send when requested by the channel. This might require inserting *stuffing samples* if the buffer gets empty. Alternatively, the buffer must be large enough so that it does not overflow with input samples.

The buffer approach is also used if the various sources are transmitting asynchronously. That is, suppose they are not always transmitting information. The sizing of the buffer requires a probability analysis, and the resulting multiplexer is know as a *statistical multiplexer (stat MUX)*. The stat MUX represents an efficient technique for multiplexing channels since a source only has a time slot when it needs it. On the negative side, since individual source messages are not occurring at a regular rate, the messages must be *tagged* with a user ID. If the channels are synchronous with samples occurring at a regular and continuous rate, the stat MUX is *not* the best approach.

The second general technique involves sub- and super-commutation. This requires that all sampling rates be multiples of some basic rate. Meeting this requirement might require sampling some of the channels at a rate higher than what you would use without multiplexing. For example, if you have two channels with required sampling rates of 8 kHz and 15.5 kHz, in order to effect that combination you might choose to sample the higher frequency channel at 16 kHz.

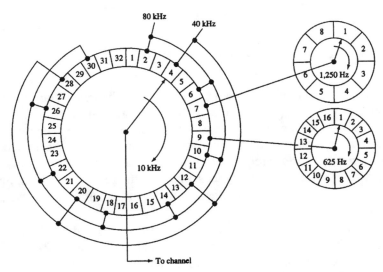

Figure 2.14 - Sub- and Super-Commutation

The concept of sub- and super-commutation is quite simple, and we illustrate it with an example. Figure 2.14 shows a *commutator wheel* with 32 slots. Suppose we wish to multiplex the following 44 channels:

1 channel sampled at 80 kHz   1 channel sampled at 40 kHz

18 channels sampled at 10 kHz   8 channels sampled at 1250 Hz

16 channels sampled at 625 Hz.

Note that all of the sampling rates are multiples of 625 Hz. Let us choose the basic rate of the commutator wheel to be 10,000 rotations per second. Therefore, each of the 18 channels which must be sampled at 10 kHz get one slot on the wheel. The channel that must be sampled at 40 kHz needs four equally-spaced slots on the wheel, so it is sampled four times during each 0.1 msec rotation of the wheel. Similarly, the 80 kHz channel needs eight equally-spaced slots on the wheel. These higher rate channels are multiplexed using *super-commutation*.

The channels that need to be sampled at less than 10 kHz only must be sampled on selected rotations of the wheel. For example, a 1250 Hz channel needs to be sampled once every eight rotations of the wheel while a 625 Hz channel needs to be sampled only once every 16 rotations. We accomplish this using *sub-commutation* wheels as shown on the figure. The eight 1250 Hz channels are commutated together with a wheel rotating at a rate of 1250 rotations per second. Each 0.1 msec, one of the channels is connected to a cell on the main commutator wheel. Similarly, the sixteen 625 Hz channels are commutated with a wheel rotating at 625 rotations per second.

Clearly a lot of design work had to go into this configuration. It's convenient when you are the textbook author and can choose nice numbers. The chosen numbers work out perfectly. Life is usually not so nice, and manipulation is needed to design the commutation system. In some ways, this process resembles finding the lowest common denominator in combining fractions, but of course it is orders of magnitude more complex than this simple algebra problem.

2.2.3. Intersymbol Interference and Crosstalk

When we talked about time division multiplexing. we made it sound too easy. We made it appear that you simply intersperse the pulses from the various channels. But life is not that good to us. Pulses in real life are never confined to a precise interval of time. The non-ideal characteristics of the channel effectively spread the pulses thus causing interference with adjacent pulses. Let's attempt to quantify these effects.

The Fourier transform of the PAM waveform was presented in Figures 2.5 and 2.9. The envelope of this Fourier transform follows a $\sin(f)/f$ curve. In fact, this envelope is the Fourier transform of a single square pulse. The first zero of this envelope is at the reciprocal of the pulse width.

The pulsed waveform may be transmitted down a coaxial cable. The signal is transmitted with little distortion provided the upper frequency cutoff of the channel is sufficiently high (at least several times the first zero of the transform).

Example 2.2: A discrete time analog signal is created by sampling a speech waveform at 10,000 samples per second. Each sampling pulse is 0.01 msec wide. This is transmitted through a channel which can be approximated by a lowpass filter with cutoff frequency at 100 kHz. Evaluate the effects of channel distortion.

Solution: Since the sampling is occurring at 10,000 samples per second, we will assume that the speech waveform has a maximum frequency below 5 kHz to avoid aliasing errors. The transmitted signal consists of pulses 0.01 msec wide. If one of these pulses forms the input to a lowpass filter with cutoff at 100 kHz, the output of the filter is as shown in Figure 2.15. Thus, although the original pulse may be confined to its assigned time interval, the filtering effects of the channel may widen the pulse to overlap adjacent intervals.

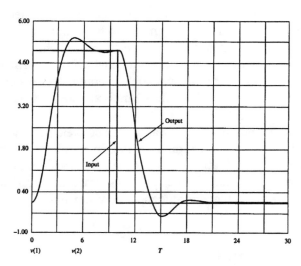

Figure 2.15 - Pulse
Through Lowpass Filter

This overlap from one time slot to adjacent time slots is known as *intersymbol interference (ISI)* or *crosstalk*. In TDM systems, adjacent pulses could be from other conversations, so the conversations can interact with each other (thus the name *crosstalk*). The term "crosstalk" also applies to the leakage of signals from one set of wires to an adjacent set of wires, as when multiple wires form part of one cable.

There are several ways of reducing the effects of intersymbol interference. Pulse spreading can be decreased by increasing the bandwidth of the system. Unfortunately, this is a luxury which requires a flexibility we don't often have. However, one parameter over which we do have control is the shape of the pulses used to transmit the sample values.

Let's begin our study of pulse shapes by assuming we transmit ideal impulse samples. Further assume that the channel can be modeled as an ideal lowpass filter. The channel shapes each impulse into a *(sin t)/t* type pulse as shown in Fig. 2.16.

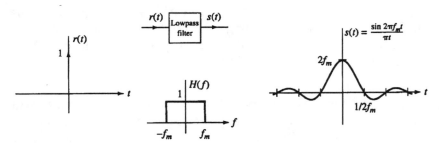

Figure 2.16 - Pulse Shaping by the Channel

Although the received waveform clearly extends into adjacent time slots, it has the very nice property that it goes to zero at the sample points. If the channel passes frequencies up to f_m, the zeros of the spread pulse are spaced by $1/2f_m$. This is the Nyquist sampling rate,

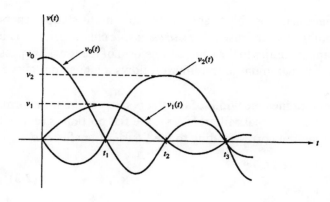

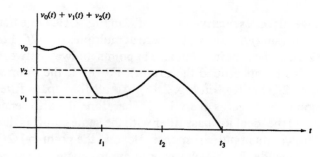

Figure 2.17 - Ideal
Lowpass Filter Shaping

so if we sample at this minimum rate, the spread pulse goes through zero at all adjacent sample points. Figure 2.17 illustrates this for three adjacent sample values.

Note that while the pulses interact with each other, at each sampling point only one $(\sin t)/t$ curve is non-zero. All of the others go through zero at that point. The sum of the various $(\sin t)/t$ curves reconstructs the original $s(t)$ since this is the final statement of the sampling theorem.

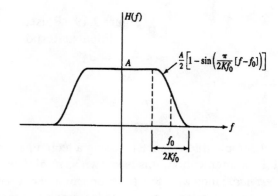

Figure 2.18 - Raised Cosine
Frequency Characteristic

The ideal bandlimited pulse shape is impossible to achieve because of the sharp corners on the frequency spectrum. A desirable compromise is the *raised cosine* characteristic. The Fourier transform of this pulse shape is similar to the square transform of the ideal lowpass filter, except that the transition from maximum to minimum follows a sinusoidal curve. This is shown in Fig. 2.18.

The value of the constant, K, determines the width of the flat portion of the transform. If $K=0$, the transform is that of the ideal lowpass filter. If $K=1$, the flat portion is reduced to a point at the origin. The time function corresponding to the Fourier transform is

$$h(t) = A\frac{\sin 2\pi f_0 t}{2\pi f_0 t} \frac{\cos 2\pi K T f_0}{1-(4Kf_0)^2} \tag{2.3}$$

This time function is sketched for several representative values of K in Fig. 2.19. Note that for $K=0$, the time function is of the form $(\sin t)/t$. This goes to zero at multiples of $1/2f_0$. For $K=1$, the response not only goes to zero at these points, but also at points midway between these values (except for the first set of points around the origin). For $K=1$, the Fourier transform has frequency content up to $2f_0$ (Set K=1 in Fig. 2.18). An ideal lowpass filter with cutoff at $2f_0$ has an impulse response with zeros every $1/4f_0$. Therefore, the difference between the raised cosine with $K=1$ and the ideal lowpass filter with the same cutoff is that the ideal filter has two additional zeros in its impulse response. Beyond the point $t=1/2f_0$, the zeros of both impulse responses coincide. It is much easier to approximate the raised cosine filter—indeed, the ideal lowpass filter cannot be built in the real world. The price we pay is intersymbol interference between *adjacent* samples (i.e., There is no interference *more than* one sample period away).

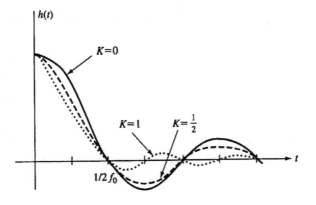

Figure 2.19 - Raised Cosine Impulse Response

We can compensate for this adjacent-sample interference by using a technique known as *partial response signaling*. In digital communication, this is known as *duobinary*. It is a form of *controlled intersymbol interference*. Since we know the proportion of one sample value that interacts with the next sample point, we can compensate by giving our receiver memory. At the risk of insulting some of you, permit me to give a classroom example.

Suppose that I (the professor) knows that student X is copying from student Y during an exam. Further suppose I know that this *interstudent interference* changes student X's grade by 10% of student Y's grade. I can completely compensate for this by lowering student X's grade by 10% This process can easily be extended to more than two students. The cancellation becomes simple if the original pulses can only take on one of two values (as in a binary digital system). Then with interference from adjacent pulses, we only have limited possibilities (actually three).

2.2.4. Pulse Width Modulation

In Section 2.2.1, we discussed PAM. We were able to transmit a discrete waveform by using the samples to vary the height of pulses. Now we'll explore varying the width of those pulses instead of the amplitude.

As in the case of PAM, we start with a signal which is a periodic train of pulses. Figure 2.20 shows an unmodulated carrier pulse train, $s_c(t)$, a representative information signal, $s(t)$, and the resulting *pulse-width modulated* (PWM) waveform.

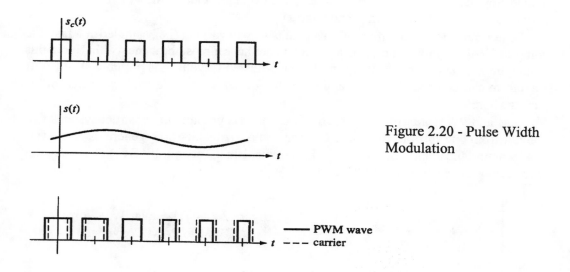

Figure 2.20 - Pulse Width Modulation

The width of each pulse varies according to the instantaneous sample value of $s(t)$. The larger the sample value, the wider the corresponding pulse.

Finding the Fourier transform of the PWM waveform is a complex computational task. Part of the reason for this complexity is that PWM is a *non-linear* form of modulation. A simple example illustrates this. If the information signal is a constant, say $s(t)=1$, the PWM wave consists of equal width pulses. This is true since each sample value is equal to every other sample value. If we now transmit $s(t)=2$ via PWM, we again get a pulse train of equal-width pulses, but the pulses would be wider than those used to transmit $s(t)=1$. The principle of linearity dictates that if the modulation is linear, the second modulated waveform should be twice the first. This is not the case as is illustrated in Fig. 2.21. (You

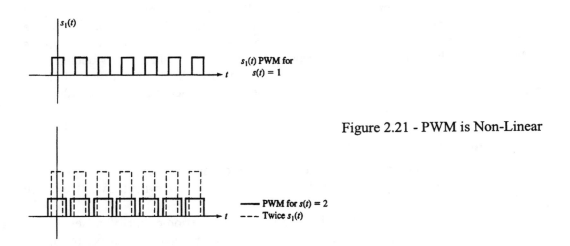

Figure 2.21 - PWM is Non-Linear

should stop here and take a moment to convince yourself that PAM, which we studied in Section 2.2.1, *is a linear* form of modulation.)

If one assumes that the information signal is slowly varying (i.e., sampled at a fast rate compared to the Nyquist rate), then adjacent pulses have almost the same width. Under this assumption, an approximate analysis of the modulated waveform is possible. The PWM waveform can be expanded in an approximate Fourier series. We shall not perform the analysis here.

PAM and PWM are related to each other, and it is possible to construct systems that convert from one form to the other. We can use a triangular-wave generator to convert between time and amplitude. (We used this principal on the counting ADC.) The triangular waveform we use is shown in Fig. 2.22.

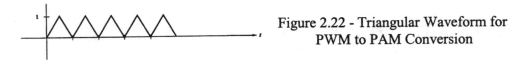

Figure 2.22 - Triangular Waveform for PWM to PAM Conversion

The process we describe here is illustrated in Fig. 2.23. Figure 2.23(a) shows a block diagram of the generator, and Fig. 2.23(b) shows typical waveforms.

We start with a normalized information signal, $s(t)$. This is put through a sample and hold circuit to yield $s_1(t)$. The triangular wave is shifted down by one unit in order to form $s_2(t)$. The sum of $s_1(t)$ and $s_2(t)$ then results in $s_3(t)$. The times for which $s_3(t)$ is positive represent intervals whose widths are proportional to the original sample values. We need only put this shifting triangular waveform into a comparator with output of one for positive input and zero for negative input. This results in $s_4(t)$, the PWM waveform. The range of pulse widths can be adjusted by scaling the original time function.

Since the heights of the pulses in PWM are constant but the widths depend on $s(t)$, the average power of the PWM waveform varies with the amplitude of $s(t)$. For example, if $s(t)$ were a musical selection, less power would be required during soft parts of the music, and more power during loud parts. This reduces the efficiency of the communication system since the pulse amplitudes would have to be chosen to assure that the *maximum* power does not exceed that permitted by the system.

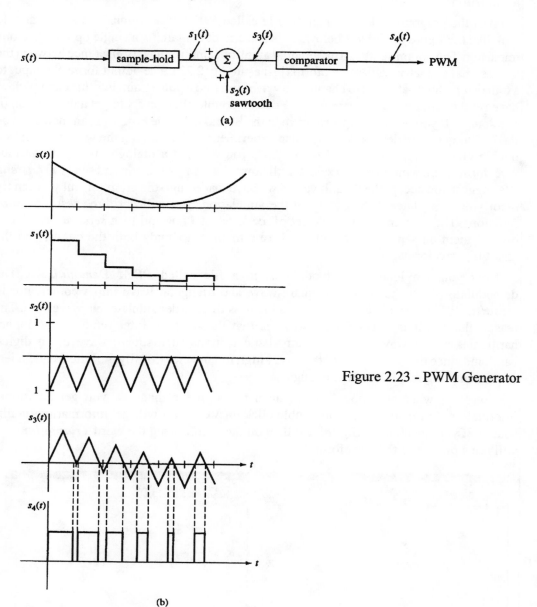

Figure 2.23 - PWM Generator

Computer Simulation Example

The CD packaged with this text includes Tina software examples. Once you load Tina (see Appendix F for instructions), you can run two simulations relating to Pulse Width Modulation.

Start the program and then open the file called **PWM Modulator**. You screen should look like the figure on the left below. This is a circuit containing a single op-amp and one transistor. The op-amp acts as a differential amplifier amplifying the difference between the input signal and a triangular waveform (just as in Fig. 2.23). The output forms the input to a transistor which saturates to change the waveform into a pulse train. For this example, we have set the input to be a square wave. You can double click on **IN** to get a description of this source. If you click on **Square wave** in the **Signal** dialogue box, you can then click on the "..." to get more details. You can also experiment with changing the waveform. If you double click on "Square wave" or on "...", you will get a dialogue box with various waveforms from which you can select. Pull down the **Analysis** menu and select **Transient**. We suggest you accept the default values we have loaded into the program, but you can try modifying them later. When you run the simulation, you will get three resulting time functions. If they print on a single set of axes, select **View** and then **separate curves** to display them on separate axes. Please take a moment to study both the circuit and the resulting waveforms.

Now you can load the demodulator, in a file called **PWM Demodulator**. The demodulator feeds the sum of a square wave and triangular wave into a comparator to create the PWM waveform. It then uses a lowpass filter to demodulate this waveform. The reason the result is not a square wave is that the lowpass filter removes the higher harmonics of the waveform. It is unrealistic to transmit a square wave using digital baseband since the waveform contains very high frequencies (ideally, infinite frequencies) so the sampling rate is not high enough.

Note that when you double click on a block in the diagram, you get additional information. For example, if you double click on VG1, you will get information on this source. If you the click on **Signal** and then on the ... following the word **Triangular**, you will see a picture of the waveform.

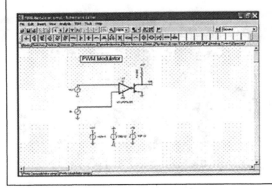
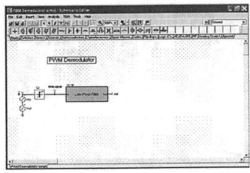

2.2.5. Pulse Position Modulation

Pulse position modulation (PPM) possesses the noise advantage of PWM without the problem of a variable average power which is a function of signal amplitude. A linear information signal, *s(t)*, and the corresponding PPM waveform are illustrated in Fig. 2.24. The larger the sample value, the more the corresponding pulse deviates from its unmodulated position. Please take a moment to study this graph. At first glance, the pulse train may appear periodic, but it's not.

A PPM waveform can be derived from a PWM waveform. The relationship between the two is that, while the position of the pulse varies in PPM, the location of the leading (or trailing) edge of the pulse varies in PWM. Suppose, for example, that we detect the location of each trailing edge in a PWM waveform (e.g., differentiate and look for large negative pulses). If we now place a constant-width pulse at each of these points, the result is PPM. This is illustrated in Fig. 2.25.

Clearly both PWM and PPM are more complex than is PAM. The justification for choosing one of these more complex systems is that they provide greater noise immunity than does PAM. In PAM, additive noise directly affects the reconstructed sample value. This disruption caused by noise is less severe in PPM and PWM, where the additive noise must affect the zero crossings in order to cause an error. Along with greater complexity, PWM and PPM have other negative properties. If you use time division multiplexing, you must be sure that adjacent sample pulses do not overlap. If pulses are free to shift around or to get wider (as they are in PPM and PWM), one cannot simply insert other pulses in the spaces and be confident that no overlap will occur. Sufficient spacing must be maintained

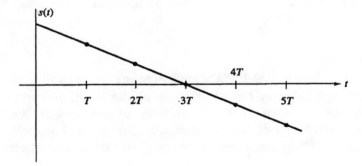

Figure 2.24 - Pulse Position Modulation

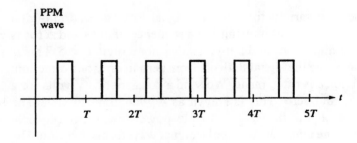

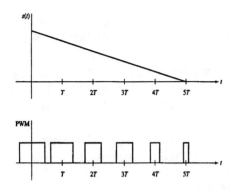

Figure 2.25 - Conversion From PWM to PPM

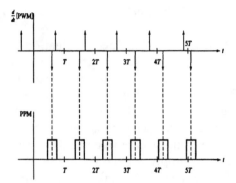

to allow for the largest possible sample value. This decreases the number of channels that can be multiplexed.

2.3. Digital Baseband

All three pulse modulation schemes, PAM, PWM and PPM are *analog* forms of communication, even though the signals can be thought of as a list of numbers. When that list of numbers is drawn from a specific set of possibilities, the communication becomes digital. Even though digital communication can consist of symbols with more than two possible values, we'll be concentrating on binary signals through much of our studies.

2.3.1. Signal Formats

We begin this section by presenting various time domain signal formats used in digital communication. One of the simplest ways of transmitting a sequence of ones and zeros is to transmit a $+V$-volt signal for a digital one and a 0-volt signal for a digital zero. This is known as *unipolar* transmission, since the signal deflects from zero in only one direction. A modification of this natural choice is often made. We send a constant of +V volts for a digital one and -V volts for a digital zero. This is known as *bipolar* transmission. The voltage amplitude, V, is chosen on the basis of available power and other physical considerations. Figure 2.26 shows the unipolar and bipolar signal waveforms that would be used to send the digital sequence 1 1 0 0 0 1 0 1. The specific signaling format illustrated in Fig. 2.26 is *NRZ*, or *non-return to zero*. The pulse stays at its value throughout the interval without returning to zero. This particular type of transmission is known as *NRZ-L*, where the voltage level corresponds to the logic level. The assignment of the higher voltage to a binary one is arbitrary. Alternatively, we could have represented a logic one with -V volts and a logic zero with +V volts.

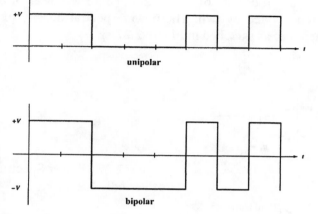

Figure 2.26 - Waveforms for Sending 11000101

Before exploring other signaling formats, we will take a few minutes to examine the Fourier transform of the NRZ signal. We need this information to know whether the channel has sufficient bandwidth to transmit this type of signal without distortion. The actual waveform depends on the binary sequence being transmitted. We consider two cases—periodic data and random data.

Suppose that the data comprise an alternating train of ones and zeros. That is, the sequence is described by

$$a_n = 0\,1\,0\,1\,0\,1\,0\,1... \qquad (2.4)$$

The bipolar NRZ-L waveform is then simply a square wave with frequency equal to half the bit rate. The Fourier series of the square wave is given by

$$s_{NRZ}(t) = \sum_1^\infty b_n \sin \pi n R_b t \qquad (2.5)$$

where R_b is the bit rate in bits per second and the Fourier coefficients are given by

$$b_n = \begin{cases} (-1)^{\frac{n-1}{2}} \dfrac{4V}{n\pi} & for\ n\ odd \\[2ex] 0 & for\ n\ even \end{cases} \tag{2.6}$$

We are interested in the power spectral density of the NRZ signal. The power of each sinusoid is the amplitude squared divided by two. The power spectral density of the periodic NRZ-L waveform is therefore as sketched in Figure 2.27.

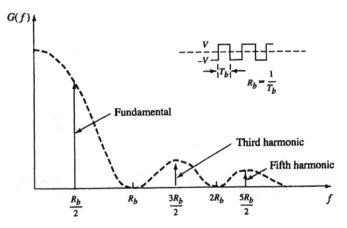

Figure 2.27 - Power Spectrum of Periodic NRZ

Note that the outline of the impulses (the dashed line in the figure) is a $(\sin^2 f)/f^2$ type curve. While the periodic data result gives us a starting point, it is not very relevant to the real world. You wouldn't go through a lot of trouble to transmit an alternating train of ones and zeros. A more meaningful spectrum results if we assume a random sequence of ones and zeros. The power spectral density of a time function, $s(t)$, can be found by taking the ratio of energy to time. That is, power is the rate of change of energy. We do this operation over a limited time interval and then take the limit as the interval length becomes infinite. Setting the time interval length to ΔT, the power spectral density is given by

$$G(f) = \left| \left\{ \lim_{\Delta T \to \infty} \frac{1}{\Delta T} \int_{-\Delta T/2}^{\Delta T/2} s(t) e^{-j2\pi ft} dt \right\} \right|^2 \tag{2.7}$$

The term in brackets in Eq. (2.7) is the Fourier transform of the truncated signal. Since $s(t)$ is NRZ bipolar, the signal is always either $+V$ or $-V$, and stays constant within an interval. As ΔT increases by jumps equal to the bit intervals, we keep adding identical terms, since the squaring operation makes the contribution of each period identical. We can therefore perform the averaging operation over one period to get

$$G(f) = \frac{1}{T} \left| \int_{-T_b/2}^{T_b/2} Ve^{-j2\pi ft}dt \right|^2$$

(2.8)

$$= V^2 T_b \left(\frac{\sin\pi fT_b}{\pi fT_b} \right)^2$$

where T_b is the bit period. This power spectral density is sketched in Fig. 2.28. Please take a moment to compare this to Fig. 2.27.

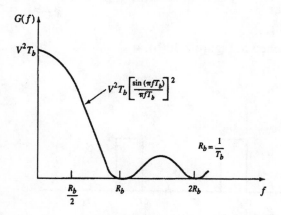

Figure 2.28 - Power Spectrum of Random NRZ

The total average power of the bipolar NRZ waveform is V^2 since the square of the NRZ waveform is a time function with this constant value. The integral of the power spectral density yields the power in any frequency band. One can show that 91% of the total power of an NRZ signal is contained in frequencies below the bit rate, R_b [just use MATLAB to integrate Eq. (2.8)].

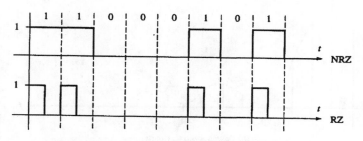

Figure 2.29 - NRZ and RZ signaling

If we went through the trouble of adding words "non return-to-zero", you can probably guess there is another format that does, in fact, return to zero. Figure 2.29 illustrates unipolar NRZ and *RZ (return to zero)* signaling for the sequence 11000101. Note that in RZ, the pulse returns to the "0" state before reaching the end of the bit interval. We are illustrating this for the case where this return occurs at the midpoint of the bit interval, though this is not the only choice for the return point. Note that the pulses in RZ are not as wide as in the NRZ case, and this change increases the bandwidth.

The power spectral density of a (unipolar) RZ signal is found much the same way as we did for the NRZ signal. We again separate the analysis into periodic data and random data. The periodic data situation gives a time function as shown in Fig. 2.30(a). Since we are interested in power, time shifting of the time function does not change the result. We therefore can shift the function $T_b/4$ to the left to yield an even time function with Fourier series given by,

$$s_{RZ}(t) = \sum a_n \cos \pi n R_b t \qquad (2.9)$$

The resulting power spectral density is sketched in Figure 2.30(b).

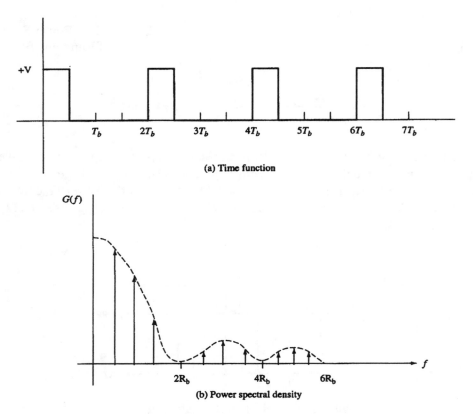

(a) Time function

(b) Power spectral density

Figure 2.30 - Power Spectrum of Periodic RZ

The outline of the impulses goes to zero at multiples of $2R_b$, exactly twice that of NRZ signaling.

The analysis for random data is more complex. The signal can be decomposed into a periodic component at the bit rate added to a random component. This decomposition is shown in Fig. 2.31.

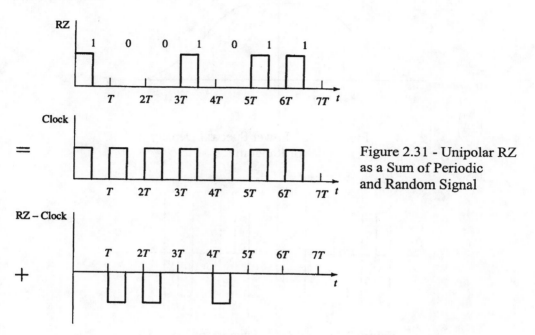

Figure 2.31 - Unipolar RZ as a Sum of Periodic and Random Signal

If we assume that ones and zeros are equally likely (i.e., each occurs 50% of the time), the power spectral density is given by

$$G(f) = \frac{\pi V^2}{8} \sum_{-\infty}^{\infty} \frac{\sin^2(n\pi/2)}{(n\pi/2)^2} \delta(f - nR_b) + \frac{V^2}{16R_b} \frac{\sin^2(\pi f/2R_b)}{(\pi f/2R_b)^2} \tag{2.10}$$

The first term is the power spectral density of the periodic clock signal. The second term is derived using the same technique as for Eq. (2.8). This power spectral density of Eq. (2.10) is sketched in Figure 2.32.

One problem associated with the transmission schemes we have presented is that the average value of the waveform is a function of the fraction of ones. This causes problems if the electronics does not transmit *dc* (i.e., is *ac* coupled). A variation that overcomes this is *alternate mark inversion (AMI)*. This is the same as unipolar RZ except that every other non-zero pulse is inverted. Figure 2.33 illustrates the NRZ-AMI signal for the binary sequence 11000101. This waveform will have an average value of zero independent of the number of ones.

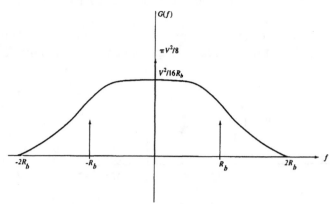

Figure 2.32 - Power Spectral Density
of Random RZ

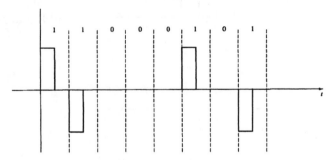

Figure 2.33 - NRZ-AMI signal for 11000101

MATLAB EXAMPLE

This simple MATLAB program plots the spectral density of the RZ signal. It plots the second term in Eq. (2.10).

```
function RZ_density()
%EXAMPLE
%Plot the spectral density of an RZ signal
clc;                                % clear command window
close all;                          % close all (previously) open figures
disp('   Plot the spectral density of an RZ signal')
disp('    ')
disp('---------------PRESS A KEY TO CONTINUE-------------')
pause
```

```
disp(' ')
disp(' ')
Rb = input('Please enter the bit rate: ');
V = input('And the voltage amplitude V: ');
Tb = 1/Rb;
f=-2*Rb:2*Rb;
G = V^2.*(sin(pi.*f/(2*Rb)).^2)./(16*Rb.*(pi.*f/(2*Rb)).^2);
plot(f,G)
xlabel('frequency, Hz'); ylabel('G(f)'); title('Spectral Density of RZ');
```

There are two particular problems associated with NRZ-L transmissions. First, when the data are *static* (i.e., no change from one bit interval to the next), there is no transition in the transmitted waveform. This causes timing problems when we try to establish bit synchronization (as we discuss in Section 2.4.3). Substituting RZ for NRZ solves this problem since transitions occur at the midpoint of each bit interval (assuming bipolar waveforms). The second problem occurs with data inversion. If the levels are reversed during transmission (i.e., $+V$ is interpreted as $-V$ at the receiver), the entire data train is inverted and every bit is in error. Inversion can occur in several ways. The most common form of inversion occurs when the information is being sent by varying the phase of a sinusoid (we will call this phase shift keying, or PSK, in Chapter 5). A time delay corresponding to a 180° phase shift results in a data inversion. Additionally, some systems include numerous electronic devices (e.g., transistors, op-amps) each of which inverts the signal. It may sometimes prove difficult to keep track of the number of inversions, or indeed, different signals may encounter different numbers of inverters. Sometimes an even number of devices is encountered so there is no overall inversion. But sometimes there is an odd number leading to inversion of the signal.

For the above reasons, we often choose *differential* forms of encoding. In such techniques, the data are represented as *changes in levels* rather than by the actual signal level. We investigate the NRZ-M and NRZ-S systems. The terminology of M and S stems from MARK and SPACE, a carryover from the telegraph days.

In the *NRZ-M* system (sometimes referred to as NRZI), a one is represented by a change in level between two consecutive bit periods while a zero is represented by no change. Figure 2.34 illustrates the NRZ-L and NRZ-M waveforms for the data sequence, 10010. Note that we have started the NRZ-M signal at $+V$ volts. We could have started the signal at $-V$ volts, and this would give us a flipped version of the waveform.

We can implement an NRZ-M encoder with the use of an exclusive OR gate and a time delay. This is shown in Fig. 2.35.

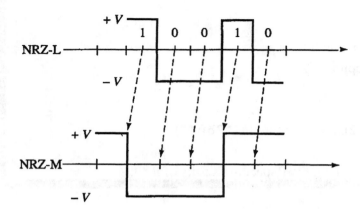

Figure 2.34 - NRZ-M
Differential Format

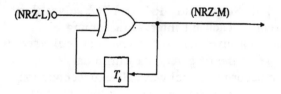

Figure 2.35 - NRZ-M Encoder

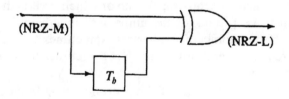

Figure 2.36 - NRZ-M Decoder

We start with the NRZ-L signal, and exclusive-OR this with the delayed NRZ-M output. If the input bit is a one, the output represents an inversion of the previous output. That is, if we take the exclusive OR operation of signal X with one, the result is the inverse of that X. Thus,

$$1 \oplus 1 = 0 \qquad\qquad 1 \oplus 0 = 1$$

The decoder for NRZ-M compares the NRZ-M signal to a delayed version of itself. If the two are identical in an interval, we know that a zero is being sent; If the two are different, a one is being sent. This decision rule describes the operation of an exclusive OR gate, so the decoder can be implemented as in Fig. 2.36.

The NRZ-S system is almost the same as the NRZ-M, except that the two outputs are reversed. A one is represented by no change in levels between two consecutive bit times, while a zero is represented by a change. Figure 2.37 shows the NRZ-L, NRZ-M and NRZ-S

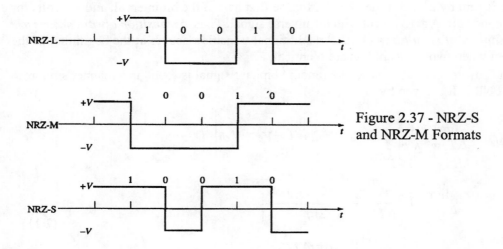

Figure 2.37 - NRZ-S
and NRZ-M Formats

representations for the same data signal we have been illustrating, 10010.

The encoders and decoders for NRZ-S are the same as those for NRZ-M except for the addition of an inverter. Derivation of these systems is left to the problems at the back of this chapter.

While the NRZ-S and NRZ-M differential systems solve the problem of waveform inversion, they do not solve the problem of loss of timing information. For example, in the NRZ-M system, a train of zeros results in a transmitted waveform without transitions. Similarly, in the NRZ-S system, a train of ones results in a loss of bit timing information.

The *biphase* (also known as *bi-ϕ* or *Manchester*) format overcomes this *static data* timing problem. Figure 2.38 shows the NRZ-L and biphase-L waveforms for the same data train we have been examining. Also shown on the figure are the basic waveforms used to send a one and a zero.

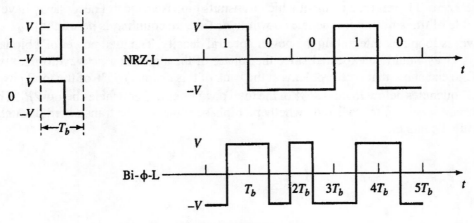

Figure 2.38 - Biphase Format

A one is sent by transmitting $-V$ volts for the first half of the bit interval, and $+V$ volts for the second half. A zero is sent with the inverse signal. Thus, *a transition always occurs are the midpoint of each bit interval*. An additional transition occurs at the beginning of the interval when two adjacent bits are identical.

The power spectral density of a random biphase signal is found in a manner similar to that of NRZ. It is given by

$$G(f) = \lim_{\Delta T \to \infty} \frac{1}{\Delta T} \left| \int_{-\Delta T/2}^{0} \pm Ve^{-j2\pi ft} dt + \int_{0}^{\Delta T/2} \mp Ve^{-j2\pi ft} dt \right|^2$$

$$= \lim_{\Delta T \to \infty} V^2 \frac{1}{\Delta T} \left| \frac{-2 + e^{j\pi f\Delta T} + e^{-j\pi f\Delta T}}{j2\pi f} \right|^2 \tag{2.11}$$

$$= \lim_{\Delta T \to \infty} \frac{V^2}{\Delta T} \left| \frac{1 - \cos\pi f\Delta T}{\pi f} \right|^2$$

$$= \lim_{\Delta T \to \infty} \frac{V^2}{\Delta T} \left| \frac{2\sin^2(\pi f\Delta T/2)}{\pi f} \right|^2$$

We now let ΔT go to infinity in jumps of the bit period to get

$$G(f) = V^2 T_b \left[\frac{\sin^2(\pi f T_b/2)^2}{\pi f T_b/2} \right]^2 \tag{2.12}$$

This is sketched in Figure 2.39. The first zero occurs at $2R_b$, just as in the case of RZ signaling. However, it is significant to note that the spectrum goes to zero as frequencies approach zero. Therefore, a channel which transmits biphase need not necessarily have to be capable of transmitting frequencies down to *dc* (i.e., *ac* coupling is permitted).

Power is found by integrating the power spectral density. The total power of a biphase signal is V^2 as would be expected since the square of the signal is a *dc* waveform at this level. One can show that (see problems at the back of this chapter) 65% of the total power lies in frequencies below R_b, and 86% of the total power lies in frequencies below $2R_b$. This is important when we try to decide whether a biphase signal can be transmitted through a bandlimited channel.

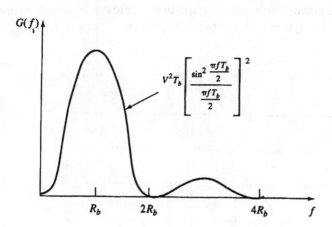

$$V^2 T_b \left[\frac{\sin^2 \frac{\pi f T_b}{2}}{\frac{\pi f T_b}{2}} \right]^2$$

Figure 2.39 - Spectral Density
of Random Biphase

Figure 2.40 shows an encoder for biphase-L. We start with NRZ-L and feed both this and a double-frequency clock signal into an exclusive OR gate. When the data logic level is zero, the output of the gate is the same as the clock. When the data logic level is one, the output of the gate is the inverse of the clock.

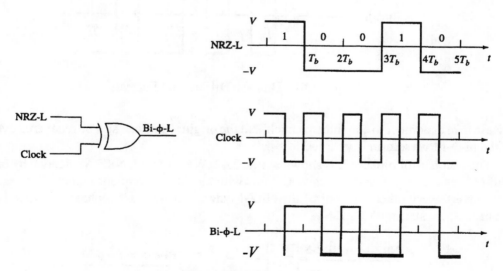

Figure 2.40 - Biphase-L Encoder

We have thus solved the timing problem, but signal inversion during transmission or reception still results in bit reversal causing unacceptable bit errors.

We can combine differential encoding with biphase encoding to solve both timing and inversion problems. The *biphase-M* and *biphase-S* formats are differential. A transition occurs at the *beginning* of every bit period, unlike the biphase-L where transitions occur in the middle of the intervals. In the biphase-M system, a one is represented by an *additional* transition in the midperiod, while a zero is represented by no midperiod transition. In the biphase-S system, the data one results in no midperiod transition, while the data zero results in a midperiod transition. These formats are illustrated in Fig. 2.41.

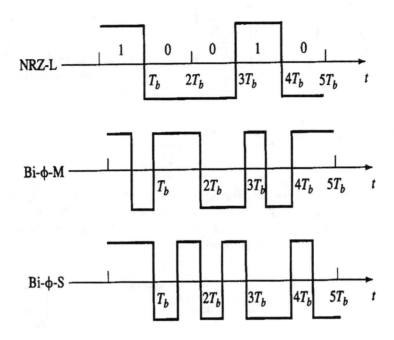

Figure 2.41 - Biphase Differential Formats

Please take a moment to convince yourself that the original data can still be recovered even if the bi-ϕ-M waveform is inverted.

The encoder for biphase-S is shown in Fig. 2.42. We start with NRZ-M, delay it by one half of the bit interval, and exclusive OR this with the double frequency clock. The result is to invert every other segment of the NRZ-M signal. If we want biphase-M instead of biphase-S, we start with NRZ-S.

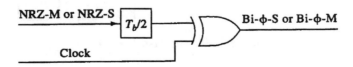

Figure 2.42 - Biphase-S Encoder

2.3.2. Pulse Code Modulation (PCM)

Pulse code modulation (*PCM*) is a direct application of the analog to digital converter. Rather than simply present it as such, we rederive the results using a different approach. The combination of the two approaches should help give you an intuitive feel for the process. We begin with a pulse amplitude modulation system.

Suppose that the amplitude of each pulse in a PAM system (see Section 2.2.1) is rounded off to one of several possible levels. Rounding off the pulse amplitudes is the same as first operating on the original continuous time function by rounding off to several possible levels. This yields a staircase type of waveform as shown in Fig. 2.43. We then sample the staircase function and transmit the samples using PAM.

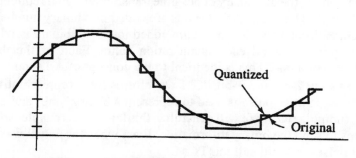

Figure 2.43 - Rounding off to Discrete Levels

The rounding off process is known as *quantization*, and it introduces an error known as *quantization noise*. That is, the staircase approximation is not identical to the original function, and the difference between the two represents an error. We discuss this error in detail when we examine performance in Section 2.5.

The *dictionary* of possible PAM pulse heights has been reduced to include only those specified quantization (round-off) levels. A received pulse is compared to the dictionary of possible transmitted pulses, and it is decoded into that entry which most closely resembles the received signal. In this way, small errors (up to one half of the quantization step size) are corrected.

The system's ability to do *error correction* is the major reason for quantizing. Suppose, for example, you wanted to transmit a signal from California to New York over coaxial cable. If you transmitted the signal via PAM, the resulting noise could make the reception unintelligible. Noise would be added along the entire transmission path with additional noise added in each amplifier. Many amplifiers would be required to cancel attenuation along the way, and the noise would get bigger and bigger with distance.

If the same signal were now sent using quantized PAM, errors could often be corrected. Suppose the amplifiers were spaced in such a way that the noise introduced between any two is less than one half of the quantization step size. Each amplifier could then restore the function to its original staircase form before amplifying and sending it on its way. That is, each amplifier would round each received pulse to the nearest acceptable level and then retransmit, thereby eliminating all but the strongest noise. Since the amplifiers do more than simply enlarge the received waveform, they are given the name, *repeater*.

The more quantization round-off levels used, the closer the staircase function resembles the desired signal. The number of levels then determines the signal *resolution*–how small a change in signal level can be detected by looking at the quantized version of the signal.

If high resolution is required, the number of quantizing levels must increase. At the same time, the spacing between levels decreases. As the levels get closer and closer together, the noise advantages start disappearing–it takes a smaller and smaller amount of noise to cause errors.

If resolution could be somehow improved *without increasing* the size of the dictionary (i.e., without moving dictionary words closer together), the error-correcting advantages could be maintained. Pulse code modulation (PCM) is a method of accomplishing this.

In a binary PCM system, the dictionary of possible transmitted signals contains only two entries, "zero" and "one". The quantization levels are coded into binary numbers. Thus, if there are eight quantization levels, the values are coded into 3-bit binary numbers. Three pulses would be required to send each quantization value. Each of the three pulses represents either a zero or a one. This is identical to the concept of A/D conversion.

Figure 2.44 shows the 2-bit and 3-bit PCM waveforms for a representative *s(t)*. For illustrative purposes, a positive pulse is used to represent a binary "one" and a zero pulse represents a "zero" (unipolar RZ). We will generalize this later in this chapter where we use different waveforms to send the binary information. Even when the actual waveforms are not in the shape of pulses, we still call this PCM.

A PCM demodulator is simply a binary digital to analog converter. The modulator and demodulator are available as LSI integrated circuits and are collectively given the name *CODEC* (for coder-decoder).

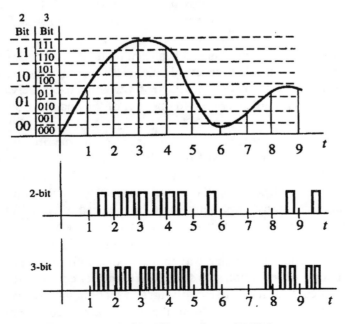

Figure 2.44 - Illustration of PCM

Non-Uniform Quantization

Figure 2.45(a) illustrates uniform quantization as an input vs. output function. The range of sample values (input) is divided into quantization regions each of which is the same size as every other region. Thus, for example, with 3-bit quantization, we divide the entire range of sample values into eight equal regions. Since we normally reconstruct a sample (i.e., digital-to-analog conversion) as the midpoint of the region, we can consider the entire process as one that transforms all sample values within a particular region into the midpoint of that region. Looked at another way, if we cascade an A/D converter and a D/A converter, we would like the two conversions to cancel each other and the output of the D/A to exactly equal the input to the A/D. However, this will not be the case. We would like the input-output relationship of the cascaded system to follow a 45-degree straight-line curve. But instead, it follows the staircase of Figure 2.45(a). The more bits of quantization we use, the finer the staircase function will be, and the more is resembles the straight line.

It is sometimes better to use quantization intervals which are *not* all the same size. That is, we replace the quantization function in Fig. 2.45(a) with that of Fig. 2.45(b). The function of Fig. 2.45(b) has the property that the spacing between quantization levels is *not* uniform, and the output levels are *not* in the center of each interval.

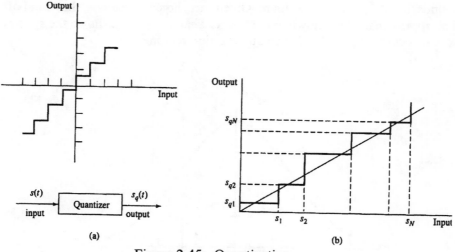

(a)

(b)

Figure 2.45 - Quantization

Let us first try to intuitively justify non-uniform quantization. Consider a musical piece where the voltage waveform ranges from -2 to +2 volts. Suppose further that 3-bit uniform quantization is used. Let's focus our attention on two of the eight regions. All voltages between 0 volt and 0.5 volt are coded into the same code word, 100, corresponding to the output reconstructed value of 0.25 volt. Likewise, all samples between 1.5 and 2 volts are coded into a single code word, 111, corresponding to an output reconstructed value of 1.75 volt. During soft music passages, where the signal may not exceed 0.5 volt for long periods, a great deal of music definition would be lost. The quantization provides the *same resolution* at high levels as at low, even though the human ear is less sensitive to changes at higher levels. The response of the human ear is non-linear. It would therefore seem reasonable to consider using small quantization steps at the lower signal levels and larger steps at the higher levels.

An alternative justification for examining non-uniform quantization applies if the signal spends a greater percentage of time at the low levels than at the higher levels. It is better to have higher resolution at these lower levels at the expense of lower resolution at the high levels. Since the signal spends much less time at the higher levels, the average quantization error may well decrease using this approach.

The *quantization error*, or *quantization noise* is a measure of the effectiveness of a quantization scheme. We examine that in detail in Section 2.5. For now, let's take a qualitative approach.

The average quantization error is a function of three things: the quantization regions (s_i's in Fig. 2.45), the round off values (s_{qi}'s in Fig. 2.45), and the probability density function of the sample values. We will show later that once quantization regions have been chosen (the s_i's in Fig. 2.45), the best choice of the s_{qi}'s is to be the center of gravity of the corresponding portion of the probability density. Figure 2.46 shows a representative example of a probability density function that resembles a Gaussian density. We have divided this into eight equal regions (this will prove to be a poor choice) and indicated these with boundaries of s_0 to s_8 on the figure. Given this choice of regions, the round off levels would be approximately as shown in the figure. That is, we have shown the s_{qi} to be at the approximate center of gravity of each quantization region.

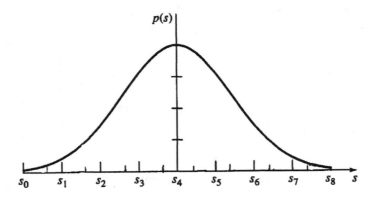

Figure 2.46 - Signal Probability Density

If in addition to the specific quantization levels we also make the actual intervals variable, a complicated optimization problem results and a complex system would be required to implement the result. For this reason, and also because systems should be applicable to a variety of input signals, sub-optimum systems are almost always used. This leads us to a study of *companding*.

Companding

The most common form of non-uniform quantization is known as *companding*. The name is derived from the words "compressing-expanding". The process is illustrated in Fig. 2.47.

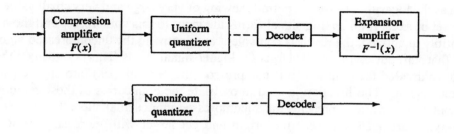

Figure 2.47 - Companding

The original signal is compressed using a memoryless non-linear device (the compression amplifier). The compressed signal is then uniformly quantized. Following transmission, the decoded waveform must be expanded using a non-linear function which is the inverse of that used in compression.

The idea of intentionally distorting a signal is common to other systems you will study. One example is *pre-emphasis* and *de-emphasis*, as used in FM. As another example, suppose that you were in a classroom trying to listen to the professor, but the light fixtures were buzzing loudly at 60 Hz. You could hang an acoustic *notch filter* on your ear, with the transfer function shown in Fig. 2.48(a). In this manner, you would significantly decrease the annoying hum. However, you would also lose portions of the professor's voice signal around 60 Hz. The professor could compensate for this by using an amplifier with the frequency characteristic shown in Fig. 2.48(b). This amplifier enhances portions of the audio around 60 Hz. If the product of the two transfer functions is a constant, you would hear an undistorted version of the lecture. The additive hum would be attenuated, and the overall transmission would improve. This concept of performing a *reversible* operation on

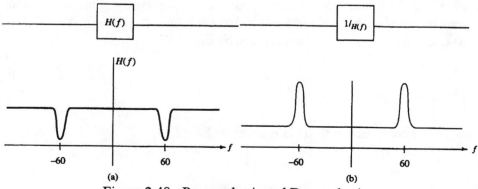

Figure 2.48 - Pre-emphasis and De-emphasis

a signal to improve some characteristic is quite common in communications.

We begin our examination of companding by analyzing the compression process. Prior to quantization, the signal is distorted by a function similar to that shown in Fig. 2.49. This operation compresses the larger values of the waveform while enhancing the smaller values, much as the logarithm is used to permit viewing of very large and very small values on the same set of axes. If an analog signal forms the input to this compressor and the output is uniformly quantized, the result is equivalent to quantizing the uncompressed signal with steps that start out small and get larger for higher signal levels. This is shown on Fig. 2.49. We have divided the output of this compressor (the vertical axis) into eight equal quantization regions. The function is used to translate the boundaries of these regions to the horizontal axis, which represents the uncompressed input signal. Note that the regions on the s-axis start out small near the origin and get larger with increasing values of the magnitude of s.

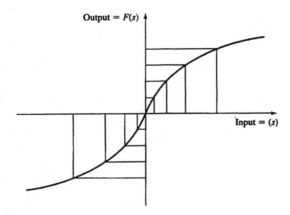

Figure 2.49 - Compression Function

If you have any hope of purchasing an off-the-shelf PCM encoder with companding, we need to have a limited number of standard compression formulae. The most common application of companding is in voice transmission. North America and Japan have adopted a standard compression scheme known as *μ-law companding*. Europe has adopted a different, but similar standard known as *A-law companding*.

The μ-law compression formula is given by Eq. (2.13).

$$F(s) = sgn(s)\frac{\ln(1+\mu|s|)}{\ln(1+\mu)} \tag{2.13}$$

The A-law compression formula is given by Eq, (2.14).

$$F(s) = sgn(s)\frac{A|s|}{1+\ln(A)} \quad for \ \ 0 \le |s| < \frac{1}{A}$$

$$(2.14)$$

$$F(s) = sgn(s)\frac{1+\ln(A|s|)}{1+\ln(A)} \quad for \ \ \frac{1}{A} \le |s| \le 1$$

The μ-law function is sketched in Fig. 2.50 for selected values of μ. If we attempted to sketch the A-law on the same curves, you would probably not notice the difference because the two laws are so similar.

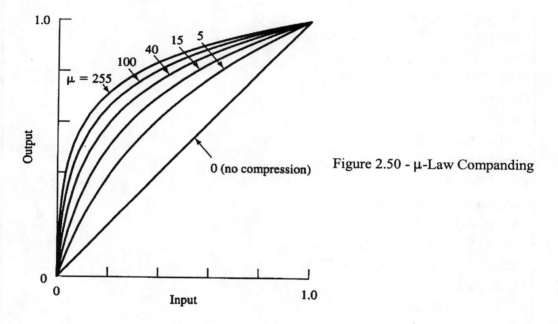

Figure 2.50 - μ-Law Companding

The parameter, μ, defines the degree of curvature of the function (i.e., the amount of compression). A commonly used value is $\mu = 255$. For the A-law companding, the commonly used value of the parameter is A=87.6.

MATLAB EXAMPLE - This MATLAB program plots mu law and A law curves. That is, it calculates Eq. (2.13) and Eq. (2.14).

clc;

close all;

disp ('Plot of Eq. (2.13) and Eq. (2.14)');

disp('The red curve indicates compression for your input value of A or \mu');

disp('The blue curve indicates compression for the standard values of A=87.6 and

```
\mu=255');
disp('The black curve is for no compression. That is, A=0 and \mu=0');
disp ('--------PRESS A KEY TO CONTINUE ---------')
pause
disp (' ')
disp(' ')
mu = input ('Please enter the value of mu: ');
A = input ('Please enter the value of A: ');
s=0:.001:1;
F=(log(1+mu*s))/(log(1+mu));
F255=(log(1+255*s))/(log(1+255));
F40=(log(1+40*s))/(log(1+40));
F100=(log(1+100*s))/(log(1+100));
Fzero=s;
subplot(2,2,1)
plot (s,F,'r',s,F255,'b',s,Fzero,'k')
xlabel('Input');
ylabel('Output')
title('\mu-law companding');
s=(1/A):.001:1;
Alaw=(1+log(A*s))/(1+log(A));
Alaw87=(1+log(87.6*s))/(1+log(87.6));
Alawzero=s;
s1=.01:.001:1/A;
Alaw1=A*s1/(1+log(A));
Alaw187=(1+log(87.6*s1))/(1+log(87.6));
Alaw1zero=s1;
subplot(2,2,2)
plot (s,Alaw,'r',s1,Alaw1,'r',s,Alaw87,'b',s1,Alaw187,'b',s,Alawzero,'k',s1,Alaw1zero,'k')
title('A-law companding');
xlabel('Input');
ylabel('Output')
```

One way to implement the μ-255 compander is to simulate a non-linear system which follows the μ-255 input/output curve. Then place the sample values through this system, and uniformly quantize the output using an 8-bit A/D converter. An alternate way is to approximate the μ-255 curve by a piecewise linear curve, as shown in Fig. 2.51. We

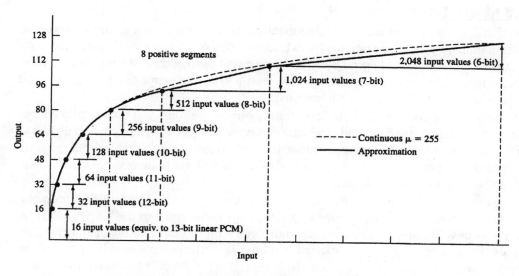

Figure 2.51 - μ-255 Piecewise Linear Approximation

illustrate the positive input portion only–the curve is an odd function. We are also showing this for 8-bit A/D.

Note that we have approximated the positive portion of the curve by eight straight line segments. We divide the positive output region into eight equal segments, which effectively divides the input region into eight *unequal* segments. Within each of these segments, we uniformly quantize the value of samples using four bits of quantization. Thus, each of these eight regions is divided into 16 subregions, for a total of 128 regions on each side of the axis. We therefore have a total of 256, or 2^8 regions which corresponds to eight bits of quantization. The specific technique for sending a sample value is to send eight bits coded as follows:

1 bit is used to give the polarity of the sample: 1 for positive, 0 for negative.

3 bits are used to identify which piecewise segment the sample lies in

4 bits are used to identify the quantization level within each sample region.

The logarithmic relationship of the μ-255 law leads to an interesting dependence among the eight segments. Each segment on the (positive) input axis is twice as wide as the segment to its left. The resolution of the first segment to the right of the origin is therefore twice that of the next segment, and so on. The sixth region to the right of the origin covers an interval on the input axis which is 1/16 of the total swing. Thus, the resolution for samples in this particular interval is the same as that of uniform quantization using 8-bits of A/D. The resolution of the region just to the left of this is the same as that of 9-bit uniform quantization. Similarly, as we move to the left, each region has the resolution of a uniform quantizer with one more bit of quantization than the previous. These levels are marked on the figure.

2.3.3 M-ary Baseband

In Section 2.3.1, we examined various baseband signal formats. As one example, we found that the nominal upper frequency for NRZ-L signals was R_b, the bit rate (Figure 2.28). As an example, if you wanted to send 8000 bits per second down a wire, the wire would have to be capable of passing frequencies up to at least 8 kHz. But suppose the channel could only pass up to 4 kHz? What are our options?

If you were not concerned about "real time" communications, you could build a large buffer to store the binary data, and send data down the channel at a rate one-half as fast as it arrives from the source. In most cases this would not be acceptable both because messages are time sensitive, and because the buffer would have to increase in size without limit as the message gets longer and longer.

If we could somehow make the pulses wider, that would lower their frequency content. But it would seem that we are limited to a maximum width equal to the bit period, T_b. A solution to this dilemma is to have each pulse carry more than one bit of information. Suppose, for example, each pulse carried two bits. Then the pulse would have to change every $2T_b$, and the frequency would be cut in half. But for the pulse to carry two bits, it must have one of four possible levels instead of two. This is the essence of M-ary communications. If each pulse carries only one bit, we call it binary. If it now carries two bits, we call it 4-ary; three bits and it is 8-ary; and so on.

Figure 2.52 shows a binary sequence, the resulting binary NRZ-L (unipolar) waveform, and the equivalent 4-ary waveform.

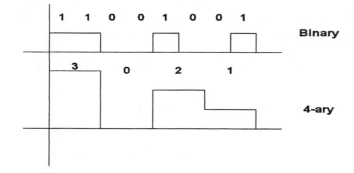

Figure 2.52 - M-ary
Baseband Communication

The minimum width of the pulses in the 4-ary case is twice that of the binary, so the nominal highest frequency of the 4-ary waveform is $R_b/2$. Had we combined bits in groups of three, we would have pulses that take on one of eight possible values, and the minimum pulse width would be three times that of binary resulting in a bandwidth reduction by a factor of three.

By this point in your education (or indeed, your life), you should know that nothing comes for free. There are always trade offs, and as one thing gets better, another thing gets worse. The price we pay for requiring additional signal heights is that the various levels get closer together. This assumes that we don't increase the maximum pulse amplitude beyond what it was for binary communications. As the levels get closer together, it will take a smaller additive disturbance to make one number turn into another, resulting in reception errors.

2.3.4. Time Division Multiplexing

To extend the concept of TDM as explored in Section 2.2.2 to digital signals, we need only modify that discussion slightly. This is true since each sample, instead of needing one pulse for transmission, now requires a number of pulses equal to the number of bits of quantization. As an example, for 8-bit PCM, each multiplexed signal requires eight pulses during each sampling period.

T-1 Carrier System

In the early 1960's, commercial and public digital communication systems were developed for voice channels. The *Bell System* (which no longer exists) already had an elaborate carrier system in operation, with cables that could handle bandwidths up to several MHz. In addition to the advantage inherent in all forms of digital communication, an additional advantage of importance to phone companies is that the signaling information required to control telephone switching operations could be economically transmitted in digital form.

The *Bell System T-1 digital communication system* was formulated to be compatible with existing communication systems. The existing equipment was designed primarily for interoffice trunks within an exchange, and therefore was suited for relatively short distances of about 15 to 65 kilometers. As the phone companies adapt their hardware to broadband data communications, the distance limitations are proving to be troublesome, particularly for Digital Subscriber Loop (DSL–don't get nervous. We will avoid discussion of DSL vs. cable!).

The system develops a 1.544 megabit/second pulsed digital signal for transmission through the channel. The signal is developed by time-division multiplexing 24 voice channels. Each channel is first sampled at a rate of 8000 samples/second. The samples are then quantized using 8-bit μ-law companding. The least significant bit is periodically (typically once every six samples) devoted to *signaling* information. This includes the necessary bookkeeping to set up the call and keep track of billing (the financial driving force behind most of today's communication systems). The 24 channels are then time-

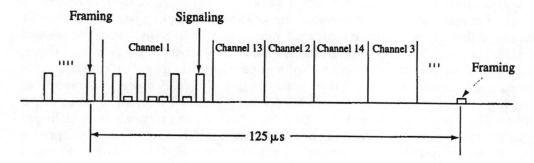

Figure 2.53 - TDM in T-1 System

division multiplexed as shown in Fig. 2.53.

The interleaved order of channels is selected because the system was designed to use existing multiplex equipment. The quantized samples from the 24 channels form what is known as a *frame*, and since sampling is done at an 8 kHz rate, each frame occupies 1/8000 = 125 μsec.

There are 192 information and signaling bits in each frame (24 × 8), and a 193rd bit is added for *frame synchronization*. This framing signal is a fixed pattern of ones and zeros in every 193rd pulse position. The product of 8000 frames per second with 193 bits per frame yields a transmission rate of 1.544 Mbits/sec.

The T-1 system can be used to send data as well as PCM since the basic system is simply a data communication system capable of transmitting 1.544 Mbps. When used for data, effort is required to properly interface with the hardware (which is expecting a structured frame of coded voice samples). Data often must be structured into 8-bit blocks, and framing pulses must be included.

T-1 as described above has been adopted by the United States, Canada and Japan. The T-1 structure for Europe is different, and follows the CCITT (Comite Consultatif Internationale de Telegraphique et Telephonique–the organization that sets voluntary international communication standards) recommendation. This system multiplexes 32 channels instead of 24. The bit rate is 2.048 Mbps instead of 1.544 Mbps.

2.3.5. Delta Modulation

In the PCM method of source encoding, each sample value is coded into a binary number. The resulting binary number must be capable of telling the receiver the corresponding sample value which ranges over the entire dynamic range. For example, if we start with a signal that ranges from -5 volts to +5 volts, the digital code must be capable of indicating sample values over a 10-volt range. The resulting resolution (and therefore round-off error) depends on this dynamic range.

If we could somehow reduce the dynamic range of the numbers to be communicated, the round-off regions would get smaller and the resolution would improve. Various techniques of source encoding operate on this principle.

Delta modulation is a simple technique for reducing the dynamic range of the numbers to be coded. Instead of sending each sample value independently, we send the *difference* between a sample value and the previous sample value. If sampling were at the Nyquist rate, this difference would have a dynamic range twice as large as that of the original samples. That is, at the Nyquist rate, each sample is independent of the previous sample. Two adjacent samples could differ by plus or minus the total range. One could be at the minimum and the other at the maximum of the signal amplitude. However, if we sample at a rate higher than Nyquist, the samples are not independent. The dynamic range of the difference between two samples is less than that of the samples themselves. Sampling at higher rates means that we must send bits more often. But if this process reduces dynamic range, we may be able to reduce the number of bits per second. Thus, the overall process could result in an improvement.

Delta modulation approximates this difference between adjacent sample values using *only one bit* of quantization. A one is sent if the difference is positive, and a zero is sent if the difference is negative. Since there is only one bit of quantization, the differences are coded into only one of two levels. We refer to these two possibilities as either $+\Delta$ or $-\Delta$. At every sample point, the quantized waveform can either increase or decrease by Δ.

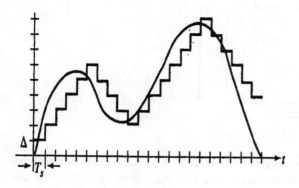

Figure 2.54 - Delta Modulation

Figure 2.54 shows an analog waveform and its quantized version. Note that all of the steps in this staircase are the same size. We attempt to fit a staircase approximation to the analog waveform. We shall examine the choice of sampling rate and step size later in this section. In fitting the staircase to the function, we need only make a simple decision at each sample point. If the staircase is below the analog sample value, the decision is to increment positively (i.e., an UP step). If the staircase is above, we increment negatively (i.e., a DOWN step). The staircase *cannot* stay at the same value since this would require sending three possible messages. The transmitted bit train for the example shown in Fig. 2.53 is

    1 1 1 1 1 1 0 0 0 0 1 1 1 1 1 1 1 0 0 0 0 0

The receiver reconstructs the staircase approximation directly from the received binary information. If a one is detected, the receiver increments with a positive step. If a zero is detected a negative step occurs. The receiver therefore can reconstruct the staircase function. (Convince yourself of this by looking only at the binary sequence above and drawing the resulting staircase function. You should find it's identical to Fig. 2.54). We then smooth this staircase (e.g., with a LPF) to get an approximation of the original function.

The above description leads to a simple implementation of the quantizer (A/D converter), using a comparator and a staircase generator. The resulting A/D converter is shown in Fig. 2.55.

The key to effective use of delta modulation is the intelligent choice of the two parameters, *step size* and *sampling rate* (*riser* and *tread width* if you are into staircase construction). These parameters must be chosen such that the staircase signal is a close approximation to the actual analog waveform. Since the signal has a definable upper frequency, we know the fastest rate at which it can change. However, to account for the fastest possible change in the signal, the sampling frequency and/or the step size must be

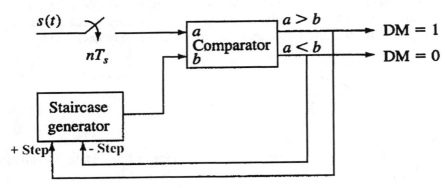

Figure 2.55 - Delta Modulator

increased. Increasing the sampling frequency results in the delta modulated waveform requiring a larger bandwidth (i.e., more bits per second must be transmitted). Increasing the step size makes the quantization error larger.

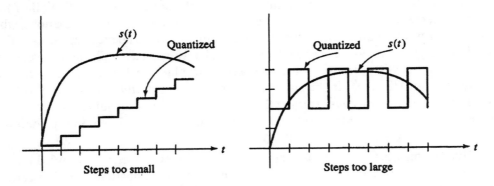

Figure 2.56 - Consequences of Incorrect Step Size

Figure 2.56 shows the consequences of incorrect step size. If the steps are too small, we experience a *slope overload* condition where the staircase cannot track rapid changes in the analog signal. That is, the maximum slope the staircase can track is $\pm\Delta/T_s$. If the signal slope is larger than this, the staircase is playing "catch up". If, on the other hand, the steps are too large, considerable overshoot occurs during periods when the signal is not changing rapidly. In that case, we have significant quantization noise, known as *granular noise*.

Delta modulation uses a source encoding process which employs memory in order to reduce dynamic range. Under certain circumstances, it is possible to achieve the same level of performance as PCM with fewer bits transmitted each second. Well, you know what's coming next....there must be a trade off consideration. However, because the system has memory, bit transmission errors propagate. In PCM, a single transmission bit error causes an error in reconstructing the associated sample value. The error affects *only* that single reconstructed sample. If a bit error occurs in delta modulation, the D/A converter in the receiver goes up instead of down (or vice versa), and all later values contain an *offset error* of twice the step size. If a subsequent bit error occurs in the opposite direction, the offset error is canceled. If the offset error poses a significant problem (often it does not due to *ac* coupling), the system can be periodically restarted from a reference level (usually zero). However, this introduces a transient response associated with each restart.

Adaptive Delta Modulation

We have shown that the appropriate step size to use in delta modulation depends on how rapidly the signal changes from one sample to the next. When the signal is changing rapidly, a larger step size helps avoid overload. When the signal is changing slowly, a smaller step size reduces the amount of overshoot, and therefore the quantization noise.

Adaptive delta modulation is a scheme which permits adjustment of the step size depending on the characteristics of the analog signal. Of course, the receiver must be able to adapt step sizes in exactly the same manner as the transmitter. If this were not the case, the receiver could not uniquely recover the original quantized signal (the staircase function). Since the transmission consists of a series of binary digits, the step size must be derivable from this bit train (unless we are willing to send a separate control signal as *overhead*).

If a given length string of bits contains an almost equal number of ones and zeros, we can assume that the staircase is oscillating about a slowly varying analog signal. In such cases, we should reduce the step size. On the other hand, an excess of either ones or zeros within a string of bits would indicate that the staircase is trying to *catch up* with the function. In such cases, we should increase the step size.

In one practical implementation, the step size control is performed by a digital integrator. The integrator sums the bits over some fixed period. If the sum deviates from what it would be for an equal number of ones and zeros, the step size is increased. In practice, the bit sum is translated into a voltage which is then fed into a variable gain amplifier. The amplification is a minimum when the input voltage corresponds to an equal number of ones and zeros in the period. The amplifier controls the step size.

Several adaptive delta modulation algorithms are simpler to implement than that discussed above. Two such rules are the Song algorithm and the space shuttle algorithm.

The *Song algorithm* compares the transmitted bit with the previous bit. If the two are the same, the step size is increased by a fixed amount, Δ. If the two bits are different, the step size is reduced by the fixed amount, Δ. Thus, the step size always changes, and it can get larger and larger, without limit, if necessary. We illustrate this for a step input function in Fig. 2.57.

A step input function represents an extreme case, and would not occur in the real world since a step function possesses infinite frequency. Note that a damped oscillation occurs following the rapid change in the signal.

If an analog signal is expected to have many abrupt transitions resembling the step function, the damped oscillations observed after the transition in the Song algorithm might be troublesome. Photographs of distinct and detailed objects in space could have many such transitions as they are scanned for transmission (i.e., a rapid change from bright to dark). The *space shuttle algorithm* is a modification of the Song algorithm which eliminates the damped oscillations. When the present bit is the same as the previous bit, the step size

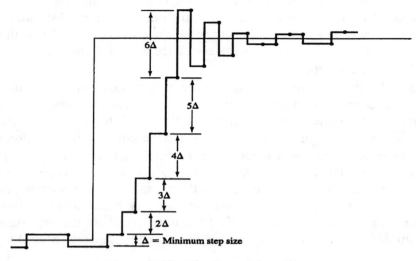

Figure 2.57 - The Song Algorithm

increases by a fixed amount, Δ. This is the same as for the Song algorithm. However, when the bits disagree, the step size reverts immediately to its minimum size, Δ. This is in contrast to the Song algorithm where the step size decreases toward the minimum at a rate of Δ every sampling period. The space shuttle algorithm is illustrated in Fig. 2.58 for the same step function input as in Fig. 2.57.

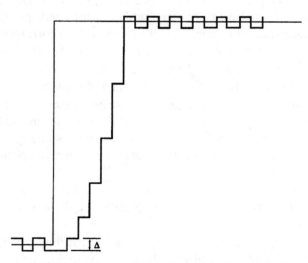

Figure 2.58 - The Space Shuttle Algorithm

2.3.6. Other A/D Conversion Techniques

In addition to PCM, DM and ADM, there are other methods for coding analog information into a digital format. The goal of each system is to send the information with maximum reliability and minimum bandwidth. We now introduce four of these methods: *Sigma-Delta modulation* (Σ-Δ), *delta pulse code modulation* (DPCM), *differential PCM* and *Adaptive differential pulse code modulation* (ADPCM). You can probably suggest others, and if your proposal has advantages for certain types of signals, you may attach your name to an algorithm (this will impress your non-engineering friends).

Sigma-Delta modulation begins by integrating the original signal and then feeding this integral into a form of delta modulator. In this process, the input to the delta modulator is actually the difference between the integral of the information signal and the integrated output pulses. Since integration is a smoothing process, we can expect changes in the integral waveform to be smaller than those of the original waveform. Integration also attenuates the higher noise frequencies (i.e., the transfer function of an integrator has a magnitude which is inversely proportional to frequency). Typical sampling rates are on the order of 64 times the Nyquist rate, so we often must send more bits than in a PCM system. However, sigma-delta modulation is able to achieve improved resolution (reduction of round-off errors). This form of modulation is part of the broader class known as *oversampling*. We discuss this in Section 2.5.3 when we analyze the round-off errors.

Prior to introducing the other analog to digital conversion techniques, we will take a moment to view delta modulation in a way different from that we used previously. This new approach will point the way to variations. In delta modulation, we approximate a continuous waveform with a staircase wave. At each sampling point, we develop an error term that is the difference between the signal and the staircase function. We quantize this error to develop a *correction term,* which is then added to the staircase function. In the case of basic DM, the quantization is done in units of one bit. In *delta PCM* (DPCM), we quantize the error into more than one bit, and add this term to the previous staircase value, as shown in Fig. 2.59. Therefore, instead of the stair steps having only one possible magnitude, they can now have two, four, eight or any power of two different magnitudes. At each sampling point, we must now send more than one bit of information–the various bits represent the PCM code for the error term. The advantage of DPCM over ordinary PCM is that, with proper choice of sampling interval, the error being quantized has a smaller dynamic range than that of the original signal. For the same number of bits of quantization, we can therefore achieve better resolution. The price we pay is complexity of the modulator. If the signal were always at its maximum frequency (the one that determines the sampling rate in PCM), DPCM would be essentially the same as PCM. However, since the signal frequencies are usually distributed over a range, adjacent samples are often correlated and it is possible to get better performance from this system than from a PCM system with the same rate of bit transmission.

Differential PCM is another technique for sending information about *changes* in the samples rather than about the sample values themselves. The differential approach includes an additional step which is not part of delta PCM. The modulator does not send the difference between adjacent samples, but instead uses the difference between a sample and its *predicted* value. The prediction is made on the basis of previous samples. This is

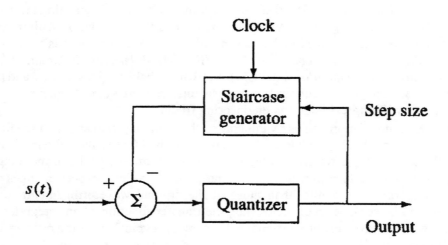

Figure 2.59 - Delta-PCM

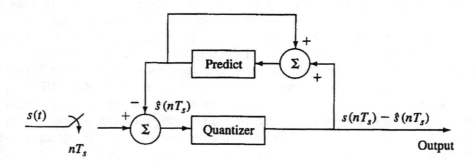

Figure 2.60 - Differential PCM

illustrated in Fig. 2.60. The symbol $\hat{s}(nT_s)$ is used to denote the predicted value of $s(nT_s)$.

The simplest form of prediction is where the estimate is a linear function of the previous sample values. Thus, using only one sample, we have

$$\hat{s}(nT_s) = As[(n-1)T_s] \qquad (2.15)$$

where A is a constant. The predict block in Fig. 2.60 is therefore a multiplier of value A.

The challenge is to choose A to make this prediction as *good* as possible. We get a measure of *goodness* by defining a prediction error as the difference between the sample and its estimate. Thus,

$$e(nT_s) = s(nT_s) - \hat{s}(nT_s)$$

$$= s(nT_s) - As[(n-1)T_s]$$

$$(2.16)$$

The mean-square value of this error is

$$mse = E\{e^2(nT_s)\} = E\{[s(nT_s) - As[(n-1)T_s]]^2\}$$

$$= E\{s^2(nT_s)\} + A^2 E\{s^2([n-1]T_s)\} - 2AE\{s(nT_s)s([n-1]T_s)\} \qquad (2.17)$$

$$= R(0)[(1+A^2) - 2AR(T_s)]$$

where $R(t)$ is the autocorrelation of $s(t)$. The error can be minimized with respect to A by setting the derivative equal to zero.

$$\frac{d(mse)}{dA} = E\{s([n-1]T_s)[s(nT_s) - As([n-1]T_s)]\}$$

$$= E\{s([n-1]T_s)e(nT_s)\} = 0$$

$$(2.18)$$

or

$$2AR(0) - 2R(T_s) = 0 \qquad (2.19)$$

This yields,

$$A = \frac{R(T_s)}{R(0)} \qquad (2.20)$$

Equation (2.18) has an interesting intuitive interpretation. It states that the expected value of the product of the error with the measured sample is zero. That is, the error has no component in the direction of the observation. The two quantities are *orthogonal*. This makes sense since, if the error *did* have a component in the direction of the observation, we could reduce that component to zero by readjusting the constant, A.

The predictor of Fig. 2.60 takes the most recent sample value (which it forms by adding the prediction to the difference term) and weights it by $R(T_s)/R(0)$. We assume that the input process has been observed sufficiently long to be able to estimate its autocorrelation.

Example 2.3

Find the weights associated with a predictor operating on the two most recent samples of a signal. Also evaluate the performance.

Solution

The prediction is given by

$$s(nT_s) = As[(n-1)T_s] + Bs[(n-2)T_s]$$

where the object is to make the best choice of A and B. For this best choice, the error has no component in the direction of the measured quantities. Thus,

$$E\{(s(n(T_s) - As[(n-1)T_s] - Bs[(n-2)T_s])\, s[(n-1)T_s]\} = 0$$
$$E\{(s(n(T_s) - As[(n-1)T_s] - Bs[(n-2)T_s])\, s[(n-2)T_s]\} = 0$$

Expanding these, we find that

$$R(T_s) - AR(0) - BR(T_s) = 0$$
$$R(2T_s) - AR(T_s) - BR(0) = 0$$

and solving for A and B yields

$$A = \frac{R(T_s)[R(0)-R(2T_s)]}{R^2(0)-R^2(T_s)}$$

$$B = \frac{R(0)R(2T_s)-R^2(T_s)}{R^2(0)-R^2(T_s)}$$

The mean-square error is then

$$mse = E\{[s(nT_s)-\hat{s}(nT_s)]^2\}$$

$$= E\{s^2(nT_s)\} - E\{\hat{s}(nT_s)s(nT_s)\}$$

$$= R(0) - \frac{R(0)R^2(T_s)+R^2(2T_s)-2R(2T_s)R^2(T_s)}{R^2(0)-R^2(T_s)}$$

As a specific example of the predictor derived in Example 2.3, suppose the autocorrelation of $s(t)$ is as shown in Fig. 2.61, and that the sampling period is 1 sec.

The equation for mse then yields mse = 1.895. For comparison, if we had forgotten to measure $s[(n-1)T_s]$ and $s[(n-2)T_s]$ and simply guessed at the mean value, or zero, the mean-square error would have been $R(0)$, or 10.

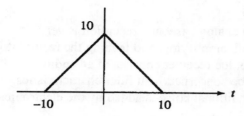

Figure 2.61 - Autocorrelation
for Example 2.3

Experiments have shown that for a speech signal a differential PCM system using a predictor operating on the most recent sample can save one bit per sample over PCM. That is, a differential PCM system can achieve approximately the same round-off error performance as the PCM system with one less bit per sample. Thus, if we think of a single voice channel as requiring 8 bits of PCM quantization, the transmission rate is 64 kbps. Differential PCM only needs 7 bits per sample, thus reducing the rate to 56 kbps and freeing the channel for other uses.

Yet another approach to A/D conversion is *adaptive DPCM*. Here, the predictor coefficients do not remain constant for the entire transmission. For each group of a given length of samples, say n, we compute a covariance matrix, $[R_{ij}]$. We then use this matrix to solve for the predictor coefficients. The predictor coefficients are then changed for the next n samples. Since the predictor coefficients are no longer constant, there must be some way to make sure that the receiver uses the same coefficients. The most common method for accomplishing this is to send the updated coefficients as *overhead*, usually multiplexed with the sample information.

We have presented four alternate techniques for encoding analog information into a digital format. There are other possibilities, many of which represent perturbations and combinations of the basic techniques. The techniques described can be coupled with companding to improve performance. In each case, we must weigh the performance improvement (i.e., decreased quantization error) against the increasing complexity of the hardware.

2.4. RECEIVERS

This section presents the design of receivers for the various forms of signal we have discussed. Following this, Section 2.5 explores the performance of the various receiver configurations.

2.4.1. Analog Baseband Reception

This is the shortest section in this book because analog baseband reception is very simple. It consists of filtering the received signal, amplifying, and feeding the result into a transducer (e.g., speaker). In the case of audio, the receiver consists of an audio amplifier. In designing receivers, one needs to be concerned with filter characteristics and noise. Noise is added in the channel, and additional noise is added by the electronics in the receiver.

2.4.2. Discrete Baseband Reception

Reconstruction of the original signal from *natural-sampled* PAM follows directly from the sampling theorem. Indeed, natural-sampled PAM is the type we encountered in our first proof of the sampling theorem in Section 1.5. Recovery of the original analog signal from its sampled version requires a lowpass filter. The natural-sampled PAM receiver is therefore as shown in Fig. 2.61. The process of converting a PAM waveform to the continuous analog waveform is known as *demodulation*.

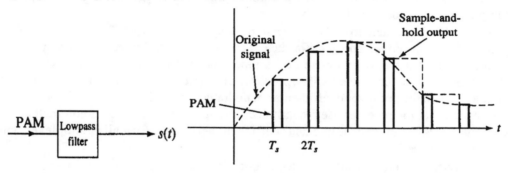

Figure 2.61 - Natural-Sampled   Figure 2.62 - Sample and Hold for PAM
        PAM Receiver   Demodulation

Demodulation of *instantaneous sampled PAM* requires a little more work. We could use a *sample and hold* circuit to recover a staircase approximation of the original waveform. The holding time is set equal to the sampling period. The result is shown in Fig. 2.62 for a representative *s(t)*.

The resulting staircase function can be lowpass filtered to get a smooth approximation to the original waveform. This form of sample and hold is an approximation to a lowpass filter.

Recall that the Fourier transform of instantaneous PAM is formed by first repeating the baseband Fourier transform at multiples of the sampling frequency, and then shaping the result with the transform of a single pulse (we did this in Section 2.2.1). The baseband portion of the PAM Fourier transform is of the form $S(f)H(f)$, where $H(f)$ is the Fourier transform of the square pulse. Therefore, *s(t)* can be recovered from the PAM waveform

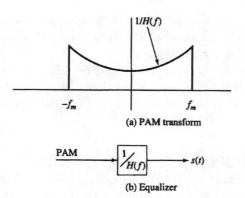

(a) PAM transform

(b) Equalizer

Figure 2.63 - Flat-top PAM demodulator

by using a shaped lowpass filter where the transfer function is the reciprocal of $H(f)$. This is shown in Fig. 2.63. The filter with transfer function $1/H(f)$ is known as an *equalizer* since it cancels the effects of the pulse shaping.

The equalizer and lowpass filter demodulators do not require that the receiver recover timing information. On the other hand, the sample and hold demodulator does require such information at the receiver. That is, the receiver must know when to sample the incoming waveform so it does not miss the original sample points. This requires *symbol synchronization*.

Reception of PWM or PPM can be viewed as a two-step process. We first convert the received waveform into PAM, and then use a PAM receiver. The conversion is accomplished using an integrator. For PWM, we simply start the integrator at the sample point and integrate the received pulse. Since the height of the pulse is constant, the integral is proportional to the pulse width. The output of the integrator is sampled prior to the next signal sampling point, and the sample generates a PAM waveform. This process is illustrated in the bottom waveform of Figure 2.64.

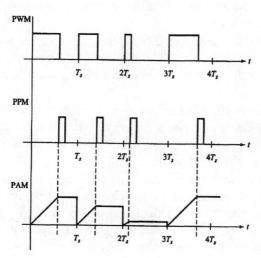

Figure 2.64 - Conversion of PWM and PPM to PAM

There are two observations you should make regarding this figure. First, we are using the form of PWM that sets the left edge of each pulse at the sampling point. Second, the resulting PAM waveform is delayed by one sampling period.

Conversion of PPM to PAM can also be illustrated in Fig. 2.64. Here, we start the integrator at each sampling point, and set it to integrate a constant. The integrator stops when the pulse arrives. Since the PPM pulse is at the trailing edge of the PWM pulse, there is no essential difference between PWM and PPM reception.

2.4.3. Digital Baseband Reception

Suppose you were given the NRZ-L baseband signal of Fig. 2.65(a) and asked to decide whether a binary one or binary zero was being transmitted in a particular interval. You might propose a system which simply samples the signal at the midpoint of the interval. These sample values could then be compared to zero. The signal is decoded as a one if the sample is positive and zero if the sample is negative. A receiver operating on this principle is illustrated in Fig. 2.65(b).

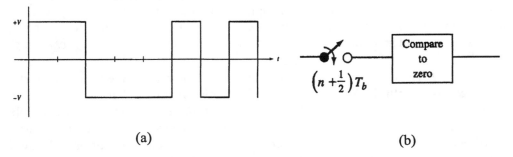

(a)   (b)

Figure 2.65 - Single-Sample Detector

Although Fig. 2.65(a) is drawn as an ideal noise free waveform, the sample value is actually a random variable because of the presence of noise.

One might suggest a modification of the receiver of Fig. 2.65(b) by adding a filter at the input. This filter reduces the noise, and could be chosen so that it does not change the mid-interval value of the signal. A *lowpass* filter might be chosen. Such a filter has a response to the signal pulse of the form $(\sin t)/t$.

We justify in Appendix C that the *matched filter* is the best type of filter to maximize the signal-to-noise ratio at the input to the decision device. Please take the time to study matched filters and the correlator implementation. Since the matched filter is "best", we shall explore the use of the matched filter as the building block of a receiver.

Let us start by ignoring the channel characteristics (i.e., assume the received signal is the same as the transmitted signal). The filter would then be matched to a perfect square pulse of $+V$ volts in order to look for the presence of a data one in the interval. The matched filter would have impulse response, $h(t)=V$ for $0<t<T_b$ (i.e., same as signal in one bit interval). Therefore, the receiver would be as shown in Fig. 2.66(a). Note that we are using the correlator implementation of the matched filter. If the signal portion of the input

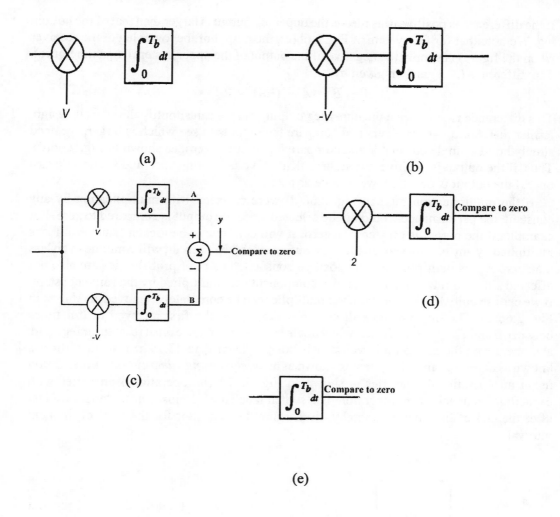

Figure 2.66 - Matched Filter Detector for +*V* Volt Pulse

to this detector is a square pulse of amplitude *V* and duration T_b, then the filter (correlator) is perfectly matched, and the output signal to noise ratio will be at its maximum value. In a sense, this filter is pulling the signal out of the background noise, but *only* if that signal is matched to the filter.

A second receiver, matched to the -*V* volt pulse would be needed to look for the data zero. This receiver is shown in Fig. 2.66(b). If the input to this receiver were a -*V* volt pulse of duration T_b, then this filter would have the effect of pulling that signal out of the background noise. So one way to determine which bit is being sent is to feed the received signal into both of the systems of Fig. 2.66(a) and 2.66(b), and observe which output is largest. Since we are looking for the largest output, an alternative technique is to take the difference between the two outputs and compare this to zero. This is shown in Fig. 2.66(c).

If the difference is positive, this means the upper leg output is larger than that of the bottom leg. We note that for the system of Fig. 2.66(c), the output of the lower leg of the receiver (*B* on the figure) is simply the negative of the output of the upper system (*A* on the figure). The difference (*A-B*) is then given by

$$A - B = A - (-A) = 2A$$

This difference is then twice the upper leg output, and we can simplify the overall design. Rather than build two receivers and compare the outputs to see which is larger, we need simply build a single correlator and compare its output to zero, as shown in Fig. 2.66(d). Thus, if the output is positive, we assume that +*V* volts is being sent, and we decode as a one. If the output is negative, we decode as a zero.

If the multiplication by 2 in Fig. 2.66(d) were replaced with multiplication by any constant, the performance would not be affected since the output is compared to zero. For example, if the output is greater than zero, it will continue to be greater than zero if it is multiplied by any positive constant. If the output is less than zero, it will continue to be less than zero if it is multiplied by any positive constant. The error probabilities are also not affected since both the signal and noise components are multiplied by the same constant. If we choose unity for the constant, the multiplier can be completely removed, as shown in Fig. 2.66(e). The diagrams are all shown operating on the first received bit for times between 0 and T_b. Once we decide which bit is being sent first, we must reset the integrators and operate on the next bit interval, which extends from T_b to $2T_b$. The reset operation is known as *dumping*, and the system is known as an *integrate and dump* circuit. A realization using an operational amplifier is shown in Fig. 2.67. The operational amplifier with capacitive feedback is an integrator. The switch labeled nT_b^+ closes instantaneously just after the end of the interval, thereby discharging the capacitor for the start of the next interval.

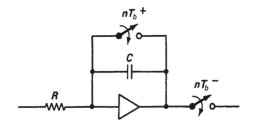

Figure 2.67 - Integrate and
Dump Detector

Binary Matched Filter Detector

In the discussion of the integrate and dump detector, we reasoned that matched filters could be effectively used in baseband receivers. We now generalize the concept of the matched filter detector to apply to any signal waveshapes. This will require a bit of work, but we feel that it is important for you to see the derivations. The results we obtain will be applicable to baseband as well as to the various forms of modulation introduced in the following chapters.

Figure 2.68 shows the *binary matched filter detector*.

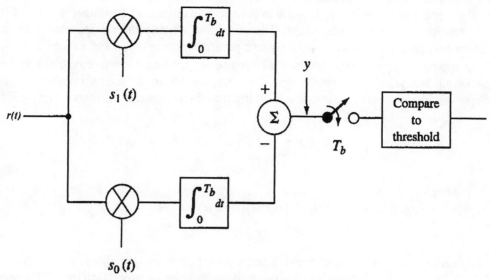

Figure 2.68 - Binary matched filter detector

We illustrate the receiver using correlators, but an exact equivalent receiver is that using matched filters to replace the multipliers and integrators (please see the discussion in Appendix C). The received signal is $r(t)$ and $s_0(t)$ and $s_1(t)$ are the signals used to transmit a zero and a one (+V and -V square pulses in NRZ-L). We assume that these two signal waveforms are known at the receiver–that is, the time functions we use are the result of passing the transmitted signals through the channel. We illustrate the receiver operating on the first bit interval, 0 to T_b. Following this decision, the receiver would then operate on the second bit interval, which extends from T_b to $2T_b$. It continues this operation for all future intervals $[nT_b$ to $(n+1)T_b]$.

The receiver compares the output of the two matched filters, one matched to signal $s_0(t)$ and the other matched to $s_1(t)$. It takes the difference of these two outputs and compares that difference to a threshold. We shall see that this threshold is often zero, in which case the detector is testing which matched filter output is largest–if the top filter output is larger than the bottom, y is positive; If the bottom output is larger, y is negative. Actually the switch is redundant in the idealized system. The output of each integrator is a number, not a time function. However, in real life, the integrators would operate over a sliding window and you would have to sample the outputs at the end of each interval.

The two signals, $s_0(t)$ and $s_1(t)$, could be baseband pulse waveshapes (e.g., NRZ, biphase), or they could be more complex waveforms of the type we study later in this text. Our study of the matched filter detector is now reduced to two challenges. First, we must decide how to choose the threshold against which the output, y, is compared. Second, we must evaluate the performance of this detector. Since choice of the threshold is considered part of receiver design, we examine that now. Performance of the detector is covered in Section 2.5.

The input to the threshold device, y, is a random variable. This is true since the output of each integrator is a *number* (NOT a time function) composed of a deterministic part (due to the signal) and a random part (due to the additive noise). The additive noise is assumed to be zero-mean Gaussian and white. Therefore the input to the comparator is Gaussian. A Gaussian density is specified by only two parameters, the mean and variance.

Suppose the input to the detector is $r(t)=s_i(t)+n(t)$, where i is either zero or one depending on which signal is being transmitted. That is, we send one of the two possible signals, and noise is added during transmission. The input to the comparator is then given by Eq. (2.23).

$$y = \int_0^{T_b} r(t)[s_1(t) - s_0(t)]dt$$

$$= \int_0^{T_b} s_i(t)[s_1(t)-s_0(t)]dt + \int_0^{T_b} n(t)[s_1(t)-s_0(t)]dt$$

(2.23)

We find the average value of y by adding together the average values of the two integrals in Eq. (2.23). The first integral is non-random, so its average is equal to itself. The average of the second integral is zero since we assume that the noise is zero-mean. Therefore, the mean value of y is

$$m_y = \int_0^{T_b} s_i(t)[s_1(t)-s_0(t)]dt$$

(2.24)

This mean value depends on which signal is being sent.

The variance of y is the expected value of the square of the difference between y and its mean. That is,

$$\sigma_y^2 = E\{[y - m_y]^2\}$$

$$= E\left\{\left[\int_0^{T_b} n(t)[s_1(t) - s_0(t)]dt\right]^2\right\}$$

(2.25)

Note that this result is independent of which signal is sent. We'll hold off expanding this variance until we get to evaluating performance in the next section. We'll come back to Eq. (2.25) at that time.

Now that we have found the mean and variance of y, we can sketch the probability density of y under the assumption that a zero or a one is sent. Figure 2.69 shows the two probability density functions, labeled $p_0(y)$ and $p_1(y)$. Note that the two densities have the same variance but different mean values.

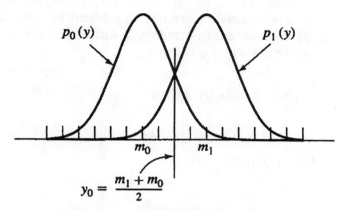

Figure 2.69 - Probability Densities of y

We are finally ready to select the threshold, which we will call y_o. If y is greater than the selected threshold, we assume that $s_1(t)$ is being sent, while if y is less than the threshold, we assume $s_0(t)$ is being sent. The correct choice of the threshold depends on the fraction of time a one is being sent, and also upon the cost of the errors. For example, if we know that ones are being sent 99% of the time, we would bias our threshold to a lower value to assure that we are more likely to decide that a one is being received (i.e., the threshold is broken more often). Similarly, if the cost of errors is not equal, we would bias the decision.

Why does cost come into the picture? Let's take a simple example. Suppose you were told that accidentally calling a transmitted "zero" a "one" costs you US$0.05, while erroneously calling a transmitted "one" a "zero" costs you a passing grade in this class. You would probably set the threshold very low so that you hardly ever decide that a zero is being received. In fact, we would not be surprised to hear you calling out the received bits as "one, one, one, one, one, one, ...one". You would want to do everything in your power to avoid accidentally missing a transmitted one.

More practical examples of costs affecting threshold exist in radar where the cost of erroneously labeling a zero as a one (a false alarm) might cost you a wasted antiballistic missile (only US$50 million), while the cost of calling a one a zero (a miss) might mean the nation's capitol is wiped out. Under such a scenario, you would certainly bias the decision toward having false alarms rather than misses (assuming your country permits unlimited budget deficits).

This intuitive discussion can be formalized under the topic of *decision theory*. We're saved from this complication because in the majority of digital communication applications, there is no reason to bias the decision. We should make the decision corresponding to the highest conditional probability. For example, if we measure $y = 2$, and $p_1(2)$ is larger than $p_0(2)$, we would decide that a one is being transmitted [1]. Referring to Fig. 2.69, this means that the threshold is chosen as the point at which the two probability density functions cross. This is labeled as y_o on the diagram. Note that due to symmetry, y_o this is the midpoint between the two mean values. It is therefore given by Eq. (2.26).

$$y_o = \frac{m_1 + m_0}{2} = \frac{1}{2} \int_0^{T_b} [s_1(t) + s_0(t)][s_1(t) - s_0(t)] dt$$

(2.26)

$$= \frac{1}{2} \int_0^{T_b} [s_1^2(t) - s_0^2(t)] dt$$

The integral of the square of the signal is the *signal energy* over the bit period. Denoting this energy as E_1 and E_0 for the two signals, we have

$$E_0 = \int_0^{T_b} s_0^2(t) dt \quad E_1 = \int_0^{T_b} s_1^2(t) dt$$

$$y_o = \frac{E_1 - E_0}{2}$$

(2.27)

If the signals have equal energy, the threshold is at the origin. In such cases, we compare the difference between the two integrator outputs to zero, and we are effectively testing which of the outputs is larger.

M-ary Baseband Reception

We have already agreed that the matched filter is the best type of detector to pull a signal out of background white noise. The matched filter detector for a flat pulse is simply an integrate and dump detector as we saw in Figure 2.66, where we must scale the output. So we can intuitively speculate that the best detector for M-ary baseband is as shown in Figure 2.71.

All we need to do is decide where to set the thresholds. For example, if the 4-ary signal of Figure 2.52 were to form the input to this detector, the integrator would be integrating one of four possible levels over the symbol period, T_s. Note that for 4-ary baseband, this signal period is twice the bit period. Suppose that the four amplitude levels were 0, A, $2A$, and $3A$. Then the output of the integrate and dump of Figure 2.70 would be either 0, AT_s, $2AT_s$, or $3AT_s$. Unless you had some reason to bias your decisions, you would set the thresholds in the middle of the intervals. Thus, if the output of the integrate and dump were less than $AT_s/2$, you would call this the symbol 00. If it were between $AT_s/2$ and $3AT_s/2$, you would call it

---

[1] If you've had a thorough course in probability, you are familiar with the terms *maximum a-posteriori* and *maximum likelihood*. We are using maximum likelihood decisions in this application.

symbol 01. Outputs between $3AT_s/2$ and $5AT_s/2$ would be called symbol 10, and outputs above $5AT_s/2$ would cause you to decide symbol 11. We will look at the performance of this system later. For now, we simply observe that the performance is not "symmetrical". For example, if you send symbol 01, negative noise can change this to 00 while positive noise can change it to 10. On the other hand, if you send symbol 00, no amount of negative noise will cause an error. The translation from symbol errors to bit errors is also non-uniform. If symbol 01 changes to symbol 10, this represents two bit errors, while if symbol 00 changes to 01, that represents only one bit error.

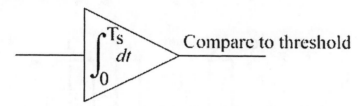

Figure 2.71 - Integrate and Dump for
M-ary Baseband Reception

Timing Recovery

Timing is an operation that significantly distinguishes a digital communication system from an analog system. Before a receiver can even begin to decide which of the various symbols it is receiving (e.g., zeros or ones in a binary system), it must establish symbol timing. In considering binary digital communication, the symbols are bits and symbol timing is the same as bit timing. We generalize this later when we introduce non-binary digital systems. So we must synchronize a local clock with the timing of the received symbols.

Once the symbol time is established and the receiver decoder is making decisions regarding which symbols are being received, it is necessary to establish other timing relationships, including character/word and block/message synchronization. It is not sufficient to decode a long string of symbols unless these symbols can be associated with the correct message. This is particularly important in time division multiplexed transmissions where the symbols must be *decommutated*.

Symbol Synchronization

The received waveform is usually in the form of an electrical signal extending over all time, so we must be able to chop the time axis into segments corresponding to the signal segments for each symbol.

Data transmission can be either synchronous, where symbols are transmitted at a regular periodic rate, or asynchronous, where spacing between words or message segments is irregular. Asynchronous transmission is often given the descriptive name, start/stop. Asynchronous communication requires that symbol synchronization be established at the start of each message segment or code block. This requires *overhead*. In the synchronous mode, symbol timing can be established at the very beginning of the transmission, and only minor adjustments (*tracking*) are needed thereafter.

The problem of bit (or symbol) synchronization is greatly simplified if a periodic component exists in the incoming symbol sequence. This is the case in some forms of signaling.

In the unipolar return-to-zero (RZ) signaling system, a transition occurs at the midpoint of every interval in which a one is being sent. We can therefore decompose the waveform into a sum of a periodic square wave plus a random pulse signal (we assume zeros are being sent randomly). We did this decomposition when we found the power spectral density of the RZ signal. Figure 2.72 is a repeat of this for a representative binary sequence.

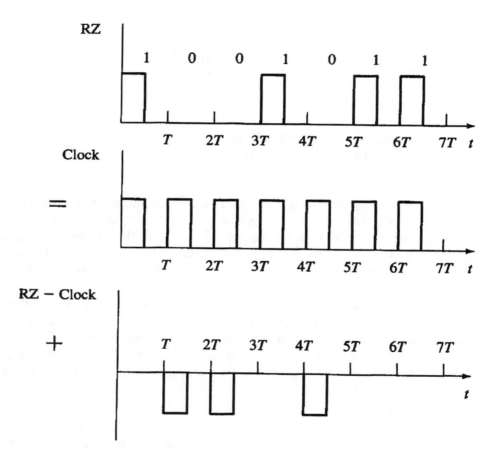

Figure 2.72 - Periodic Component in RZ Signal

Because the RZ signal contains a periodic component at the bit rate, its power spectrum contains a series of discrete spectral lines at the bit rate and its harmonics. A narrow bandpass filter tuned to the vicinity of the bit rate can extract the clock information. Alternatively, a phase lock loop (PLL) can be used to lock on to the periodic component which is at the bit rate.

The RZ system can be used in those cases where simplicity is important, and signal to noise ratio is high. The system has poor power efficiency since the transmitted signal is zero a significant fraction of the time. We therefore suffer a performance degradation when compared to NRZ or biphase signaling (we shall prove this in the next section). However, with NRZ or biphase transmission and random data, the received signal does *not* contain a periodic component at the bit rate. We must therefore resort to more sophisticated synchronization approaches. We now explore three of these techniques: the maximum a-posteriori, the early-late gate and the digital data transition tracking loop.

In the *maximum a-posteriori* (*MAP*) approach to bit synchronization, we observe the received signal over a finite length interval, and decode the signal using an assumed locally generated clock. The decoded signal is then correlated with a stored replica of the known symbol sequence, and the clock is adjusted to achieve maximum correlation. Perfect correlation is usually not obtainable due to bit errors which occur in transmission. Note that this scheme requires that part of the transmitted signal be dedicated to this function. That is, the receiver must know what it is looking for, so signal information cannot be transmitted during this synchronization period. *MAP* synchronization can be performed in serial or parallel, but problems are associated with either technique. In the serial mode, adjustments are made in the local clock, and the next group of received symbols is correlated with the stored sequence. This requires a long acquisition time as the known sequence must be transmitted more than once. Alternatively, the receiver could try different clock adjustments simultaneously using parallel processing. This is expensive in terms of hardware.

Because of the these shortcomings, practical approaches have been devised which approximate the *MAP* synchronizer. These approaches include adaptive control loops which lock on to the desired timing by minimizing an error term. The *early-late gate* is one such approach. A block diagram of the system is shown in Fig. 2.73.

The error signal is defined as the difference between the square of the area under the early gate and the square of the area under the late gate. This error goes to zero when the early and late gates coincide. In this case, the correct timing is at the midpoint of the gate periods. When the error is non-zero, the timing changes in a direction to cause the error to go to zero. Note that when the signal is zero, both the early and late gates have zero output and no adjustments occur.

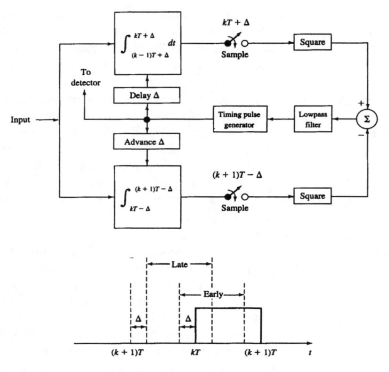

Figure 2.73 - Early-Late Gate

The *digital data transition tracking loop* (DTTL) is shown in Fig. 2.74. When this loop is in lock, the upper portion integrates the input pulse over periods corresponding to the bit

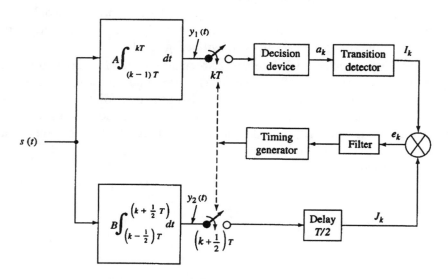

Figure 2.74 - DTTL

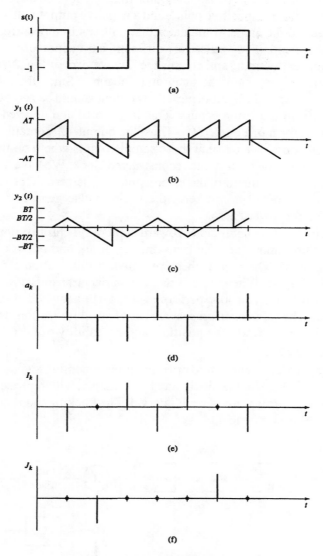

Figure 2.75 - Typical DTTL Waveforms

period, while the lower portion integrates over intervals spanning transitions. This is best understood by examining typical waveforms when the loop is in lock. Figure 2.75 illustrates a typical set of these waveforms.

Figure 2.75(a) shows a representative NRZ-L baseband waveform and Figs. 2.75(b) and (c) show the output waveforms for the in-phase and quadrature integrators. Note that the quadrature integrator dumps (sets to zero) at the midpoints of the intervals.

The remainder of the system compares integrator outputs from one interval to the next. Figure 2.75(d) shows the output of the decision device, which is simply a sampler and hard limiter. It tests the output of the in-phase integrator to see if it is positive or negative. We indicate a positive output with a positive pulse, and a negative output with a negative pulse. If two adjacent pulses are the same, it indicates that no transition occurred in the original waveform. On the other hand, if two pulses are different, a transition occurs. The transition detector compares adjacent outputs, and creates the I_k waveform of Fig. 2.75(e). In the case where no transition occurs, we view the quadrature channel. Since the integrator for this channel is offset by one half of the symbol period, its output should be zero when a transition occurs, and non-zero if no transition occurs. This is sampled and delayed by one-half bit period to form the waveform of Fig. 2.75(f). Thus, the output I_k contains a pulse at each transition point, and J_k contains a pulse at those sample points where no transition occurs. These two waveforms combine to form the reconstructed clock. We have analyzed the loop in the lock condition. In this situation, the error signal, e_k, is zero. This is found by multiplying I_k by J_k. If this product is not zero, the timing pulse generator is adjusted in a direction to force the error toward zero.

In analyzing the control loop approaches toward symbol synchronization, it is important to be aware of acquisition times (i.e., the transient response), and of the chance of falsely locking at the wrong point. The approaches presented in this section effectively recover timing in the absence of noise. When noise is present, or data transmission is at high speeds relative to loop acquisition time, the loop performance must be analyzed to assure that system specifications are met. The references present curves of synchronization error as a function of signal to noise ratio, mean time to acquisition, and probability of false lock.

Non-Linear Clock Recovery

Non-linear filtering approaches attempt to operate on the incoming signal in a manner that produces a periodic component at the clock rate. For example, with biphase-L coding, there is always a transition at the midpoint of each interval. The tracking loop shown in Fig. 2.76 can be used to extract this component.

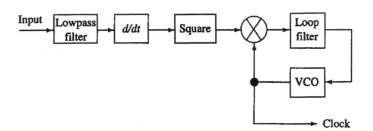

Figure 2.76 - Non-Linear Tracking Loop

The system first differentiates the filtered waveform, producing a sharp pulse (ideally, an impulse) at the center of each interval. Some of these pulses will be positive and others negative depending on the direction of the original transition. A squaring device is used to make each of these pulses positive, and therefore generate a periodic element within the signal waveform. This periodic element is extracted with a narrowband filter tuned to a frequency close to the bit rate, or by a phase lock loop.

Frame Synchronization

When using time division multiplexing the timing of the overall frame must be established. The receiver must be able to associate individual bits with the proper transmitting source. The decommutator must recognize when a new frame begins.

Frame synchronization can be accomplished by transmitting a very distinctive voltage waveform at the start of each frame, much as a *comma* or *period* is used to separate phrases or sentences. If we are restricted to two possible signal waveforms (i.e., binary), our flexibility is highly limited and we cannot use a distinctive voltage waveform. Frame synchronization should therefore be accomplished as an integral part of symbol transmission. This is usually done by sending a specified sequence, or *prefix*, at the start of each frame.

As a first approach, let us assume that we send a binary one at the beginning of each frame. Thus, if the words are k bits long, a one would be inserted as every $(k+1)^{th}$ bit. The receiver would add together bits separated by k positions. The addition could be performed by an integrator. A shift-register arrangement, as shown in Fig. 2.77 performs this addition for five words.

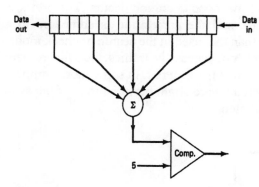

Figure 2.77 - Addition of
Five Framing Bits

When the system is locked on to the one bit marking the beginning of each frame, the output of the summing device is five. The probability of achieving a sum of five for any other position would be $(0.5)^5$, or 1/32, assuming that the bits within words are equally likely to be zero or one. This false synchronization probability could be reduced by including more frames in the sum. In any case, this system would continually check the sum and adjust if necessary. Therefore, if the system goes out of synchronization, it will usually re-achieve sync within $k+1$ input bits. The reliability of this technique can be improved by enlarging the prefix code to be longer than one bit in length.

Before examining prefixes in more detail, let's take a moment to formalize the strategies used to establish and maintain frame synchronization. Synchronization occurs in three modes: Search, check and lock. In some systems, the search and check modes are combined into a single *acquisition* mode.

Suppose that the prefix is N bits long. Then in the search mode, the frame synchronizer correlates each successive N data bits with a locally-stored reference word. A threshold is set (T_S) depending upon system noise (bit error rate), and if the correlation exceeds this threshold, the search mode is exited, and the *check* mode is entered (See Fig. 2.78). The selected length of the prefix and the number of errors tolerated in the search mode depend upon the noise level and critical nature of the signal. For example, in the   *T1 voice communication system*, the frame consists of 192 bits and only one additional prefix bit. In other systems (particularly military, and/or low SNR applications) the frame might be several thousand bits long and the prefix might be in the order of 20 bits in length. In such cases, the tolerable errors during search may range from zero (in a high SNR case) to several bits (in a low SNR case). There is a tradeoff between acquisition time and probability of false lock. If the threshold is set too low, the loop spends very little time in the search mode, but the probability of locking on to the wrong point in the frame increases.

In the *check* mode, the synchronizer looks at the next frame and correlates the prefix with the stored word. In some cases, more than one additional frame is examined. A threshold is set (T_C,), and if that threshold is exceeded, the lock mode is entered. If the threshold is not reached, we assume that the proper synchronization point has not been obtained, and the search mode is reentered. Note that the threshold for the check mode is not necessarily the same as that used in the search mode.

The *lock* mode is the normal mode in which we hope operation occurs. The prefix is continually correlated against the stored word, and the correlation is compared to a threshold (T_L). This threshold is usually lower than that used in the search or check mode. If the threshold is not broken in a specified number of successive frames, the synchronizer is considered to be out of lock, and the search mode is reentered. In some more complex synchronizers, a modified search mode is entered where slight perturbations of the sync position are tried before the full search is reinitiated.

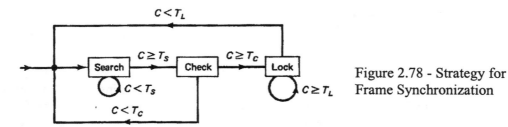

Figure 2.78 - Strategy for Frame Synchronization

Codes for Synchronization

The distinctive symbol sequence marking the beginning of each frame is known as a *marker*, or *prefix*. Let us assume that this sequence is m bits long. The receiver then correlates each m-bit received sequence with a stored version of the known prefix. If the correlation exceeds a specified threshold, the search mode is exited.

We desire a prefix sequence for which there is a minimum probability of falsely establishing frame synchronization at the wrong location. For example, suppose we choose the 7-bit prefix sequence, 0001101. The received sequence will be of the form

    ....x x x x x x 0 0 0 1 1 0 1 x x x x x....

where "x" is used to denote a portion of the information message. We consider the x's to be randomly distributed with equal probability of each being zero or one. Let us define the *correlation* between the stored sequence and the portion of the received sequence as the difference between the number of bit agreements and the number of bit disagreements. Thus, the correlation ranges from +7, when all bits are identical in the two sequences, to -7, when all bits are different. When synchronization is achieved, in the absence of bit errors, the correlation is 7. Suppose we compare the two sequences and that the timing is incorrect by one bit period. We then would compare

x 0 0 0 1 1 0

with

0 0 0 1 1 0 1

The correlation is either +1 or -1 depending upon whether x is zero or one, respectively (please check this yourself). In the other direction, we compare

0 0 1 1 0 1 x

with

0 0 0 1 1 0 1

Again, the correlation is either +1 or -1 depending upon whether the value of x is zero or one respectively.

Clearly, in design of prefix sequences, we wish to obtain minimum correlation of the sequence with truncated shifted versions of itself.

The truncated correlation, for a shift of k positions, is given by

$$C_k = \sum_{i=1}^{m-k} x_i \, x_{i+k} \qquad (2.28)$$

where x_i is either +1 or -1, and m is the length of the prefix.

Among the codes for use in synchronization are the *Barker code*, for which the correlation is limited to a magnitude of one for $k \neq 0$. Barker codes have been discovered for lengths of 3, 7, 11, and 13. Thus, for example, with a length of 13, the correlation for a zero shift is 13, while for any non-zero shift, the correlation is bounded in magnitude by 1. Unfortunately, Barker codes do not exist for lengths greater than 13 (Actually, it can be proven that none exist for lengths between 14 and 6084). The known Barker codes are given by

1 1 0

1 1 1 0 0 1 0

1 1 1 0 0 0 1 0 0 1 0

1 1 1 1 1 0 0 1 1 0 1 0 1

In a very noisy environment, we may require prefixes longer than 13 bits in length. *Neuman-Hofman* codes are designed to minimize the maximum value of cross-correlation between the prefix pattern and the incoming data stream. As an example, the Neuman-Hofman code of length 24 is

0 0 0 0 0 1 1 1 0 0 1 1 1 0 1 0 1 0 1 1 0 1 1 0

The correlation, with zero shift, is 24. The maximum magnitude correlation for other shifts occurs at a shift of 10, and results in a correlation of -4. The reason we are concerned with negative correlations (i.e., magnitude) is that these codes are often used in differential systems where bit reversals may occur. Thus, rather than look for the maximum positive correlation to declare frame synchronization, we look at the magnitude of the correlation.

Example 2.4

A frame synchronization scheme is to be designed with a search mode, a check mode and a lock mode. Each frame contains 250 bits of information and a preamble of length N bits. The frame must be transmitted within one second. The bit error rate is given by

$$P_e = 0.5 - 0.5e^{-\frac{f}{2500}}$$

where f is the number of bits per second being transmitted.

(a) If the preamble is 10 bits long and frame synchronization must be acquired within 10 frames (with 99% probability), find the threshold to be used in the search mode.

(b) Once the system is in lock, we wish it to remain in this condition with probability 99.9% Find the lock threshold value.

Solution: Before solving this problem, we note that the given bit error rate is a function of the bit transmission rate. As the bit rate increases, the time allocated to each bit decreases. This decreases the transmitted energy per bit. We shall see in Section 2.5 that the decrease in energy per bit has the effect of increasing the bit error rate. Therefore adding additional framing bits is not without cost.

(a) With a 10 bit preamble length, the bit error rate is given by

$$P_e = 0.5 \left(1 - e^{-\frac{260}{2500}}\right) = 0.0493$$

Let us denote the search mode threshold as T_S. That is, if there are at least T_S matches out of the 10 prefix bits, the search mode is exited. The probability of exactly T_S matches out of 10 bits is

$$\binom{10}{T_S}(1 - 0.0493)^{T_S}(0.0493)^{10 - T_S}$$

where the first term is the number of ways that T_s matches can be distributed among 10 bit positions. The probability of 10 matches is $(0.9507)^{10}=0.603$. The probability of nine matches is 0.916. The specifications in the problem are "loose" enough that we need look no further. We are asked to acquire synchronization within 10 frames with a probability of

99% If we require the full 10 bits to agree, the probability of this happening at least once during 10 frames is given by

$$1 - (1 - 0.603)^{10} = 0.9999$$

We have calculated this quantity by taking the difference between one and the probability of less than 10 matches occurring 10 times in a row. Thus, the probability of acquiring frame sync within 10 frames is 99.99% assuming we require 10 matches. We therefore set T_S at 10.

(b) Once the search mode is terminated, we want to stay in synchronization with a probability of 99.9% Let us assume that the check is made over only one frame at a time (this is a high signal to noise ratio environment). We then need to check the formula given above to find out how many agreements are required to meet the specification. The probability of at least seven agreements is given by

$$\sum_{n=7}^{10} \binom{10}{n} (1 - 0.0493)^n (0.0493)^{10-n} = 0.999$$

Therefore, the threshold during lock should be set at 7 matches.

Design Example

Suppose you are asked to design a system that must transmit 1 million information bits per second. The information is arranged in frames of 1,000 bits each. The environment is extremely noisy, and the application is highly sensitive. You must design the frame synchronization system to have a probability of less than 10^{-6} of falsely locking at the wrong point. Frame synchronization must be acquired within five frames with a probability greater than 99.9 percent. These are relatively "tight" specifications, and they might apply to a military system.

We are at quite an early point in our study of digital communications to tackle a design problem of this magnitude. However, we shall use it to establish parameters and motivate the further studies.

A very important quantity that the problem has not specified is the bit error rate. We discuss this in detail in Section 2.5, but for now, you need only know that the bit error rate is the probability of making an error (i.e., mistaking a transmitted one for a zero, or vice versa). The problem states that the environment is very noisy, so we could expect bit error rates on the order of 10^{-2} or more. As the bit transmission rate increases, less time is available to send each bit, and the pulse spreading (intersymbol interference) increases. Thus, as the transmission rate increases, we would expect the bit error rate to also increase. This is an important observation, since lengthening of the frame synchronization prefix must increase the transmission rate. The problem specifies that each frame contains 1,000 information bits, and the information must be transmitted at 1 Mbps. Thus, the bit transmission rate will be $(1,000 + N)/1,000$ times 1 Mbps.

The actual relationship between bit error rate and transmission speed depends upon the noise power, the bandwidth of the channel, channel factors such as whether there is multipath, the transmission scheme, and the detector design. Since we need this information to solve the problem, and many of these topics are not covered until later in the text, let us (arbitrarily) assume that the bit error rate is related to the transmission rate by the graph of Fig. 2.79. The key values are summarized in the following table and graph:

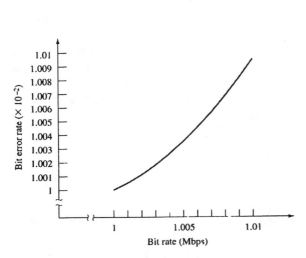

Transmission rate (Mbps)	BER($\times 10^{-2}$)
1.000	1.0000
1.001	1.0005
1.002	1.0011
1.003	1.0018
1.004	1.0026
1.005	1.0035
1.006	1.0045
1.007	1.0056
1.008	1.0068
1.009	1.0081
1.010	1.0095

Figure 2.79 - Error Rate vs. Bit Rate

We are now ready to try various framing techniques. To start the analysis, let us assume that a single framing bit is added to each frame and that the receiver looks for that bit over a single frame. This is clearly not acceptable, since the probability of false lock would be 50 percent. That is, the probability that any information bit would be the same as the transmitted frame synchronization bit. Even if the receiver scans over five adjacent frames (as permitted in the specifications), the probability of the information having any specified five-bit pattern is 2^{-5}, or 3.1×10^{-2}. The specifications call for a false-lock probability of less than 10^{-6}. Clearly, we must go to longer frame synchronization sequences.

We will start by concentrating on the probability of acquiring synchronization in order to gain facility with the probability calculations. Suppose we decide to try a Barker code of length 3, that is, 110. Suppose further that we set the threshold at the correlation value 3. That is, we look for perfect agreement between the received prefix and the locally-generated replica. The probability of locking in a single frame will then be the probability that no bit errors occurred in the prefix, or

$$PR(lock) = (1 - BER)^3 = (1 - 0.010018)^3 = 97\%$$

where BER is the bit error rate. This does not meet the specifications, so there is no need to examine the probability of falsely locking onto the wrong point. We must investigate either lowering the threshold or viewing more than a single frame at a time.

Suppose we decide to lower the correlation threshold to 2. Then the probability of locking in a single frame is still the probability of there being no bit errors. This is true because a single bit error will reduce the correlation by 2. Thus, the probability of locking in a single frame still does not meet the specifications. We will not even waste time trying to lower the required correlation threshold to 1, since it should be clear that the probability of false locking under this condition would also be unacceptable.

The alternative is to increase the number of frames over which we look for agreement. Suppose we examine the received sequence over two frames and look for a total correlation of at least 4. We can then permit a single bit error over the six bits, so the probability of acquisition becomes

$$(1 - BER)^6 + 6 \times BER \times (1 - BER)^5 = 99.85\%$$

This still does not meet the specifications. If we wish to permit two bit errors over the two frames (i.e., look for a total correlation of at least 2), we will meet this specification, but not be close to the false-lock specification. Hence, we must look over at least three frames. Suppose we do this and require a total correlation of at least 5. That is, we are examining nine bits and are permitting up to two bit errors. The probability of acquisition is then

$$(1 - BER)^9 + 9 \times BER \times (1-BER)^8 + 36 \times (1 - BER)^7 = 99.99\%$$

We have therefore met the first specification and must now examine the probability of false lock. Unfortunately, this is where we get into trouble. The probability of falsely locking is the probability of getting at least seven agreements between the synchronization sequence (three 3-bit transmissions over three frames) and the data. If the data are considered to be random, the probability of false lock is

$$(0.5)^9 + 9 \times (0.5)^8(0.5)^1 + 36 \times (0.5)^7(0.5)^2 = 46 \times (0.5)^9 = 8.98 \times 10^{-2}$$

which clearly does not meet the specification. Even if we were to look over the full five frames permitted by the specifications and require perfect agreement (doing so could not achieve the required probability of acquisition), the probability of false lock would be

$$(0.5)^{15} = 3 \times 10^{-5}$$

This does not meet the specification of 10^{-6} maximum probability of false lock. Obviously, we need a longer synchronization sequence.

Suppose we go to the 7-bit Barker sequence. Let us approach the analysis by examining the acquisition probability as a function of the required correlation. Note that the bit error rate increases to 1.0056×10^{-2} since we must transmit at a faster rate. If we use the full five frames, we are transmitting 35 synchronization bits. The probability of errors among these 35 bits is found from the binomial distribution and is as follows:

$$Pr(0 \; errors \; in \; 35 \; bits) = (1 - BER)^{35} = 0.702$$

$$Pr(1 \; e \; in \; 35 \; bits) = 35 \times (BER)^1(1 - BER)^{34} = 0.2496$$

$$Pr(2 \; errors \; in \; 35 \; bits) = \binom{35}{2} \times (BER)^2(1 - BER)^{33} = 4.31 \times 10^{-2}$$

$$Pr(3 \; errors \; in \; 35 \; bits) = \binom{35}{3} \times (BER)^3(1 - BER)^{32} = 4.82 \times 10^{-3}$$

$$Pr(4 \; errors \; in \; 35 \; bits) = \binom{35}{4} \times (BER)^4(1 - BER)^{31} = 3.914 \times 10^{-4}$$

$$Pr(5 \; errors \; in \; 35 \; bits) = \binom{35}{5} \times (BER)^5(1 - BER)^{30} = 2.465 \times 10^{-5}$$

Since the specifications require a 99.9-percent probability of locking within five frames, we see that permitting up to three errors among the 35 synchronization bits would meet the specifications. That is, if three errors are permitted, the probability of not locking is the probability of four or more errors, which is approximately 4×10^{-4}. Thus, the probability of acquisition is approximately 99.96 percent, which meets the specifications.

We require a correlation of at least 29 [35 - 2 × 3)] to declare that frame synchronization is achieved. It is now necessary to check that the probability of false lock is within the specification. The probability of achieving agreement between 32 of the 35 bits, assuming random data, is

$$Pr(false \; lock) = 1 \times (0.5)^{35} + 35 \times (0.5)^{34}(0.5)^1$$

$$+ \binom{35}{2}(0.5)^{33}(0.5)^2 = \binom{35}{3}(0.5)^{32}(0.5)^3$$

$$= 2.09 \times 10^{-7}$$

This meets the specification, so we have been successful at configuring one possible frame synchronization scheme.

We have ignored other design trade-off considerations, including the entire issue of hardware and implementation. We will be ready to tackle these issues after learning about specified transmission techniques.

2.4.4 Intersymbol Interference Revisited

Before performing a mathematical analysis of intersymbol interference, we shall begin with a pictorial approach in order to introduce the concept of *eye patterns*. Suppose that binary information is transmitted using a baseband digital waveform. Let's assume a 1-V pulse is used to send a binary one and a 0-V pulse for a binary zero (i.e., unipolar baseband). When the waveform goes through the system, it is distorted. Any sharp corners of the wave are rounded since practical systems cannot pass infinite frequencies. Therefore, the values in previous sampling intervals affect the value within the present interval. If, for example, we send a long string of ones, we would expect the channel output to eventually settle to being a constant, one. Similarly, if we send a long string of zeros, the output should eventually settle toward zero. If we alternate ones and zeros, the output might resemble a sine wave, depending on the frequency cutoff of the channel.

If we now plot all possible waveforms within the interval (assuming all possible past transmissions), including those for a one and those for a zero, we get a pattern that resembles a picture of an eye. Figure 2.80 shows some representative transmitted waveforms and the resulting received waveform.

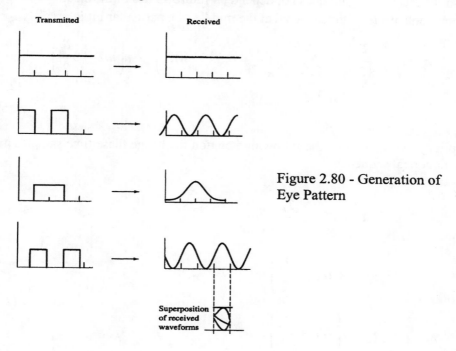

Figure 2.80 - Generation of Eye Pattern

The partial eye pattern is sketched in the figure. The number of individual waveforms contributing to the eye pattern depends on the memory of the system.

Eye patterns are easy to generate in the laboratory. Suppose a random NRZ waveform forms the input to a system and the output is observed on an oscilloscope. Suppose the oscilloscope is synchronized to the sampling interval. The eye pattern is then observed on the screen. The eye opening (blank space in the center) is significant. If the eye is "open", it indicates that all waveforms for a transmitted one are separated from those of a transmitted zero, and accurate decisions can be made. If the eye closes, errors will be made in receiving certain sequences. Note that for the ideal distortionless channel, the eye consists of two horizontal lines, one at the top and one at the bottom of the interval.

Now suppose that we transmit a particular bipolar signal, and the square pulses that are transmitted are distorted by the channel. Suppose the received waveform due to one transmitted pulse is $s(t)$. The signal at the receiver is then,

$$r(t) = \sum_{n = -\infty}^{K} a_n s(t - nT_b) + n(t) \tag{2.29}$$

In Eq. (2.29), $n(t)$ is the additive channel noise, a_n is either +1 or -1 depending on the bit transmitted, and K is the largest integer less than t/T_b. We are assuming the channel is causal, so the received signal does not depend on future values of the data.

We now look at the received signal at the middle of a particular bit interval.

$$r\left(mT_b + \frac{T_b}{2}\right) = \sum_{n=-\infty}^{m} a_n s\left(\frac{T_b}{2} + (m - n)T_b\right) + n\left(mT + \frac{T_b}{2}\right) \tag{2.30}$$

For simplicity, we will adopt the following notation to change these time samples into a discrete numerical sequence.

$$t_i = iT_b + \frac{T_b}{2}$$

$$s_i = s\left(iT_b + \frac{T_b}{2}\right)$$

$$r_i = r\left(iT_b + \frac{T_b}{2}\right) \tag{2.31}$$

$$n_i = n\left(iT_b + \frac{T_b}{2}\right)$$

Using this notation, Eq. (2.29) becomes,

$$r_m = \sum_{n=-\infty}^{m} a_n s_{m - n} + n_m \tag{2.32}$$

The intersymbol interference is represented by the summation term in Eq. (2.32) where the n=m term in the sum is the desired signal component, and the other terms represent the interference from past binary information. Separating out the correct term, we have

$$r_m = a_m s_0 + \sum_{n=-\infty}^{m-1} a_n s_{m-n}$$

(2.33)

$$= a_m s_0 + I_m$$

I_m is the intersymbol interference, and it depends on the specific shape of the signal, $s(t)$, and on the particular sequence of binary digital transmitted. The peak value of the interference can be found by assuming that the particular sequence is such that all terms in the sum are positive. Therefore,

$$I_m(peak) = \sum_{n=-\infty}^{m-1} |s_{m-n}|$$

(2.34)

The mean-square value of interference is now calculated.

$$\overline{I^2} = E\left\{ \left[\sum_{n=-\infty}^{m-1} a_n s_{m-n} \right]^2 \right\}$$

(2.35)

$$= \sum_{j=-\infty}^{m-1} \sum_{i=-\infty}^{m-1} E\{a_i a_j\} s_{m-i} \, s_{m-j}$$

We now assume that the a_i are independent of each other, and that ones and zeros are equally likely. The cross products go to zero, and the double summation reduces to

$$\overline{I^2} = \sum_{n=-\infty}^{m-1} s_{m-n}^2$$

(2.36)

Example 2.5

Find the peak and mean-square intersymbol interference for a signal resulting from sending ideal impulse samples through a channel with a triangular passband characteristics, as shown in Fig. 2.81.

Solution: We first find the impulse response of the channel.

$$h(t) = \frac{T}{n^2 t^2} \sin^2\left(\frac{\pi t}{T}\right)$$

From this result, we find the sample values:

$$s_0 = \frac{4}{n^2 T}$$

$$s_1 = \frac{4}{\pi^2 3^2 T}$$

.

.

$$s_n = \frac{4}{\pi^2 (2n + 1)^2 T}$$

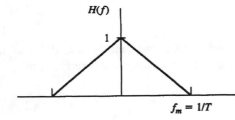

$H(f)$

$f_m = 1/T$

Figure 2.81 - Channel Characteristics
for Example 2.5

The peak and mean-square intersymbol interference are found directly from the equations.

$$I_m(peak) = \frac{4}{\pi^2 tT} \sum_{n=-\infty}^{m-1} \frac{1}{|2(m - n) + 1|^2}$$

$$\overline{I_m^2} = \left(\frac{4}{\pi^2 T}\right)^2 \sum_{n=-\infty}^{m-1} \frac{1}{|2(n - m)|^4}$$

What is the Best Pulse Shape?

Suppose that a waveform, *s(t)*, is used to transmit a binary one, and that *-s(t)* is used to send zero. The question then arises as to the best choice of *s(t)*. We start by assuming that the transmitted pulses are generated using an impulse fed through an ideal low-pass filter. This is shown in Fig. 2.82. Why do we start with this assumption? You may think that the impulse is the best shape for avoiding intersymbol interference. But the channel cannot pass infinite frequency, so it will cut off signal content above a certain frequency. If we limit the frequencies with a lowpass filter, the channel may allow the entire signal to pass undistorted.

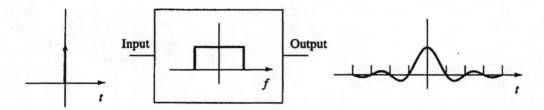

Figure 2.82 - Pulses formed by Ideal Lowpass filter

The assumed pulse is therefore of the form,

$$\frac{\sin(\pi t/T)}{\pi t/T} \tag{2.37}$$

Let's further assume that *T* in Eq. (2.37) is the sampling period. This pulse has the property that it goes through zero at every sampling point except for the point at time zero. Therefore, in the intersymbol interference analysis, s_m is zero for all non-zero *m*, and we have no intersymbol interference. Figure 2.83 illustrates this for three pulses.

This observation is actually a restatement of the sampling theorem. If the channel passes frequencies up to $1/2T_b$, then the individual samples values are independent.

The ideal bandlimited pulse is difficult to achieve because of the sharp corners in the frequency spectrum. Even if we could achieve this channel characteristic, the relatively significant sidelobes of the impulse response place demands on the detection system. If sampling is not done at the precise multiples of T_b, intersymbol interference becomes a significant factor. Thus, synchronization must be precise, and timing jitter must be kept to a minimum.

A desirable compromise is the *raised cosine* characteristic. The Fourier transform of this pulse is similar to the square transform presented above, but the transition from maximum to minimum follows a sinusoidal curve. The transform is shown in Fig. 2.84.

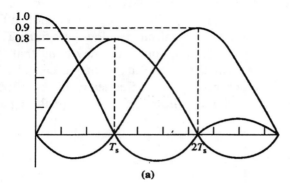

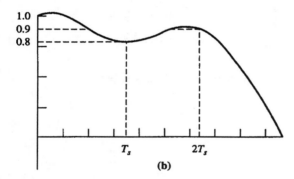

Figure 2.83 - Sequence of Pulses
Formed by Ideal Lowpass Filter

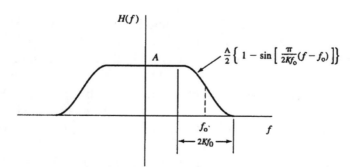

Figure 2.84 - Raised Cosine Fourier Transform

The value of K determines the width of the constant portion of the transform. If $K=0$, the transform becomes that of the ideal bandlimited pulse. If $K=1$. the flat portion is reduced to a point at the origin and the bandwidth is $2f_o$. Because of the rounded corners, we can approximate this characteristic to a far greater degree of accuracy than is possible for the square transform.

The time function corresponding to the raised cosine transform is given by

$$h(t) = A \, \frac{\sin 2\pi f_o t}{2\pi f_o t} \, \frac{\cos KT2\pi f_o}{1 - \dfrac{8K^2 f_o t^2}{\pi}}$$

(2.38)

This function is sketched for several representative values of K in Fig. 2.85.

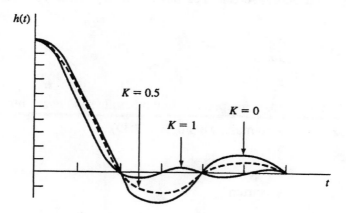

Figure 2.85 - Raised Cosine Time Function

Note that at $K=1$, the response goes to zero not only at the zeros of the $sin(t)/(t)$ function, but also at points midway between these sample values. It is therefore possible to sample at the same rate as for the ideal channel, with no resulting intersymbol interference. The price we pay is increased bandwidth.

2.5. Performance

Now that you are an expert on receiver design, it's time for the bad news. Well, maybe it's not all bad. If receivers performed perfectly, there would be little job security for communication engineers. Thankfully, they do not operate perfectly. In this section, we examine the performance of the various receivers we designed in Section 2.4.

In *analog* communication, we normally wish the output to be as close to the input waveform as possible. In *digital* communication, the normal measure of performance is the *rate at which bit errors occur*.

2.5.1 Analog Baseband

The output signal to noise ratio of a baseband analog receiver depends on the input signal to noise ratio, the characteristics of the receiver, and the noise added by the electronics in the receiver.

The signal to noise ratio at the input to the receiver depends on the characteristics of the channel and of the noise that adds during transmission. We normally consider the additive noise to be white Gaussian, so the total power of the added noise is the product of the system bandwidth with the noise power per hertz.

The noise added by the receiver is characterized by the *noise figure*. *Thermal noise* is produced by the random motion of electrons in a medium. The intensity of this motion increases with increasing temperature and is zero only at a temperature of absolute zero.

If the voltage across a resistor were examined using a sensitive oscilloscope, a random pattern would be displayed on the screen. The power spectral density of this random process is of the form,

$$G(f) = \frac{A\,|f|}{e^{B|f|} - 1} \qquad\qquad (2.39)$$

A and *B* are constants that depend on temperature and other physical constants.

MATLAB EXAMPLE - In this example, we plot Eq. (2.39).

```
function thermal_noise()
%EXAMPLE
%Plot thermal noise frequency spectrum
%G(f) = A*|f|/exp(B*|f|-1)
%Where parameters A and B are user inputs

clc;                                    % clear command window
close all;                        % close all (previously) open figures
disp('   Plot thermal noise frequency spectrum')
disp('    G(f) = A*|f|/exp(B*|f|-1)')
disp('      Where parameters A and B are user inputs')

disp('     --------------------------------------------------')
%pause
disp(' ')
disp(' ')
fm = input('Please enter desired highest frequency on x-axis. Example: 1E9: ');
A = input('Please enter parameter A: Example: 1E-8: ');
B = input('And parameter B: Example: 1E-7: ');

f= 1:fm/10000:fm;                                            % Freq.
Bandwidth
```

```
G = A*abs(f)./(exp(B*abs(f))-1);                          % Noise Power Spectrum Density

%plot(f,G)
semilogx(f,G)
xlabel('frequency, hz'); ylabel('G(f)'); title('Power Spectral Density of Thermal Noise');
```

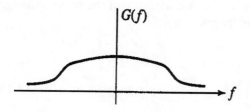

Figure 2.86 - Power Spectral Density of
Thermal Noise

Figure 2.86 shows the curve of Eq. (2.39). For frequencies below the knee of the curve, $G(f)$ is almost constant. If we operate in this frequency range, we can consider thermal noise to be white noise. In fact, thermal noise appears to be approximately white up to extremely high frequencies, on the order of 10^{13} Hz. For frequencies within this range, the mean-square value of the voltage across the resistor has been shown to equal

$$\overline{v^2} = R(0) = 4kTRB \tag{2.40}$$

where k is Boltzmann's constant (1.38×10^{-23} $J/^\circ K$), T is the temperature in $^\circ K$, R is the resistance value in ohms, and B is the observation bandwidth in Hz. This means that the height of the two-sided spectral density over this constant region is $2kTR$.

Of more practical concern is the actual power generated by a resistor. That is, if a resistor is connected to additional circuitry, how much noise power is generated in that additional circuitry? We know from basic circuit theory that this depends on the impedance of the external circuit. The power transferred is a maximum when the load impedance matches the generator impedance (if reactance is present, the load impedance is the complex conjugate of the source impedance). This yields the *maximum available power*, which (using a voltage-divider relationship) is

$$N = \frac{\overline{v^2}}{4R} = kTB \tag{2.41}$$

with a corresponding power spectral density of

$$G_n(f) = \frac{kT}{2} \qquad (2.42)$$

Equation (2.42) is the power spectral density of the available noise power from a resistor.

If we have a system with a number of noise-generating devices within it, we often refer to the system *noise temperature*, T_e, in °K. This is the temperature of a single noise source that would produce the same total noise power at the output.

If the input to the receiver contains noise, the receiver then adds its own noise to produce a larger output noise. The system *noise figure* is a measure of the noise added by the system (adjusted for system gain). It's actually given by

$$\frac{N_{out}}{N_{in}} \frac{C_{in}}{C_{out}} \qquad (2.43)$$

where C_{in} and C_{out} are carrier input and output powers respectively. You'll learn about carriers in the next chapter. For now, you can think of C_{out}/C_{in} as the gain. The noise figure is expressed in decibels. For example, a noise figure of 3 dB indicates that the system is adding an amount of noise equal to that which appears at the input, so the output noise power is twice that of the input.

We now turn our attention to the channel. The technique we use to analyze the channel additive noise is typical of the way we will approach broadcast communication systems in the following chapters. Let us assume a channel noise model as shown in Fig. 2.87.

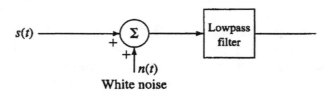

Figure 2.87 - Additive Noise in Baseband Channel

The noise, *n(t)*, is added to the signal, *s(t)* and the sum is lowpass filtered. We assume the lowpass filter passes frequency up to a maximum, f_m. This filter is normally part of the receiver. We include it in this model since, without it, the noise has significant components outside the frequency band of interest.

We assume that the additive noise is white with two-sided power spectral density of $N_o/2$. That is, the noise has a power of N_o watts/Hz. The noise power at the output of the filter is then $N_o f_m$ watts, and the signal to noise ratio is

$$SNR = \frac{P_s}{N_o f_m} \qquad (2.44)$$

In Eq. (2.44), P_s is the signal power.

Baseband analog receiver performance also depends on nonlinearities in the electronics. This is expressed in measures such as dynamic range and harmonic distortion. The term, *dynamic range* usually refers to the ratio (in decibels) of the strongest to the weakest signal that a receiver can process without noise or distortion exceeding acceptable limits. Although this sounds like a simple concept, application to a practical transmission is quite complex. For example, the behavior of a receiver when a single sinusoid forms the input may be quite different from that when the input is a complex sum of many signal components. Keep in mind that we are discussing nonlinear effects, so the actual receiver dynamic range may depend upon input signal characteristics.

Harmonic distortion is normally measured by letting the receiver input be a pure sinusoid. Nonlinearities in the receiver change this sinusoid to a periodic function with harmonics. The ratio of power of the harmonics to power of the fundamental is a measure of harmonic distortion.

2.5.2. Discrete Baseband

The signal to noise ratio of a PAM signal depends on the form of the receiver. If the receiver simply samples the received waveform at periodic points in time, the sample values are

$$r(nT_s) = s(nT_s) + n(nT_s)$$

where $r(t)$ is the received signal and $n(t)$ is the additive noise. The signal to noise ratio is then

$$\frac{S}{N} = \frac{\overline{s^2}}{\overline{n^2}} \qquad (2.45)$$

The numerator of Eq. (2.45) is the average signal power while the denominator is the average noise power. If the channel can be modeled as an ideal lowpass filter, the noise power is simply $N_o B$ where N_o is the noise power per Hz and B is the system bandwidth.

Calculation of the signal to noise ratio for PPM and PWM systems is more difficult. Since we can easily convert a PWM signal to a PPM signal, we shall calculate the SNR only for PPM. Figure 2.88 shows an *ideal* square pulse and an approximation to a *practical* square pulse.

The job of the PPM receiver is to locate the trailing edge (or leading edge) of this pulse. One way to do this is with a comparator or threshold detector. A threshold is set and when the signal breaks through this threshold, we assume we have located the trailing edge. The

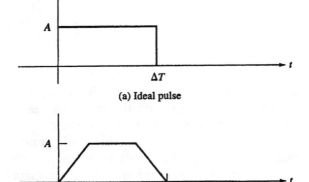

(a) Ideal pulse

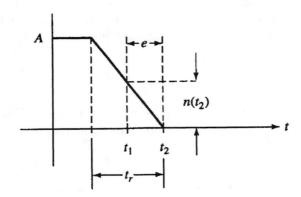

(b) Approximation to practical pulse

Figure 2.88 - Ideal and
Practical Square Pulse

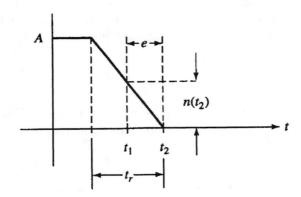

Figure 2.89 - Noise
Affecting Location
of Trailing Edge

threshold value would normally be set at the midpoint, $A/2$, and this value is known as the *slicing level*. In the *ideal* square pulse case, as long as the additive noise waveform never exceeds the slicing level in magnitude, the location of the trailing edge will not be affected.

If we now look at the practical approximation to the square pulse, and add noise to this, we have the situation shown in Fig. 2.89. We are focusing on the trailing edge of the pulse. The rise time is shown as t_r. We have added enough noise, $n(t_2)$, to move the threshold crossing from t_1 to t_2. How much does a change in amplitude affect the time at which the signal crosses the threshold? Some simple geometry (similar triangles) can be used to derive the relationship,

$$\frac{e}{n(t_2)} = \frac{t_r}{A} \tag{2.46}$$

Remember that A is the amplitude of the pulse, t_r is the pulse rise time, and $n(t_2)$ is the additive noise at the time when the perturbed signal crosses the slicing level. We solve for the timing error to obtain

$$e = \left(\frac{t_r}{A}\right) n(t_2) \tag{2.47}$$

The mean square value of the error is then given by

$$\overline{e^2} = \left(\frac{t_r}{A}\right)^2 \overline{n^2(t_2)} \tag{2.48}$$

Let us assume that the mean square value of the noise is given by the noise power per Hz (N_o) multiplied by the bandwidth (B). This assumes additive white noise. The error is then

$$\overline{e^2} = \left(\frac{t_r}{A}\right)^2 N_o B \tag{2.49}$$

The pulse rise time is related to the system bandwidth. The wider the bandwidth, the faster the signal can change. If we assume the maximum possible slope (assume a square pulse through a lowpass filter), we get the relationship

$$t_r = \frac{1}{2B} \tag{2.50}$$

Equation (2.50) is related to the sampling theorem (Can you see how the Nyquist rate comes into this equation?). We also would like to relate the results to the pulse energy instead of the amplitude since many systems are *energy limited*. This energy is approximately given by

$$E_p \approx A^2 \Delta T \tag{2.51}$$

Recall from Fig. 2.88 that ΔT is the original pulse width. The approximation of Eq. (2.51) improves as the rise time decreases (i.e., The practical pulse gets closer to the ideal pulse). Combining these results yields

$$\overline{e^2} = \frac{\Delta T}{4BE_p} N_o \tag{2.52}$$

Equation (2.52) is the desired result. It shows that the mean squared timing error is inversely proportional to the system bandwidth. In order to convert the PPM waveform to an analog signal, we change the pulse location to a pulse amplitude and then take the ratio of the mean square value of the signal samples to the mean square value of the error. The noise power is therefore related to the mean square timing error of Eq. (2.52).

2.5.3. Digital Baseband

Now that we have explored analog baseband (both continuous time and discrete), we are ready to turn our attention to digital baseband. There are two distinct sources of error in a digital communication system. The first arises from the A/D conversion process where sample values are rounded off. This produces a round-off error (quantization noise). Even if the transmission and reception of ones and zeros is perfect, we will still be left with this round-off error. The second source of error is the actual transmission process and the effects of distortion and additive noise. This causes bit errors where transmitted ones are received as zeros and vice versa. We examine these two sources of error separately.

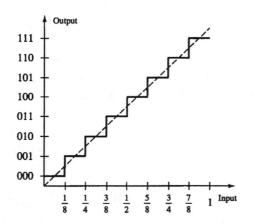

Figure 2.90 - Input-Output Relationship of Quantizer

Quantization Noise - Uniform Quantization

Let's begin our study of quantization noise in PCM by re-examining the quantization input-output relationship of Fig. 2.90. The input has been normalized (scaled and shifted to range from 0 to 1), and we illustrate 3-bit A/D. Looking at the smallest interval, all samples between 0 and 1/8 are coded into 000 and reconstructed as 1/16. Then all samples between 1/8 and 2/8 are coded into 001 and reconstructed as 3/16. This process continues for all 8 regions. Quantization noise, or error, is defined as the time function which is the difference between $s_q(t)$, the quantized waveform, and $s(t)$, the original waveform. The error is given by Eq. (2.53).

$$e(nT_s) = s(nT_s) - s_q(nT_s) \qquad (2.53)$$

Figure 2.91(a) illustrates a representative time function, $s(t)$, and a quantized version of this time function, $s_q(t)$. To simplify the diagram, we are illustrating 2-bit A/D and showing a waveform that never goes negative. ΔS is the width of the round-off region (e.g., 1/8 in Fig. 2.90).

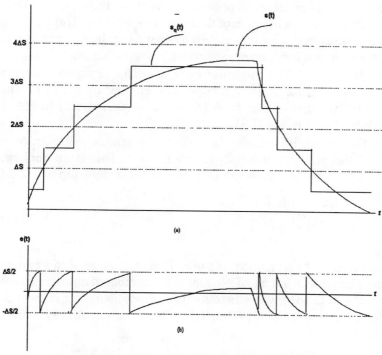

(a)

(b)

Figure 2.91 - Quantization Error

While we are illustrating time functions, it is important to note that it is the *sample values* that are rounded off, not the analog time function. Therefore, the only important values of $s_q(t)$ are those values at the time sample points, nT_s. Figure 2.91(b) shows the quantization error, $e(t)$, as the difference between $s(t)$ and $s_q(t)$. Again note that we are only interested in the values of this error function at the sample points. The magnitude of the error term never exceeds one half of the spacing between quantization levels (provided, of course, that the waveform does not leave the assumed range of zero to $4\Delta S$).

The actual error for a specific time function is of little use to use in designing systems. Instead, we need to find the average statistics of the error. To do so, we must first find the probability density function of the error.

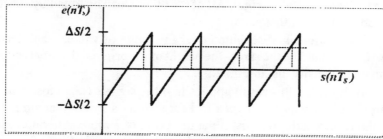

Figure 2.92 - Error as a Function of Sample Value

The error at any sample point depends only on the value of the sample at that point. Figure 2.92 illustrates the error as a function of the input sample value. This can be derived from an input-output relationship such as Fig. 2.90 by observing the distance between the ramp function and the staircase function at each point along the horizontal axis.

The error curve starts at $-\Delta S/2$ at the lower edge of each quantization interval, and increases linearly to reach $+\Delta S/2$ at the top edge. The number of cycles of this ramp function depends on the number of bits of A/D. We show four cycles in Fig. 2.92 to correspond to the 2 bits of A/D in Fig. 2.91.

If we now know the probability density function of the sample values, it becomes a simple matter to find the probability density function of e. This is an application of *functions or a random variable* (see Appendix C). The result is shown in Eq. (2.54).

$$p(e) = \sum_i \frac{p(s_i)}{|\, de_i/ds_i\, |} \tag{2.54}$$

The s_i are the various values of s corresponding to e. That is, if we set e to a certain value in Fig. 2.92, there are a number of values of s (equal to the number of quantization regions) that yield this value of e (i.e., the inverse function is multi-valued). We show this as the dashed line in the figure. The magnitude of the slope of the function is always unity, so Eq. (2.54) can be simplified to

$$p(e) = \sum_i p(s_i) \tag{2.55}$$

If you look at the intervals to the right of the origin, we can solve for the values of s corresponding to e as

$$s_0 = \frac{e + \Delta S}{2}$$

$$s_i = i\Delta S + \frac{e + \Delta S}{2} \tag{2.56}$$

s_o is the first value to the right of the origin. All of the other s_i can be found by successively adding and subtracting ΔS from this value.

If the samples were uniformly distributed over the range of values, each term in the summation of Eq. (2.55) would be the same–the height of the original density function. The result is a uniform error density, as in Fig. 2.93.

We would like to show that this result applies to other signals (i.e., those that *do not* have a uniform probability density). To get a feel for this, let's try another simple density in the hopes that the results will point us in the direction of a more general observation. Suppose now that the samples have a triangular density as shown in Fig. 2.94.

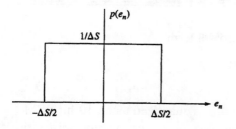

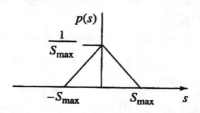

Figure 2.93 - Density of Error

Figure 2.94 - Triangular Error
Density

The result is still the uniform error density of Fig. 2.93. You can convince yourself of this by taking some representative values. In fact, as e increases, all of the s_i of Eq. (2.55) increase by the same amount. In a triangular density, for every value in the sum of Eq. (2.55) that is decreasing, another value is increasing by the same amount. The sum does not change.

A similar argument can be used for any density that is approximately linear over the range of a single quantization region. For this reason, the uniform density of Fig. 2.93 is assumed to be approximately correct over a wide range of input signals.

Now that we know the density of the error, we can find the average of its square, or the *mean square error* using simple probability theory.

$$mse = \int_{-\infty}^{\infty} p(e_n)e_n^2 de_n$$

$$= \frac{1}{\Delta S} \int_{-\Delta S/2}^{\Delta S/2} e_n^2 de_n = \frac{\Delta S^2}{12}$$

(2.57)

This result gives the average of the square of the error in a single sample of the time function. To assess the *annoyance* caused by this error, it is necessary to compare this value to the average of the square of the unperturbed time sample. This yields the very important *signal to quantization noise ratio*. Thus,

$$SNR = \frac{s^2}{e_n^2} = \frac{12s^2}{\Delta S^2} = \frac{12P_s}{\Delta S^2}$$

(2.58)

where P_s is the average signal power. You should somehow mark or highlight this equation. It is an extremely important result that will be used many times in the work to follow.

Example 2.6

Consider an audio signal comprised of a single sinusoidal term,

$$s(t) = 3\cos500\pi t$$

(a) Find the signal to quantization noise ratio when this is quantized using 10-bit PCM.
(b) How many bits of quantization are needed to achieve a signal to quantization noise ratio of at least 40 dB?

Solution: (a) Equation (2.58) is used to find the signal to quantization noise ratio. The only parameters which need to be evaluated are the average signal power and the quantization region size. The total swing of the signal is six volts, so the size of each interval is $6/2^{10} = 5.86 \times 10^{-3}$. The average signal power is $3^2/2 = 4.5$ watts. The signal to quantization noise ratio is then given by

$$SNR = \frac{12 \times 4.5}{(5.86 \times 10^{-3})^2}$$

$$= 1.57 \times 10^6$$

If we wish to express this in dB, we take the log of this and multiply by 10. Therefore,

$$SNR = 10 \times \log(1.57 \times 10^6) = 62 \ dB$$

(b) The minimum signal to noise ratio is specified as 40 dB. This corresponds to a ratio of 10^4. We use Eq. (2.48) where ΔS is unknown (since ΔS depends on the number of bits of quantization). Therefore,

$$\frac{12 \times 4.5}{\Delta S^2} > 10^4$$

and solving for ΔS yields,

$$\Delta S < 7.35 \times 10^{-2}$$

We now note that the step size is given by $\Delta S = 6/2^N$ where N is the number of bits of quantization. We need to choose an N such that ΔS is no larger than 7.35×10^2. Thus,

$$\frac{6}{2^N} < 7.35 \times 10^{-2}$$

and
$$2^N > 81.6$$
We could use logarithms to solve this equation, but that really should not be necessary. If N is 6, the left side is 64. If N is 7, the left side is 128. Therefore, we require 7 bits of quantization to achieve a signal to noise ratio of at least 40 dB. [Do NOT give the answer that $N=6.350497247$ as you would obtain using a calculator! N is the number of bits of quantization, and as such it must be an integer.]

It should be clear that each additional bit of quantization reduces the step size, ΔS, by a factor of two. This increases the signal to quantization noise ratio by a factor of four. A factor of four corresponds to 6 dB since $10 \log 4 \approx 6$. Hence, each additional bit of quantization increases the SNR by 6 dB.

While Eq. (2.58) clearly shows how to find the signal to quantization noise ratio as a function of the signal and the quantization step size, it is useful to get one general result to use as a starting point in system design. Suppose the signal, $s(t)$, is uniformly distributed between $-S_{max}$ and $+S_{max}$, as shown in Fig. 2.95.

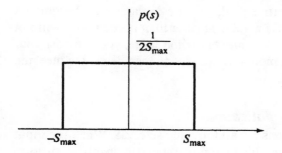

Figure 2.95 - Uniformly Distributed Signal

For this particular case, Eq. (2.58) reduces to a very simple form. We need only two quantities to solve this equation–the average signal power and the step size. The power is found from basic probability theory.

$$P_s = \int_{-\infty}^{\infty} s^2 p(s)\,ds = \frac{1}{2S_{max}} \int_{-S_{max}}^{S_{max}} s^2\,ds = \frac{S_{max}^2}{3} \tag{2.59}$$

The step size is given by

$$\Delta S = \frac{S_{max} - (-S_{max})}{2^N} = \frac{2S_{max}}{2^N} = \frac{S_{max}}{2^{N-1}} \tag{2.60}$$

Equation (2.58) then becomes

$$SNR = \frac{4S_{max}^2}{S_{max}^2/2^{2N-2}} = 2^{2N} \tag{2.61}$$

Does this result surprise you? The signal to noise ratio doesn't depend on S_{max}! As S_{max} changes, *both* the signal power and the quantization noise power change by the same factor.

We can convert the signal to noise ratio of Eq. (2.61) into decibels with the result,

$$SNR_{dB} = 10\log(2^{2N}) = 20N\log(2) \approx 6N \; dB \tag{2.62}$$

This result represents a good starting point even when the signal is not uniformly distributed. Looking back to part (b) of Example 2.7, you were asked to specify the number of bits of quantization to achieve an SNR of at least 40 dB. In order for $6N$ to be greater than 40, N must be at least 7 which is the same answer we got in the example. But be cautious in application of Eqs. (2.61) and (2.62). Signals in real life are   *not* uniformly distributed, and these equations *only* apply to the uniformly distributed case. If you incorrectly apply Eq. (2.62) to a non-uniform signal, you are in danger of designing a system with the wrong value of N. If you use a value smaller than needed, you will not meet the specifications for SNR. If you use a value that is too large, you can cost your employer large sums of money by requiring more bits per second than needed to meet the specifications.

Quantization Noise - Non-Uniform Quantization
In uniform quantization, the round-off regions are all the same size, ΔS. When the input samples are *not* uniformly distributed, it is possible to improve signal to quantization noise ratios by using *non-uniform quantization*. We begin by assuming that the samples are distributed according to a probability density, *p(s)*, as shown in Fig. 2.96.

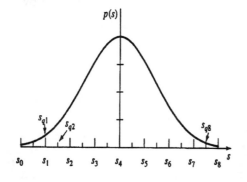

Figure 2.96 - Probability Density
of Samples

Although this appears to be approximately Gaussian, it is meant to be a representative probability density function and the results we obtain will not depend upon any specific shape. We have illustrated three bits of quantization, with the 8 regions marked by their boundaries, s_i, and by the round off levels, s_{qi}. The mean square quantization error of the samples is given by Eq. (2.63).

$$mse = E\{[s(nT_s) - s_q(nT_s)]^2\}$$

$$= \int_{-\infty}^{\infty} (s - s_q)^2 p(s) ds$$

$$= \sum_i \int_{s_i}^{s_{i+1}} (s - s_{q_i})^2 p(s) ds$$

(2.63)

In Eq. (2.63), the s_{qi} are the various quantization round-off levels, and $p(s)$ is the probability density function of the signal samples. We shall return to this equation in a few moments when we examine companded systems. For now, we use the expression to prove a statement made earlier about the best location for the round-off levels. Let us assume that the regions have been specified (the s_i are given) and we wish to find the optimum location of the round-off values, s_{qi}. We interpret the word "optimum" to mean those values that minimize the mean square error. To do so, we differentiate Eq. (2.63) with respect to the s_{qi} and set the derivative to zero. This yields,

$$\int_{s_i}^{s_{i+1}} (s - s_{q_i}) p(s) ds = 0 \quad \text{for all } i$$

(2.64)

Equation (2.64) indicates that once the quantization regions have been chosen (i.e., the s_i), the round-off levels are selected to be the center of gravity of the corresponding portion of the probability density. Thus, each quantization level, rather than being in the center of the interval, is skewed toward the higher probability end of the interval. This is intuitively satisfying since more samples can be expected to fall in this region.

Example 2.7

Assume that the density function of $s(t)$ is a zero-mean Gaussian density with variance, σ^2, of 1/9. Since the probability of a sample exceeding a magnitude of one is less than 1% (i.e., the 3σ point), assume that you quantize the region between -1 and +1 (i.e., values above a magnitude of one will saturate at either 000 or 111). Three-bit quantization is used.

(a) Find the mean square quantization error assuming uniform quantization is used.

(b) A scheme is proposed wherein the quantization regions are chosen to yield *equal area* under the probability density function over each region. That is, the probability of a sample falling within any particular interval is the same as that for any other interval. Choose the best location for the round-off values, and find the corresponding mean-square error.

Solution: (a) We use the approximate formula of Eq. (2.57) to find the mean square error in the uniform case. The size of each interval is 2/8 = 1/4. The error is therefore given by,

$$mse = \frac{\Delta S^2}{12} = \frac{1}{192} = 5.2 \times 10^{-3}$$

(b) We must first find the boundaries of the quantization regions. We are dividing this into 8 equal area segments, so the area under the density within each region must be 1/8. Reference to a table of error functions indicates that the s_i are given by

    -1, -0.38, -0.22, -0.1, 0, 0.1, 0.22, 0.38, 1

Equation (2.54) is now used to find the round-off levels, s_{q_i}. The equation reduces to

$$s_{q_i} = \frac{\int_{s_i}^{s_{i+1}} sp(s)ds}{\int_{s_i}^{s_{i+1}} p(s)ds}$$

$$= 8 \int_{s_i}^{s_{i+1}} sp(s)ds \quad for\ all\ i$$

This can be approximated or evaluated in closed form. The resulting s_{qi} are given by

    -0.54, -0.3, -0.16, -0.05, 0.05, 0.16, 0.3, 0.54

Finally, the mean square error is found from Eq. (2.63) to be

    mse = 5.3 x 10⁻³

It appears that the uniform quantizer is better than this particular non-uniform quantizer. However, with the Gaussian density and only 3-bits of quantization, Eq. (2.47) is not a very good approximation for the mean square error. Recall that this equation required the density to be approximately piecewise linear over the various regions. The exact answer to part (a) could be found by applying Eq. (2.63). The result is 6.2 x 10⁻³, and therefore, the non-uniform quantizer *does* provide an improvement in performance.

This example suggested one possible algorithm for selecting the quantization regions. In fact, this is not the best algorithm, and it even causes degradation of performance when compared to uniform quantization for some signal probability densities.

The equation for mean square error weights the probability by the square of the deflection from the quantized value before integration. In general, the problem is to minimize the error of Eq. (2.63) as a function of two variables, s_i and s_{qi}. The s_{qi} are constrained to satisfy Eq. (2.64). Unless the probability density can be expressed in closed form, this problem is computationally difficult.

The *mean value theorem* can be applied to Eq. (2.63) to get an approximation which improves as the number of bits of quantization increases. Using this approximation, the following rule for choosing the quantization regions results:

Choose the quantization regions to satisfy the uniform moment constraint,

$$(s_{i+1} - s_i)^2 p(midpoint) = constant \tag{2.65}$$

Companded Systems

We introduced companding in Section 2.3.2. Our goal in the current discussion is to derive an approximate expression for the performance improvement (or degradation) offered by a particular compression function as compared to uniform quantization. The result is approximate, and the approximation improves as the number of bits of quantization increases—that is, as the quantization regions get smaller. We begin by assuming that the round-off values are in the middle of each interval. This is the best choice only if the density can be assumed to be a constant over the width of the interval. Further assuming that the density function can be approximated over the interval by its value at the round-off level, Eq. (2.63) reduces to

$$mse = \sum_i \int_{s_i}^{s_{i+1}} (s - s_{q_i})^2 p(s) ds$$

$$= \sum_i p(s_{q_i}) \left. \frac{(s - s_{q_i})^2}{3} \right|_{s_i}^{s_{i+1}} \tag{2.66}$$

$$s_{q_i} = \frac{s_i + s_{i+1}}{2} \tag{2.67}$$

We now use the approximation that s_{qi} is in the middle of the interval, Equation (2.66) then becomes

$$mse = \sum_i p(s_{q_i}) \frac{(s_{i+1} - s_i)^2}{12} \tag{2.68}$$

It is comforting to occasionally check our work against the known result for uniform quantization. In fact, if uniform step sizes of ΔS are substituted for $(s_{i+1} - s_i)$ in Eq. (2.68), the result reduces to $\Delta S^2 / 12$ as we found earlier for uniform quantization [If you simply looked at Eq. (2.68) and said, "Oh sure, he is right", you are fooling yourself. Please take a moment now to prove this statement]. If this were not the case, we would have to recheck the derivation to find the error(s).

We can relate the interval size, $s_{i+1}-s_i$, to the slope of the compression curve. That is, if the compressed output is uniformly quantized with a step size of ΔS, the corresponding step sizes of the uncompressed waveform are approximately given by Fig. 2.49, which we repeat here.

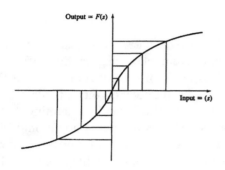

Figure 2.49 (repeated)

We need to take the limit of the summation of Eq. (2.68) as the intervals get smaller and smaller. To do so, we separate the square of the interval from the cube term and rewrite this squared term using the compression function relationship of Eq. (2.69).

$$s_{i+1}-s_i \approx \frac{\Delta S}{F'(s_i)} \tag{2.69}$$

The remaining interval multiplier becomes the differential term. Thus,

$$mse = \frac{1}{12}\sum_i p(s_{q_i})(s_{i+1}-s_i)^2(s_{i+1}-s_i)$$

$$= \frac{1}{12}\sum_i \frac{p(s_{q_i})\Delta S^2}{[F'(s_i)]^2}(s_{i+1}-s_i) \tag{2.70}$$

In the limit as the interval widths approach zero, this becomes

$$mse = \frac{\Delta S^2}{12}\int_s \frac{p(s)}{[F'(s)]^2}ds \tag{2.71}$$

The mean-square error for a uniform quantizer, $\Delta S^2/12$, appears explicitly in Eq. (2.71). If the integral in this equation is less than unity, the compander has provided an improvement over the uniform quantizer (i.e., it has reduced the mse).

Equation (2.71) can be used to evaluate a companding scheme given any particular probability density of the signal samples. You will have ample opportunity to fine tune your analysis skills in the problems at the back of the chapter. For now, we will present one set of results in a slightly different way.

We wish to compare the companded and uniform systems. For this comparison, we will choose 8-bit quantization since that is common in voice transmission. If we assume that the signal samples are uniformly distributed, we achieve a signal to quantization noise ratio of 48 dB ($6N$ dB).

Suppose now that the maximum signal power decreases, but the quantization levels are not changed (i.e., we do not redesign the A/D). As long as the signal fills at least one quantization region ($2S_{max}/256 = S_{max}/128$), the quantization noise varies over its entire range ($-\Delta S/2$ to $+\Delta S/2$), and the average noise power remains unchanged. Therefore, as the signal power decreases, the numerator of the SNR equation decreases while the denominator stays constant. Thus, the SNR decreases in the same proportion. We can plot SNR as a function of input signal power as shown in the linear curve of Fig. 2.97. Zero dB input signal power refers to the level for which the A/D converter was designed. At this level, the signal spans all 256 round-off regions and the signal to quantization noise ratio is 48 dB (i.e., $6N$).

Now looking in the other direction, suppose we make the signal bigger. As soon as the signal increases beyond the range of the quantization levels (i.e., overloads), the noise power increases rapidly. This is true since larger samples saturate the system, and the error *is no longer limited in magnitude* to $\Delta S/2$.

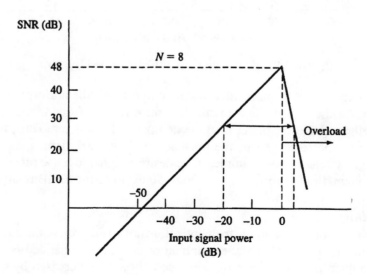

Figure 2.97 - SNR as a Function of Signal Power

Now suppose that the design specifications for your system contain a specified minimum SNR. The portion of the curve in Fig. 2.97 above that level represents the *dynamic range* of the quantizer. For example, if we desire a SNR of at least 28 dB, the acceptable input signal range is from -20dB to about +3dB relative to full-load (a dynamic range of 23 dB), as indicated by the double-ended arrow on Fig. 2.97. If the input extends outside this range, we can no longer guarantee that the system will meet the specification.

The companded system performs better than the uniform quantization system for small signals. This is true since the intervals get smaller as the sample size decreases. We can evaluate the companded system performance and compare it to that of the uniform quantizer. Figure 2.98 does this for a uniform signal density and μ-law companding (for various values of μ, including μ-255). The non-companded curve of Fig. 2.97 is repeated in this figure (see μ=0 curve) for comparison.

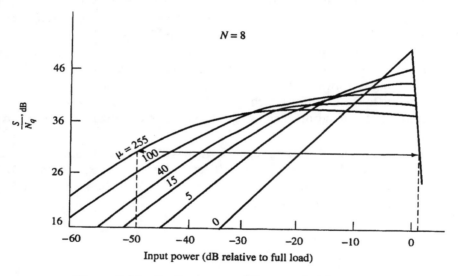

Figure 2.98 - Performance of Companded Systems

Note that, as expected, the companded system outperforms the uniform quantizer for low signal levels. As the signal starts to fill the entire range of quantization intervals, the uniform quantizer performs better than the companded quantizer. Thus, the companded system enhances performance in the presence of low-level signals, as expected. The dynamic range increases. For example, if we desire a signal to noise rate of at least 28 dB in a μ-255 system, the dynamic range is from -50 dB to about +3 dB relative to full-load, as indicated on the diagram.

Oversampling

We have been examining the signal to quantization noise ratio where the quantization noise power is the mean square value of the quantization error. In companded systems, we found that signal to quantization noise ratio could sometimes be increased by quantizing with steps that are not of uniform size. We now explore whether there are other techniques for improving the signal to quantization noise ratio.

One place we can look for help is in the frequency domain. A noise signal can be reduced by filtering out some of its frequency components. In a sampled system, a lowpass filter is used to reconstruct $s(t)$ from its samples. Wouldn't it be nice if that filter were to cut off some of the quantization noise without distorting the reconstructed signal? If so, it would improve the signal to noise ratio.

To analyze the effects of the filter, we need to know something about the frequency distribution of the quantization noise. The noise consists of samples at the sampling points, and we can assume that each sample is independent of all of the others. That is, knowing the value of quantization noise at one sampling point tells us nothing about the value at the next point. This assumption is valid as long as the signal sample values change across many quantization regions between adjacent sample points. In contrast, if there were very few quantization regions (e.g., one bit of quantization) and we sampled much higher than the Nyquist rate, we would expect the quantization noise values to change relatively slowly from sample to sample.

If the noise sample values are independent of each other, the noise power spectral density is constant up to one-half of the sampling rate. We ask you to explore this in the problems in the back of this chapter. Intuitively, you can think of this as a reverse application of the sampling theorem: "a signal specified by independent samples at a rate of $2f_m$ samples/sec describes a waveform with a maximum frequency of f_m Hz."

Now returning to the original problem, we see that if we sample at the Nyquist rate, the noise power is limited to frequencies below f_m, and all of the noise goes through the reconstruction lowpass filter. The signal to noise ratio is therefore calculated as we found earlier and

$$SNR = \frac{12P_s}{\Delta S^2} \tag{2.72}$$

Suppose we now sample at a rate higher than the Nyquist, say $2Kf_m$, where K is greater than one. The noise now has frequencies ranging up to Kf_m Hz, and the lowpass filter reduces the noise power by the ratio of

$$\frac{f_m}{Kf_m} = \frac{1}{K} \tag{2.73}$$

The signal to quantization noise ratio is now

$$SNR = \frac{12KP_s}{(\Delta S)^2} \tag{2.74}$$

As an example, suppose a speech waveform with maximum frequency of 4 kHz is sampled at 8 kHz and quantized using 8 bits. If we assume the speech is uniformly distributed (a poor approximation), the signal to noise ratio is $6N=48$ dB. Now suppose that we raise the sampling rate to 32 kHz, a factor of *4 times oversampling* . The signal to noise ratio improves by a factor of 4 to 54 dB. Oversampling is used in compact audio discs. (You might find it interesting to take note of the advertisements which assume the general public understands oversampling).

Quantization Noise in Delta Modulation

Turning our attention now to delta modulation, we once again define the quantization error as the difference between the original signal and the quantized (staircase) approximation. Thus,

$$e(t) = s(t) - s_q(t) \tag{2.75}$$

We assume that the sampling rate and step size are chosen so as to avoid overload. Under this assumption, the magnitude of the quantization noise never exceeds the step size. If we make the simplifying assumption that all signal amplitudes are equally likely, we conclude that the error is uniformly distributed over the range between $-\Delta$ and $+\Delta$, as shown in Fig. 2.99.

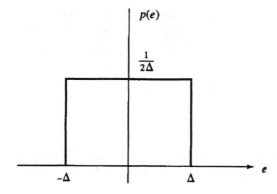

Figure 2.99 - Uniform Quantization Error Distribution in DM

The mean square value of the quantization noise is given by

$$mse = \frac{1}{2\Delta} \int_{-\Delta}^{\Delta} e^2 de = \frac{\Delta^2}{3} \tag{2.76}$$

In designing digital communication systems, how do we decide whether to use PCM or DM as the source encoding technique? As a designer, you would be concerned with many factors. Among these are complexity (i.e., cost), transmission bit rate (i.e., required system bandwidth), reliability, quantization noise, and the effects of transmission errors. It would be very nice if we could derive a simple formula that relates SNR for PCM to that of DM.

Unfortunately, many specific assumptions must be made to arrive at such a result. The bottom line is that under certain circumstances, DM will provide the same SNR as PCM with a lower transmission bit rate. Under other circumstances, the reverse is true. Adaptive delta modulation adds yet another parameter to the analysis. This is a good application for computer simulations.

We start our analysis by solving for the mean square quantization error at the output of the delta modulation receiver. The demodulator includes a lowpass filter to smooth the staircase approximation to generate a continuous curve. To find the quantization noise power at the filter output, we must first find the frequency characteristics of the quantization noise. This is not a simple analytical task, and it requires that a specific form be assumed for $s(t)$.

Let us assume that the original $s(t)$ is a sawtooth-type waveform. This is the simplest example of a waveform that is uniformly distributed. The waveform, $s(t)$, its DM quantized version, $s_q(t)$, and the resulting quantization noise, $e(t)$, are shown in Fig. 2.100. Notice that the noise function is *almost* periodic with period T_s, the sampling period. The noise would be exactly periodic with a period equal to that of the sawtooth waveform if that period were an integral multiple of T_s. . We are assuming that the step size and sampling period are chosen to avoid overload, and in this case, to give perfect symmetry. The power spectral density of $s_q(t)$ can be found precisely. It is of the form $\sin^4 f/f^4$ since the Fourier transform of a sawtooth follows a $\sin^2 f/f^2$ shape. The first zero of the power spectral density of the triangular waveform is at $f=1/T_s$. The lobes beyond this point are attenuated as the

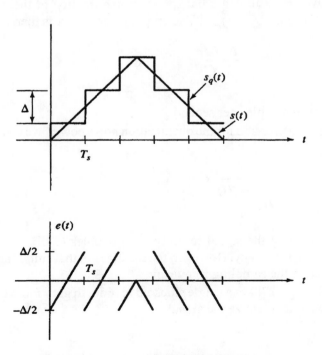

Figure 2.100 - Delta Modulation of
Sawtooth Waveform

fourth power of f^1. Therefore, there is relatively little power outside of the main lobe, so let's assume that all of the power is concentrated in a low frequency band stretching to $f=1/T_s$. Now since we are assuming that delta modulation sampling takes place at a rate well above Nyquist, the first zero of the spectrum at $f=1/T_s$ is at a much higher frequency than f_m. The lowpass filter, which cuts off at f_m, only passes a relatively small portion of the main lobe of the noise power spectrum. This is shown in Figure 2.101.

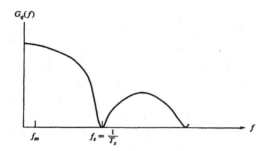

Figure 2.101 - Quantization Noise
Power Spectrum

To get an approximate result, let's assume that the spectrum is essentially flat over the range between 0 and f_s. The total power of the noise is the mean square error we found earlier, $\Delta^2/3$. Since we are assuming a flat spectrum, the fraction of the power that gets through the lowpass filter is $T_s f_m$, or f_m/f_s. The output noise power is then given by

$$N_q = \frac{\Delta^2}{3} \frac{f_m}{f_s} \tag{2.77}$$

where f_s is the number of samples per second.

Now that we have an expression for the quantization noise power, we can find the signal to noise ratio as

$$\frac{S}{N_q} = \frac{3f_s}{\Delta^2 f_m} \overline{s^2(t)} \tag{2.78}$$

Equation (2.78) implies that the signal to noise ratio doubles (3 dB increase) for each doubling of the sampling frequency. However, the result is not that simple since in DM the step size, Δ, is related to the sampling frequency.

We will now apply Eq. (2.78) to a single sinusoid in order to get a more comprehensive result. Suppose we wish to quantize the signal

$$s(t) = a\cos 2\pi f_m t \tag{2.79}$$

The maximum slope of the function is

$$\left|\frac{ds}{dt}\right|_{\text{max}} = 2\pi f_m a \qquad (2.80)$$

The maximum slope of the delta modulation staircase approximation is

$$slope_{\text{max}} = \frac{\Delta}{T_s} = f_s \Delta \qquad (2.81)$$

To avoid slope overload, the maximum slope of the staircase approximation must be equal to or greater than the maximum slope of the signal. Therefore,

$$f_s \Delta \geq 2\pi f_m a$$

$$f_s \geq \frac{2\pi f_m a}{\Delta} \qquad (2.82)$$

Now we put all the pieces together using the minimum slope in the signal to noise calculation of Eq. (2.78). This yields,

$$\frac{S}{N_q} = \frac{3f_s}{\Delta^2 f_m}\overline{s^2(t)} = \frac{6\pi a}{\Delta^3}\frac{a^2}{2} = 3\pi\left(\frac{a}{\Delta}\right)^3 \qquad (2.83)$$

We can obtain an alternate form of this relationship in terms of the frequencies by solving for a/Δ in Eq. (2.82) and using this in Eq. (2.83). This yields,

$$\frac{S}{N_q} = \frac{3}{8\pi^2}\left(\frac{f_s}{f_m}\right)^3 \qquad (2.84)$$

Equation (2.84) shows that for every doubling of the sampling frequency, the signal to noise ratio increases by a factor of 8 (9 dB). In order to avoid incorrect application of Eq. (2.84), it is important that you understand the assumptions made in its derivation. The most critical of these is the assumption that $f_s\Delta$ is the minimum value required to avoid slope overload.

Example 2.8

An audio signal of the form,

$$s(t) = 3\cos 1000\pi t$$

is quantized using DM. Find the signal to quantization noise ratio.

Solution: We must first choose a step size and a sampling frequency for this waveform. The Nyquist rate is $f_s = 1000$ samples/sec. Suppose for purposes of illustration, we choose 8 times this rate, that is, $f_s = 8000$ samples/sec. The maximum amount that the function can change in 1/8 msec is approximately one volt. If a step size of one volt is chosen, the staircase will not overload. The quantization noise power is given by,

$$N_q = \frac{\Delta^2 f_m}{3 \ f_s} = 41.7 \ mW$$

The signal power is $3^2/2$ or 4.5 watts. The signal to noise ratio is given by

$$SNR = \frac{4.5}{0.042} = 107 \quad or \quad 20.3 \ dB$$

This is far less than what is achieved using 8-bit PCM in this example.

Bit Error Rate

Now that we have examined the errors due to the A/D converter, it is time to turn our attention to errors made during the transmission of the ones and zeros. We begin by evaluating the performance of the single-sample detector. The design of this detector was presented in Section 2.4.3.

Each sample in the single-sample detector is composed of a signal ($\pm V$) plus additive noise. We assume that the additive noise is a sample of a Gaussian zero-mean noise process. We therefore find that the probability density of the sample follows one of two Gaussian curves as shown in Fig. 2.102(a).

The sample follows the curve labeled $p_0(y)$ if a zero is being sent, and it follows $p_1(y)$ when a one is transmitted. We call the received bit a one if the sample is positive, and a zero if it is negative. The error probabilities can therefore be found by integrating the tail of the curves. The probability of mistaking a transmitted one for a zero is the cross-hatched area labeled P_M on the figure. The subscript M is a carryover from radar where mistaking a target return (one) for no return (zero) is considered a   *miss*. The other transition probability is labeled as P_{FA}, the probability of false alarm. We can see from the symmetry of the diagram that these probabilities are equal. The model of this system is therefore a *binary symmetric channel*. This type of channel is often represented by Figure 2.102(b). The horizontal lines indicate transmission without error, and the diagonal lines represent bit errors.

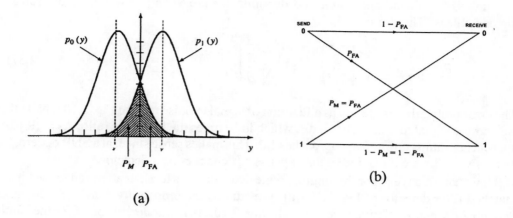

(a)

(b)

Figure 2.102 - Single-Sample Detector Performance

The variance of the noise is proportional to the bandwidth, f_m, where the proportionality constant is the noise power per Hz, N_o. Therefore, the probability of bit error is given by either P_{FA} or P_M (they are the same)[3] as

$$P_e = \frac{1}{\sqrt{2\pi N_o f_m}} \int_0^\infty \exp\left(\frac{-(y+V)^2}{2N_o f_m}\right) dy \qquad (2.85)$$

We make a change of variable to put this in the form of an error function (See Appendix C) to find

$$P_e = \frac{1}{2} erfc\left(\sqrt{\frac{V^2}{2N_o f_m}}\right)$$

$$= Q\left(\sqrt{\frac{V^2}{N_o f_m}}\right) \qquad (2.86)$$

We give answers both in terms of the error function and the Q-function. The resulting probabilities of error are identical, and either approach can be used depending on which particular table is available at the time. We can identify $V^2/N_o f_m$ as the signal to noise ratio.

---

[3]If they were not the same, the probability of bit error is the sum, $P_M Pr(1) + P_{FA} Pr(0)$ where $Pr(1)$ and $Pr(0)$ are the fraction of times a one or zero is sent.

The units of both the numerator and denominator are volts^2, or watts, so the ratio is dimensionless. Therefore,

$$P_e = \frac{1}{2} \, erfc\left(\sqrt{\frac{S}{2N}}\right) = Q\left(\sqrt{\frac{S}{N}}\right) \tag{2.87}$$

The error probability is plotted as a function of signal to noise ratio in Fig. 2.103. Note that we express signal to noise ratio in dB, which is 10 times the log of the ratio. As SNR (dB) approaches minus infinity, P_e approaches 1/2. This makes sense since, at a SNR of zero, we may as well flip a coin and save the expense of constructing a detector.

Now that we have P_e for the single sample detector, let's turn our attention to the binary matched filter detector of Fig. 2.104(a). We studied the probability density functions in Section 2.4 when we derived the curves of Fig. 2.70. As in our derivation of for the single sample detector, the error is given by the area under the tail of the Gaussian density. This is shown in Fig. 2.104(b).

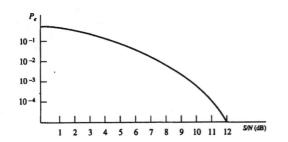

Figure 2.103 - Bit Error Rate for Single-Sample Detector

We shall derive the error for the special case where the two signals have equal energy. In the problems at the back of this chapter, you will show that our result also applies to the case of unequal signal energies. Since $P_M = P_{FA}$, the error is given by either quantity. Using P_{FA}, we have

$$P_e = \frac{1}{\sqrt{2\pi}\sigma} \int_0^\infty \exp\left[\frac{(y-m_0)^2}{2\sigma^2}\right] dy \tag{2.88}$$

To get this into the form of an error function, we change variables.

$$u = \frac{y-m_0}{\sqrt{2}\sigma} \tag{2.89}$$

$$P_e = \frac{1}{2} erfc\left(\frac{-m_0}{\sqrt{2}\sigma}\right)$$

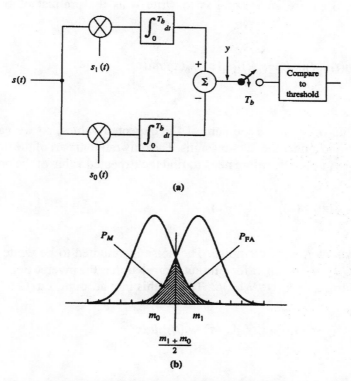

Figure 2.104 - Errors in Binary Matched Filter Detector

where m_0 is the mean of $p_0(y)$ and σ is the standard deviation. These were introduced in the previous section and they are repeated below.

$$m_0 = \int_0^{T_b} s_0(t)[s_1(t) - s_0(t)]dt$$

$$\sigma_y^2 = E\{[y - m_y]^2\} \tag{2.90}$$

$$= E\left\{\left[\int_0^{T_b} n(t)[s_1(t) - s_0(t)]dt\right]^2\right\}$$

When an integral is squared, it is not sufficient to simply square the integrand. Doing so would ignore all of the cross products. You can see this by looking at a sum.

$$\left(\sum_i x_i\right)^2 \neq \sum_i x_i^2$$

The proper way to square an integral is to write it as the product of two integrals. Therefore,

$$\sigma_y^2 = E\left\{\int_0^{T_b}\int_0^{T_b} n(t)n(\tau)[s_1(t)-s_0(t)][s_1(\tau)-s_0(\tau)]dtd\tau\right\} \tag{2.91}$$

The expected value of a sum is the sum of the expected values, so we can move the expected value symbol under the integral signs. The only random part of the integral is that containing the noise, n. We therefore need to find the expected value of the noise product,

$$E\{n(t)n(\tau)\} = R_n(t-\tau) \tag{2.92}$$

This is the *autocorrelation* of the noise. The noise is assumed to be white with power spectral density, $G_n(f)=N_o/2$. Therefore, its autocorrelation is the inverse Fourier transform of the power spectrum, or $R_n(t)=N_o\delta(t)/2$. Taking this into account, Eq. (2.91) becomes

$$\sigma_y^2 = \int_0^{T_b}\int_0^{T_b}\frac{N_o}{2}\delta(t-\tau)[s_1(t)-s_0(t)][s_1(\tau)-s_0(\tau)]dtd\tau \tag{2.93}$$

which, by the sampling property of the impulse, is equal to

$$\sigma_y^2 = \frac{N_o}{2}\int_0^{T_b}[s_1(t)-s_0(t)]^2 dt \tag{2.94}$$

We need now simply plug Eqs. (2.90) and (2.94) into Eq. (2.89) to find the bit error rate. The result can be simplified considerably and also related to physical quantities by defining the following two parameters. We define E as the *average energy* and ρ as the *correlation coefficient*.

$$E = \frac{E_0+E_1}{2}$$

$$\rho = \frac{\int_0^{T_b} s_0(t)s_1(t)dt}{E} \tag{2.95}$$

E_1 and E_0 are the individual signal energies given by,

$$E_1 = \int_0^{T_b} s_1^2(t)dt$$

$$E_0 = \int_0^{T_b} s_0^2(t)dt$$

Since we have assumed equal signal energies, we have

$$E_1 = E_0 = E$$

$$m_0 = \int_0^{T_b} s_0(t)[s_1(t)-s_0(t)]dt = \rho E - E$$

(2.96)

$$\sigma_y^2 = \frac{N_o}{2}\int_0^{T_b}[s_1(t)-s_0(t)]^2 dt = N_o E(1-\rho)$$

Equation (2.89) becomes

$$P_e = \frac{1}{2}erfc\left(\frac{E - E\rho}{\sqrt{2N_o E(1 - \rho)}}\right)$$

(2.97a)

$$= \frac{1}{2}erfc\left(\sqrt{\frac{E(1-\rho)}{2N_o}}\right)$$

In terms of the Q function, we have

$$P_e = Q\left(\sqrt{\frac{E(1-\rho)}{N_o}}\right)$$

(2.97b)

Equations (2.97) are the desired results. They show how to calculate the bit error rate from three parameters: The average energy per bit, the correlation coefficient of the two signals, and the noise power per Hz. [*Please mark/highlight these equations as we will use them many times in the work to follow.*]

Let's start with some general observations about Eq. (2.97). Both the complementary error function and the Q function decrease for increasing argument. Therefore, the probability of error decreases as the average energy increases (since E is in the numerator), the correlation decreases (since $-\rho$ is in the numerator), or the noise power decreases (since N_o is in the denominator). The observation about signal energy and noise power should be obvious, but that concerning correlation needs more investigation.

Let's take a moment to prove that the magnitude of the correlation coefficient, $|\rho|$, is bounded by one (that's why it's called a *coefficient*). That is, ρ can't get bigger than 1 nor less than -1. We begin with the observation that

$$\int_{-\infty}^{\infty} [s_1(t) \pm s_0(t)]^2 dt \geq 0$$

This inequality is true since the integrand is non-negative. Expanding and substituting we have,

$$\int_{-\infty}^{\infty} s_1^2(t) dt + \int_{-\infty}^{\infty} s_0^2(t) dt \pm 2 \int_{-\infty}^{\infty} s_1(t) s_0(t) dt \geq 0$$

$$E_1 + E_0 \pm 2E\rho \geq 0$$
$$2E \pm 2E\rho \geq 0$$
$$\mp \rho \leq 1$$

The final inequality shows that the magnitude of ρ is bounded by one. We shall get some practice in calculating ρ in the examples. For now, we consider two extremes. First assume that $s_0(t) = s_1(t)$. The correlation coefficient is then $\rho = 1$ [prove it], and Eq. (2.97) becomes

$$P_e = \frac{1}{2} erfc(0) = Q(0) = \frac{1}{2} \tag{2.98}$$

Please note that an error probability of one-half is the *worst possible situation*. You may think that an error probability of unity is worse, but this is not true. If an error were made *every* time a bit was transmitted, we could achieve perfect communication by simply adding an inverter to the receiver. So the result in Eq. (2.98) is intuitively satisfying since the transmission is supplying no information. The same signal is used to transmit both the zero and the one, so the receiver can only guess at the information. In such situation, you could save your employer money by substituting a coin for the matched filter detector. Repeatedly flip the coin to make a decision regarding the received sequence, and the probability of bit error is 1/2.

At the other extreme, if $s_1(t) = -s_0(t)$, the correlation coefficient is $\rho = -1$, and the probability of error is a minimum at

$$P_e = \frac{1}{2}erfc\left(\sqrt{\frac{E}{N_o}}\right) = Q\left(\sqrt{\frac{2E}{N_o}}\right)$$ (2.99)

Figure 2.105 shows the bit error rate as a function of the signal to noise ratio, E/N_o, for three values of correlation, -1, 0 and 1. Note that the abscissa is in dB, which is 10 times the log of E/N_o.

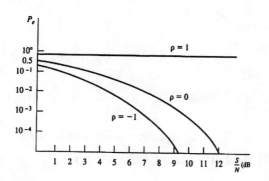

Figure 2.105 - Bit Error Rate for
Matched Filter Detector

MATLAB EXAMPLE

This MATLAB program develops a plot of bit error probability as a function of signal to noise ratio for the value of ρ that you input. For comparison purposes, the output also contains performance curves for ρ of -1, -0.5, 0, +0.5, and +1.

```
function berr()
%EXAMPLE
%Plot Bit error rate for matched filter detector
%Pe = (1/2)*erfc(sqrt((E/No)*(1-rho)/2)
%Where parameter rho = correlation coefficient and E/No = SNR
clc;                                  % clear command window
close all;                            % close all (previously) open figures
disp('  Plot Bit error rate for matched filter detector')
disp('   Pe = (1/2)*erfc(sqrt((E/No)*(1-rho)/2)')
disp('   Where parameter rho = correlation coefficient and E/No = SNR')
disp('      -------------------PRESS A KEY TO CONTINUE--------------------')
pause
```

```
disp(' ')
disp(' ')
snr_db = 0:1:16;                         % SNR (dB) on the x-axis
snr = 10.^(snr_db./10);            % Convert SNR-dB to linear form
rho = input('Please input the value of \rho, (-1< \rho <1): ');
Pe = (1/2).*erfc(sqrt(snr.*(1-rho)/2));  % Bit Error Rate Computation
axis([0 16 0 0.6]);                     % set x-y axes limits
hold on
plot(snr_db, Pe)
line_ = line(snr_db, Pe);                   t_ = text(8, 0.4,'Your input curve in
Magenta color');       set(line_,'LineWidth', 2);
set(line_, 'Color', 'Magenta'); set(t_, 'Color', 'Magenta');                set(t_,
'FontSize', 8);
rho = -1;                                    % Correlation coefficient
Pe = (1/2).*erfc(sqrt(snr.*(1-rho)/2));  % Bit Error Rate Computation
plot(snr_db, Pe)
line0 = line(snr_db, Pe);            t0 = text(0.2, 0.04, '\rho = -1');
set(line0,'LineWidth', 2);
set(line0, 'Color', 'Blue');     set(t0, 'Color', 'Blue');                set(t0,
'FontSize', 8);
rho = -0.5;                                  % Correlation coefficient
Pe = (1/2).*erfc(sqrt(snr.*(1-rho)/2));  % Bit Error Rate Computation
plot(snr_db, Pe)
line1 = line(snr_db, Pe);                  t1  =  text(0, 0.1, '\rho = -0.5');
set(line1,'LineWidth', 2);
set(line1, 'Color', 'Red');            set(t1, 'Color', 'Red'); set(t1, 'FontSize', 8);
rho = 0;                                     % Correlation coefficient
Pe = (1/2).*erfc(sqrt(snr.*(1-rho)/2));  % Bit Error Rate Computation
plot(snr_db, Pe)
line2 = line(snr_db, Pe);                  t2 = text(0.8, 0.15,'\rho = 0');
set(line2,'LineWidth', 2);
set(line2, 'Color', 'Green');    set(t2, 'Color', 'Green');                set(t2,
'FontSize', 8);
```

```
rho = 0.5;                                          % Correlation coefficient
Pe = (1/2).*erfc(sqrt(snr.*(1-rho)/2));  % Bit Error Rate Computation
plot(snr_db, Pe)
line3 = line(snr_db, Pe);                    t3 = text(2, 0.2,'\rho = 0.5');
set(line3,'LineWidth', 2);
set(line3, 'Color', 'Black');    set(t3, 'Color', 'Black');               set(t3,
'FontSize', 8);
rho = 1;                                            % Correlation coefficient
Pe = (1/2).*erfc(sqrt(snr.*(1-rho)/2));  % Bit Error Rate Computation
plot(snr_db, Pe)
line4 = line(snr_db, Pe);                    t4 = text(8, 0.51,'\rho = 1');
set(line4,'LineWidth', 2);
set(line4, 'Color', 'Blue');    set(t4, 'Color', 'Blue');           set(t4, 'FontSize',
8);
xlabel('SNR, dB'); ylabel('Pe'); title('Bit Error Rate for Matched Filter Detector');
hold off
```

Example 2.9

Binary information is transmitted using baseband signals of the form shown in Fig. 2.106.

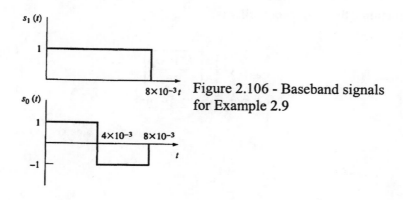

Figure 2.106 - Baseband signals for Example 2.9

Design a matched filter detector and find the probability of bit error if the additive noise has power of 10^{-3} watts/Hz.

Solution: The matched filter detector is shown in Fig. 2.107(a). Since the signal energies are equal, the threshold against which the output is compared is zero. Although the detector is correctly designed as shown, we note that for the first half of each bit interval, $s_1(t)=s_0(t)$, and the signal portion of the output is zero. We can therefore simplify the detector to that shown in Fig. 2.107(b). We now note that the bottom leg (multiplier and integrator) produces an output which is the negative of that of the top leg. Since the lower output is subtracted from the upper, the effect is the same as doubling the upper signal. We therefore further simplify the detector to that shown in Fig. 2.107(c). Finally, since the output is compared to zero, the multiplication by 2 is superfluous (and would cost extra money), so we get the simplified (integrate and dump) detector of Fig. 2.107(d).

The performance of all four detectors is identical, so we can calculate performance based upon the original problem statement. We need only find E and ρ. The energy of each signal, and therefore the average energy, is given by 8×10^3. The correlation coefficient is

$$\rho = \frac{\int_0^{T_b} s_1(t)s_0(t)dt}{8 \times 10^{-3}} = 0$$

Therefore, the probability of error is

$$P_e = \frac{1}{2}erfc\left(\sqrt{\frac{E(1-\rho)}{2N_o}}\right) = \frac{1}{2}erfc(2) = 2.3 \times 10^{-3}$$

Using Q functions, the error probability is

$$P_e = Q\left(\sqrt{\frac{E(1-\rho)}{N_o}}\right)$$
$$= Q(2\sqrt{2}) = 2.3 \times 10^{-3}$$

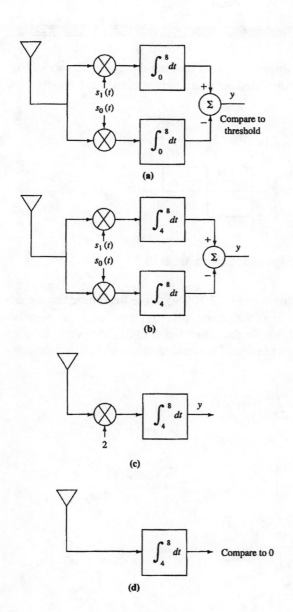

Figure 2.107 - Matched Filter Detector
for Example 2.9

Example 2.10

Design a matched filter detector to choose between the two baseband signals shown in Fig. 2.108. This is a biphase code. Assume that white Gaussian noise of power spectral density $N_o/2$ is added, and that $E/N_o = 3$.

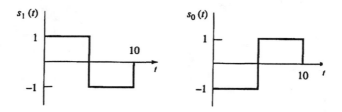

Figure 2.108 - Baseband Signals for Example 2.10

Solution: The matched filter detector is shown in Fig. 2.109(a). Note that since the signals have equal energy, the output is compared to zero. Since the lower leg produces the negative of the upper leg signal, we can remove it and double the upper leg signal. However, since the output is compared to zero, the doubling is not needed. This results in the simplified detector

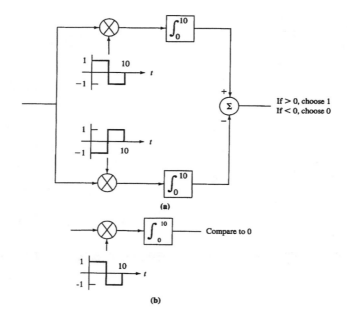

Figure 2.109 - Matched Filter Detector for Example 2.10

of Fig. 2.109(b).

Since the signals are the opposite of each other, the correlation coefficient is equal to -1. The performance is then given by

$$P_e = \frac{1}{2} erfc \left(\sqrt{\frac{E}{N_o}} \right) = Q \left(\sqrt{\frac{2E}{N_o}} \right) = 0.006$$

Thus on the average, we can expect six bit errors out of every 1000 transmitted bits.

Example 2.11

Design a matched filter detector for the two baseband signals shown in Fig. 2.110.The additive noise has a power of 0.1 watts/Hz. Find the bit error rate.

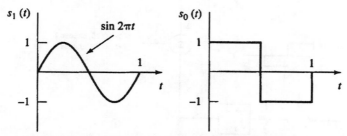

Figure 2.110 - Baseband Signals for Example 2.11

Solution: The matched filter detector is shown in Fig. 2.111. Since the signal energies are not equal, the output is compared to a non-zero threshold.

That threshold is given by

$$y_o = \frac{E_1 - E_0}{2} = \frac{0.5 - 1}{2} = -0.25$$

The average energy is given by

$$E = \frac{E_1 + E_0}{2} = 0.75$$

The correlation coefficient is

$$\rho = \frac{2 \int_0^{1/2} \sin 2\pi t \, dt}{0.75} = 0.849$$

Note that this is a very high correlation coefficient because the two signals are quite similar. This represents a very poor design. Of course, noise of 0.1 watts/Hz is enough to obliterate most life on earth.

The probability of error is then given by

$$P_e = \frac{1}{2} erfc \left(\sqrt{0.57}\right) = Q(\sqrt{1.14}) = 0.14$$

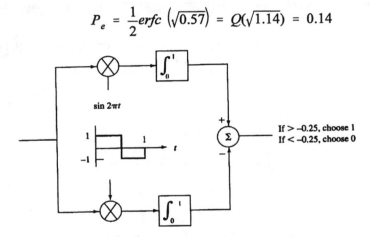

Figure 2.111 - Detector for Example 2.11

M-ary Baseband Performance

We now examine the performance of the integrate and dump M-ary detector of Figure 2.71. The output of this detector is

$$\int_0^{T_s} [s(t) + n(t)] dt = s_{out} + n_{out} \tag{2..100}$$

This is Gaussian with a mean value of $A_i T_s$, where A_i is the amplitude of the square pulse used to send the ith symbol. The variance is found the same way we did for the matched filter detector:

$$E\{[n_{out}^2]\} = E\left\{\int_0^{T_s}\int_0^{T_s} n(t)n(\tau)d\tau dt\right\}$$

(2.101)

$$=\int_0^{T_s}\int_0^{T_s} R(t-\tau)d\tau dt = \int_0^{T_s}\int_0^{T_s}\frac{N_o}{2}\delta(t-\tau)d\tau dt = \frac{N_o T_s}{2}$$

We can identify conditional probabilities representing symbol errors. Let's denote these as

$$Pr(R/S_i)$$

The index, i, is equal to 0, 1, 2 or 3 for a 4-ary system. R represents the received symbol and S represents the sent symbol. Thus, for example, $Pr(R_1/S_0)$ would be the conditional probability of say that symbol one was received conditioned on symbol zero being sent. Figure 2.112 shows the conditional probabilities for signal amplitudes of zero, A, $2A$ and $3A$.

You can see from the figure that adjacent symbol errors are equally likely, and

$$Pr(R_0/s_1)=Pr(R_1/S_0)=Pr(R_1/S_2)=Pr(R_2/S_1)=Pr(R_2/S_3)=Pr(R_3/S_2)$$

$$= \frac{1}{2}erfc\left(\sqrt{\frac{A^2 T_s}{4N_o}}\right)$$

(2.102)

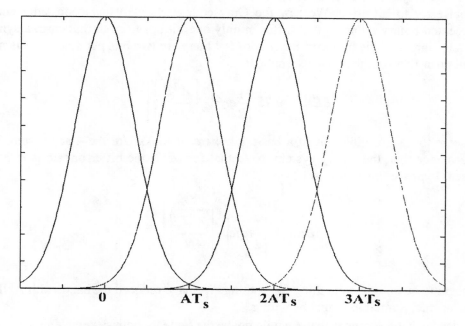

Figure 2.112 - Conditional Probabilities for M-ary Baseband

Symbol errors can also occur where the received symbol is   *not* adjacent to the transmitted symbol. For example, if symbol zero is transmitted, it could be erroneously received as symbol two or symbol three. But because the tails of the Gaussian density decrease rapidly, the probability of a non-adjacent symbol error is much smaller than the probability of an adjacent error. We shall ignore this probability and assume that all symbol errors that occur result in erroneously picking the *adjacent symbol*. The symmetry that exists for binary communication does not exist for M-ary where M>2. Note that if we send either symbol zero or symbol three, an error can only result in one way. Symbol zero can change to symbol one and symbol three can change to symbol two. On the other hand, if we transmit symbol one, it can change to either symbol zero or symbol two. A similar situation exist if we transmit symbol two. Because of this observation, the probability of a symbol error when we transmit symbol one or symbol two is twice that given in Eq. (2.102).

If we assume the symbols are equally probable, then the probability of any symbol is 1/M. The overall symbol error probability is then

$$P_e = \frac{2(M-1)}{M}\left[\frac{1}{2}erfc\left(\sqrt{\frac{A^2T_s}{4N_o}}\right)\right]$$
(2.103)

For *M*=4, this becomes

$$P_e = 1.5\left[\frac{1}{2}erfc\left(\sqrt{\frac{A^2T_s}{4N_o}}\right)\right]$$
(2.104)

This is the probability of symbol error. To compare this to the binary case, we need to relate symbol error to bit error. We saw the Gray code earlier. In that code, adjacent symbols represented binary numbers that differ in only one bit position. In that case, a symbol error translates into a single bit error. Since for *M*=4 there are two bits per symbol, this means that the bit error rate for the above example, is

$$BER = 0.75\left[\frac{1}{2}erfc\left(\sqrt{\frac{A^2T_s}{4N_o}}\right)\right]$$
(2.105)

We seem to have a violation of a basic law here. It looks like the 4-ary system has better performance than the binary system. Note that for M=2, the bit error rate is (please take a moment to prove this to yourself)

$$BER = \left[\frac{1}{2}erfc\left(\sqrt{\frac{A^2T_s}{4N_o}}\right)\right]$$
(2.106)

We seem to be getting something for nothing. We have reduced the bandwidth AND reduced the error rate. Something must be wrong.

Indeed, what is wrong is that *A* would not be the same for both systems. If it were, the binary pulses would get as large as *A* while the 4-ary pulses would get as large as 3*A*. Sure, if we are willing to increase the amplitude (and therefore the maximum power), we can appear to violate the basic trade off rules of engineering. But to compare these systems fairly, we would have to cut A by a factor of three when we switch from binary to 4-ary.

2.6 Applications

In this section, we examine three applications of baseband communications. We begin with two analog applications: the local loop in traditional telephone voice service, and closed-circuit television. This is followed with one digital application, that being recording techniques such as those used in compact audio discs.

2.6.1. Telephone - Local Loop

The telephone system is conceptually very simple. We give this very brief description because it forms the backbone of the analog communication system and is, by far, the most widespread application of analog baseband communication.

The telephone switching network is composed of many wire connections, some of which carry a single telephone conversation, and others which multiplex many simultaneous conversations. A line dedicated to a single user is known as a *loop*, while shared channels are known as *trunks*.

The *subscriber loop* is the pair of wires which connects an *end office* (also known as a *central office*) with an individual subscriber. Although the many pairs of wires are combined in a *paired cable*, each subscriber has a unique assigned pair. Lengths are on the order of several kilometers. Both directions of communication appear on the same pair of wires, so simultaneous talking by both parties causes superimposing of the signals on the wires.

The local loop communicates in much the same manner as an intercom. A transducer (microphone) converts an acoustic voice signal into an electrical voltage. The simplest form of transducer is a carbon microphone consisting of granules of carbon. The sound pressure varies the contact resistance between the granules. The receiver consists of an earphone (a diaphragm controlled by a magnet which is varied by a winding on a permanent magnet).

The telephone set includes circuitry to control *sidetones*–the coupling of the sender's voice into the sender's receiver. When you talk into a telephone, you do not want to hear your voice in the earpiece, although some coupling is needed for natural conversation. The telephone set also includes the necessary hardware to produce and receive control signals (e.g., tones for push-button dialing, ringing, and switches to indicate that the phone has been answered).

Since subscriber loops can vary greatly in distance, the attenuation also varies. This would lead to unacceptable variation in volume levels. For this reason, the telephone set contains equalizers (using varistors).

Typical telephone circuits have a passband that extends from several hundred Hz to approximately 3200 Hz. As long as the loops do not exceed several km in length, this bandwidth is easily achieved.

2.6.2. Closed-Circuit Television

Although closed-circuit television need not follow the standards established for broadcast TV, we present the concepts using these standardized numbers. In this manner, we will not have to change these numbers when we describe broadcast television in the next chapter.

The challenge of television is to reduce a moving image with sound to an electrical signal which can be transmitted through a channel. There are varying standards for picture quality and for the manner in which color is included. We describe the North American *broadcast* standard in this example. Once this is understood, the extension to other standards is relatively simple.

A Picture is Worth a Thousand Words?

Nonsense! A picture can be equivalent to far more than a thousand words. While we could rigorously define the information content of a picture using concepts from the science of information theory, we shall not do that here. Instead, we will subdivide the picture into small components similar to words.

Suppose we divide a picture into squares where each square is a certain shade. The number of squares in any given area determines the *resolution*. For example, Fig. 2.113(a) shows a picture of the letter "A" where 81 squares have been used to define the picture. In Fig. 2.113(b), the same letter is shown where the number of squares (elements) has increased to 342. The result is improved resolution.

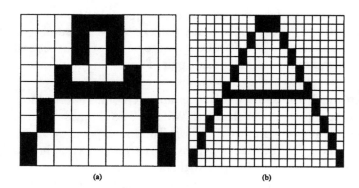

Figure 2.113 - Picture Elements

(a)   (b)

To get a feel for some reasonable numbers, we shall use commercial TV as an example. The number of elements used in United States television is set by the Federal Communications Commission (FCC) at 211,000 picture elements (*pixels*). This is divided into 426 elements in each horizontal line and 495 visible horizontal lines in each picture. The job of television transmitters is then to step through each of these 211,000 picture elements and send an intensity value for each one. The receiver interprets these transmissions and reconstructs the picture from the 211,000 intensity levels. This information must be updated rapidly to simulate motion. We are confining our attention to monochrome (black and white) television.

A conventional (*not* flat-screen) TV receiver is not much different from a conventional analog laboratory oscilloscope. A beam of electrons is shot toward a screen and bent by use of deflection plates (magnets). When a negative charge is placed on a plate, the electron beam is repelled. In the oscilloscope, we apply a sawtooth waveform to the horizontal deflection plates to sweep the beam from left to right (creating a time axis), and then more rapidly back to the

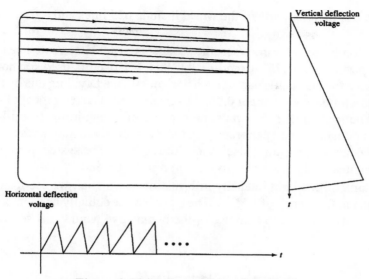

Figure 2.114 - Generation of the Raster

left. This traces a line. In TV receivers, we add a second dimension to the sweep. While the beam is rapidly sweeping from left to right on the screen, it is less rapidly sweeping from top to bottom on the screen. The net result is a series of almost horizontal lines on the screen, as sketched in Fig. 2.114. This is known as the TV *raster*.

We are describing traditional analog scanning TV. Digital television has the option of controlling the choice of picture elements using digital counting circuitry. Picture elements are not necessarily phosphors bombarded with an aimed electron beam, but could be individual LED or LCD elements arranged in a matrix and digitally scanned. Nonetheless, conventional TV remains the norm for the present. Even in the future, it will represent an educational historical study of how a significant technological (perhaps also social) problem was solved.

The screen has 495 visible horizontal lines in order to comply with the FCC regulation. The beam then returns to the top of the screen, taking the equivalent time of an additional 30 lines to do so. We can therefore think of the picture as having 525 lines, 30 of which occur during the *vertical retrace* time.

We now assign timing to this process. The human eye requires a certain picture rate (number of times the entire screen is refreshed each second) to avoid seeing flicker of the type seen in early motion pictures. The minimum number is somewhere near 40 per second. United States TV decided to use 60 frames per second (color TV actually uses a number slightly less–approximately 59.94 Hz). This matches the frequency of household current and is chosen that way to minimize the effects of the video equivalent of 60-Hz hum. That is, if the 60-Hz power signal is not completely filtered out of the video, it causes a slight gradation of brightness over the height of the picture. If this gradation is stationary, the eye probably will not notice it. However, with a frequency mismatch the gradations will exhibit a migration in the vertical direction (a rolling), and the likelihood of human detection increases. We emphasize here that the numbers being presented are for the U.S. standard (this applies to the U.S., Canada, Netherlands, Brazil, Colombia, Cuba, Japan, Mexico, Peru, Surinam and Venezuela). Many of the standards in use in other parts of the world include 625 lines and a frame frequency of 50 Hz.

At 525 lines/frame and 60 frames/sec, the product is 31,500 lines/sec. The reciprocal of this yields the time per line as 31.75   μsec/line. Of this, 5.1 μsec is used for horizontal retrace, leaving 26.65 μsec for the visible part of each horizontal line. Dividing this by the 426 elements in a line gives us a time per element of 0.0625 μsec/element. The reciprocal of this is 16 million elements/sec. The system therefore must be capable of transmitting 16 million independent shades (black, white, gray, etc.) per second. In the worst case, we may wish to display a perfect checkerboard design of alternating black and white squares. The system would then jump from the darkest to lightest and back again 16 million times a second.

If we now think of this light intensity information as a signal, we see from Fig. 2.115 that it has a fundamental frequency of 8 MHz. The square wave actually has a maximum frequency of infinity, but if we round it to a sinusoid, the human eye would still see the checkerboard design.

Figure 2.115 - Intensity Signal for Checkerboard

Here, the long arm of the law steps in again, and the FCC mandated a maximum video signal bandwidth of 4.2 MHz. Alas, how can these seemingly contradictory specifications (60 frames/sec to avoid flicker, 211,000 picture elements for proper resolution, 4.2 MHz maximum bandwidth) be met?

Engineers had observed many decades of development of motion pictures (traditional film, not videotape) in which a similar predicament occurred. In standard motion pictures, only 24 different pictures are shown each second. But 24 flashes/sec on the screen would appear to flicker considerably. Contemporary motion picture projectors flash each image twice (the film moves into position and the shutter opens and closes twice before the film moves again–a mechanical engineer's dream). Thus, the frame rate is 48/sec, though the rate of presenting new pictures is 24/sec.

Television's founders decided to play a similar trick. But in the early years of television, they didn't have inexpensive fast-access memory available to store a picture while it is scanned twice. They cut the signal frequency in half by cutting the number of lines per second in half, from 31,500 to 15,750 lines/sec. However, it was necessary to fill the entire screen each 1/60 sec to avoid flicker. The technique for cutting the line frequency in half without changing the frame frequency is known as *interlaced scanning*. During the first 1/60 sec, the odd numbered lines are traced (ending with 1/2 line). The beam then returns to the top center of the screen to trace the even-numbered lines in the next 1/60 sec. Thus, while it takes 1/30 sec to send the frame consisting of all 211,000 picture elements, the screen is scanned twice (each scan is called a *field*) during this period. The eye fills in the missing rows and detects no flicker. There is, of course, some loss of resolution on fast-moving objects, but this was deemed appropriate for conventional TV. When interlaced scanning is not used, the result is called *progressive scanning*.

Signal Design

We now translate this information into an electrical signal format. If we plot light intensity as a function of time, a staircase function results. Figure 2.116 shows an example of the letter "T" in dark black followed by the punctuation "period" in light gray. The associated signal is shown where, for simplicity, interlaced scanning has not been included and the number of lines has been drastically reduced.

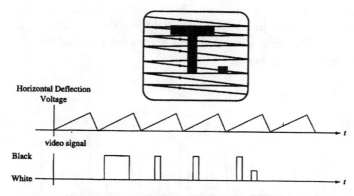

Figure 2.116 - Video Signal for Message, "T"

The information (video) signal would be a similar staircase function where the minimum step width is 0.125 μsec. In the actual video signal, the voltage corresponds to the light intensity and the receiver uses this voltage to control the electron gun. The higher the voltage applied to a grid placed between the gun and screen, the fewer electrons hit the screen and the darker the spot. As an additional modification, the staircase function is smoothed to reduce bandwidth. The eye cannot tell the difference between a smooth or rapid transition in the signal during 0.125 μsec. After all, this corresponds to only 1/426 of the width of the TV screen.

There is an additional aspect to this electrical video signal known as *blanking*. While the beam on the cathode ray tube is retracing, we want the electron stream to be turned off so that the retrace is not seen as a line on the screen.

Taking all this into account, the video signal corresponding to the picture shown in Fig. 2.116 is redrawn as Fig. 2.117.

Synchronization

The transmitter rapidly traces from left to right and from top to bottom many times each second. It sends a record of light intensity as a function of time. The receiver must be sure it is placing the transmitted intensity element in the same location on the screen as intended. If the beam in

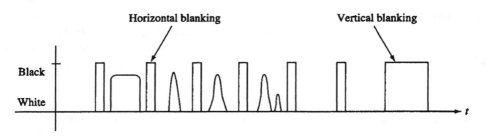

Figure 2.117 - Video Signal Including Blanking

the receiver doesn't start a scan at the same instant that the received waveform does so, the picture will appear split at best, and totally scrambled at worst. We need a method for synchronizing the two sweeping operations. This is done by means of synchronization pulses added to the video signal. The pulses are added during the blanking intervals, thereby not affecting what is seen on the screen. Figure 2.118 shows the signal of Fig. 2.117 modified with the addition of synchronization pulses.

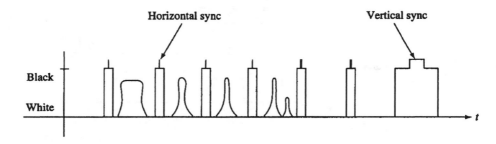

Figure 2.118 - Addition of Synchronization

Two types of synchronizing pulses are shown in the figure. The narrow pulses are horizontal synchronizing pulses, and the wide pulses are vertical synchronizing pulses.

The receiver separates these pulses from the remaining signal by means of a threshold circuit. The horizontal and vertical pulses are then distinguished by means of a single *RC* integrator circuit. The integral of the wider vertical sync pulses is larger than that of the narrower horizontal sync pulses. The separate pulses are then used to synchronize (trigger) the horizontal and vertical oscillators.

Block Diagram of TV Monitor

We are now in a position to examine the overall block diagram of a black and white TV monitor. A representative block diagram of a traditional monitor is shown in Fig. 2.119. The video signal is amplified and the amplified signal is fed into the control grid of a cathode-ray tube. This controls the brightness of the dot on the front surface of the phosphor-coated screen. Since the synchronization pulses are above the level of the video signal, they can be separated using a clipping circuit (a simple diode and battery can accomplish this). The vertical sync is separated from the horizontal sync using a simple integrator circuit. Since the vertical pulses

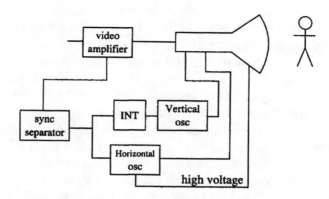

Figure 2.119 - Block Diagram of Monochrome TV Monitor

are much wider, the area is much larger. The two sets of synchronization pulses trigger the two sawtooth generators controlling the vertical and horizontal sweeping motion.

As an added bonus, the retrace portion of the horizontal deflection voltage (with a very large slope due to 5.1 μsec retrace time) is used to generate the very high voltage (over 20 kV) required on the CRT anode to pull the electrons in a straight line toward the face of the tube. This high voltage is generated by differentiating the retrace ramp using what is known as a *flyback transformer* (an inductor, so voltage is proportional to the derivative of current).

Color Television

Color TV takes advantage of the theory behind the color wheel. All colors, including black and white, can be formed as combinations of the three primary colors. The theory of resolving colors into components dates back to Sir Isaac Newton (a scientist), Johann Wolfgang von Goethe (a poet), Thomas Young (a physicist), and Hermann von Helmholtz (a physicist). Perhaps such interdisciplinary teams could help solve some of today's technological dilemmas? The term *primary colors* can be interpreted in several ways. The *primary colors of the spectrum* are red, green and blue, while the *primary colors of paints* are red yellow and blue. Television chose the primary colors of the spectrum since phosphors that glow these colors when bombarded with an electron beam are readily available.

The color TV receiver can be thought of as three separate TVs, one generating red, one generating blue, and one generating green. By mixing these in varying strengths, any color can be formed.

Classical color CRTs actually contain three separate electron guns, one for each color. The transmitted signal must therefore be capable of generating three separate video signals to control each of these guns.

But certainly you are asking where the two additional signals can be squeezed since we are operating with the same bandwidth limitation imposed on black and white TV. Monochrome transmission already uses all of the bandwidth allowed for TV. We need a system to transmit all three video signals without increasing the bandwidth.

The answer to this dilemma lies in a frequency analysis of the video signal. If we examine the Fourier transform of a black and white video signal, we find it resembles a train of impulses. An actual signal for one frame contains 525 horizontal traces. Since most pictures contain some form of *vertical continuity*–that is, the content of one horizontal line closely resembles that of the next horizontal line–the video signal is almost periodic with a fundamental frequency of 15,750 Hz (the number of lines per second).

If the video signal were exactly periodic, its Fourier transform would be a train of impulses at multiples of the fundamental frequency. Since it is almost periodic, the Fourier transform consists of pulses (not impulses) centered around multiples of the line frequency. The more periodic the time signal, the sharper the pulses in frequency.

Since the Fourier transform approaches zero between multiples of the line frequency, additional information can be placed in these spaces by use of a form of multiplexing. The additional information needed for color transmission is carried by a sinusoid which is midway between two multiples of the line frequency. This is assured by using a signal with frequency that is an odd multiple of half of the line frequency. The figure used is 3.579545 MHz, which is the 455th harmonic of half of the line frequency. The two additional signals are combined and transmitted.

If we were simply designing a color monitor with no other constraints, we could transmit the three primary color signals. However, as broadcast television moved from monochrome to color, one requirement was to make the signal compatible. That is, those viewing a color signal on a black and white monitor should see the total picture, not just one color component. For this reason, the three signals transmitted are *not* the three primary colors. Instead, the basic system transmits brightness (known as *luminance*), position in the color spectrum (*hue*), and how close the color is to a pure single frequency (*saturation*). The hue and saturation are combined with the luminance in a matrix operation which has the goal of making it easy to reproduce the three separate colors at the receiver and at the same time matching characteristics of our human color perception. The system does not provide perfect separation, and if the picture lacks vertical continuity over a span of time, the colors will interact. The combination scheme used in the United States is known as the *NTSC* (National Television Systems Committee) *standard*. This standard is used by many countries in the world. Some other parts of the world combine the colors according to different standards, notably the *PAL* (phase-alternation line), and *SECAM* (sequential color and memory). The bandwidth of the luminance signal is limited to 4.2 MHz, while the bandwidth of the other two signals are 1.5 MHz and 500 kHz. Again, this is a compromise and results in some loss of performance. For example, if you ever watched a movie on TV that used red for the titles, you probably noticed a decrease in resolution. The letters became fuzzy. The sharp corners of the letters represent frequency components above the cutoff of the signal sending this information.

2.6.3. Digital Recording - CDs

The compact disc is an excellent example of the application of baseband digital techniques. It clearly points out the advantages of digital over analog. It does not, however, represent a complete communication system in the context of our approach. The signal transmission from transmitter (studio) to receiver (purchaser) is digital, but the channel is really the disc itself which is physically carried from one location (the store) to the other (one's house). In the receiver, the signal is processed and converted to analog, and then transmitted through wires to the speakers. Another reason this differs slightly from the presentation in the text is that the signal is not carried in the form of an electrical waveform, but in a mechanical format on the disc.

In spite of these differences, compact disc technology serves as an example of the manner in which the theory is reduced to a practical situation.

We begin by describing the A/D process. The audio signal (actually two signals for stereo) is sampled at a rate of 44.1 kHz. This is more than sufficient to reproduce audio signals with frequencies up to 20 kHz. The samples are each converted to a 16 bit binary number, thus producing a 16-bit PCM signal. This yields a nominal signal-to-quantization noise ratio of 96 dB ($6N$).

The above paragraph represents a simplified approach. The system employed by CD recorders includes four areas of more complex processing: Reduction of bit rate, oversampling, block coding to reduce transitions, and forward error correction. Companding and dithering are two other features than can be applied to compact discs. We now briefly expand on each of these areas.

Although the basic CD standards call for a 16 bit A/D, actual systems employ a 14-bit A/D. The more bits of quantization used, the larger the number of bits per second that must be recorded. As this number increases, the maximum recording length of a disc reduces. Thus, a trade-off decision is required. The reduction from 16-bit to 14-bit PCM would lower the SNR by 12 dB. However, because of the signal to quantization noise ratio improvement afforded by oversampling, this reduction in bit rate causes a relatively small degradation in SNR.

We saw earlier in this chapter that sampling at higher rates has the advantage of spreading the quantization noise over a wider band the frequencies. The lowpass filter in the receiver then reduces the noise. For example, if a signal with a maximum frequency of 20 kHz is sampled at 40 kHz, the quantization noise fills the band between dc and 20 kHz. If the signal is instead sampled at 80 kHz, the quantization noise has a power spectral density extending from dc to 40 kHz. Filtering at 20 kHz cuts the noise in half and increases the signal to quantization noise ratio by 3 dB. Suppose now that a compact disc system samples at 4 times the nominal rate, or 176.4 kHz. This would increase the signal to noise ratio by 6 dB. Thus a 14-bit system would achieve 90 dB signal to noise ratio instead of 84 (6 times 14). This SNR improvement is not without cost. Sampling at four time the rate, if performed during the recording process, would increase the number of bits that must be stored on the disc by a factor of 4. Discs that can now store 75 minutes of music would only hold less than 20 minutes!

In fact, the sampler at the recorder still only operates at 44.1 kHz. The extra three samples per sampling interval are inserted at the receiver. Since these sample values are not needed to reproduce the original waveform, the receiver inserts three samples of identically zero value. For example, in a three-bit system, suppose the first 3 samples were

101 010 111

The receiver interprets these as

101 000 000 000 010 000 000 000 111 000 000 000

In addition to coding for error control, an additional block code is inserted to improve performance. The signaling technique can be considered as unipolar NRZ. The compact disc surface consists of *pits* and *lands*. Pits are portions of the disc that have been removed (depressions) and lands are the areas between the pits. A laser beam searches for the transitions between pits and lands. Due to the finite width of the beam and the mechanics of creating the discs, transitions are not instantaneous. In fact, if we look at the overall transducer system which converts the disc surface to an electrical system, we can see such effects as intersymbol interference, and we can generate eye patterns. One must also consider vibrations and the aging that occurs in the drive systems. This ultimately limits the speed of recording and the number of bits that can be placed on a disc. If the transitions are spaced further apart, the demands on the system are relaxed.

To space transitions as far as possible, a block code is used. Each block of 8 data bits is encoded in a 14-bit code word in a manner which decreases the number of transitions. For example, the bit sequence 10101010, which contains 7 transitions, is coded into 11100011100011, which only contains 4 transitions. Thus, although bits occur at a faster rate (by a factor of 14/8), transitions are more widely spaced using the *eight-to-fourteen modulation* (EFM).

Compact discs employ error correction codes (these are not discussed in this text). The actual type used is the *cross-interleaved Reed-Solomon code* (CIRC). This is a linear code composed of two interleaved codes. The first, more simple code, detects error in blocks of data. If errors are detected, the second code performs the more difficult task of error correction. If no error is detected in a particular block, the second code need not be implemented.

Companding and dither might be used. Companding was discussed in detail earlier in this chapter. *Dither* is a technique where a small amount of controlled noise is added to reduce distortion for low-level inputs. Suppose that for a portion of the signal, amplitudes are very small as in a soft portion of a musical selection. If we think of a sine wave traversing only two quantization regions, the resulting quantized waveform is a periodic square wave, as shown in Fig. 2.120

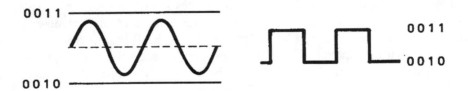

Figure 2.120 - Quantized Small Signal

This square wave consists of a fundamental and odd harmonics. The ear is not sensitive to the harmonics–they might add tonal quality to the sound, but they will not change the perceived frequency. However, for high enough harmonics, the frequencies exceed 1/2 of the sampling rate. The aliasing effect transforms these higher harmonics to lower frequencies that are not harmonically related to the signal frequency. For example, suppose the signal frequency is 1.3 kHz. The 19th harmonic occurs at 24.7 kHz, which aliases as 2.65 kHz [24.7 kHz minus one half of the sampling frequency of 44.1 kHz]. Although this aliased signal component is at a relatively low level, since the signal is also low level, the ear could perceive this as interference and distortion.

Suppose now that a low-level noise signal (on the order of 1/3 of quantization spacing) is added. This will add additional transitions as the noisy signal is pulled back and forth across the dividing line between the two round-off levels. When this quantized, it might resemble Fig. 2.121(b). We have repeated the undithered signal as Fig. 2.121(a).

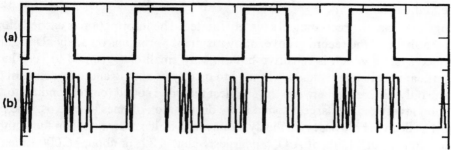

Figure 2.121 - Coding of Dithered Signal

The effects of the noise can be removed with filtering. However, the significant thing to note is that the Fourier transform of the signal of Fig. 2.121(b) does not contain the harmonic content of that of 2.121(a). These Fourier transforms are sketched in Fig. 2.122 where the top sketch is the transform of the undithered signal and the bottom is the transform of the dithered signal. Note that some of the third harmonic remains, but the higher-order harmonics are greatly attenuated. We have therefore successful eliminated the negative effects of aliasing.

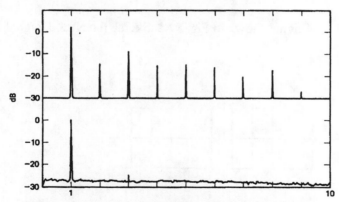

Figure 2.122 - Fourier Transform of Dithered Signal

Advantages of Digital Recording

The compact disc provides advantages when compared with classical forms of analog recording (tapes and analog discs). These advantages relate to wearing and damage of the medium, and to dynamic range.

As an analog disc wears out, the audio content experiences distortion. The transducer is normally in the form of a needle which senses shape variations in a groove in the record. Any change in shape translates to a change in the reproduced audio. In the case of a compact disc, the change would have to be sufficient to make a pit look like a land, or vice versa. Since there is no mechanical contact between the pickup mechanism (the laser beam) and the disc, wear is kept at a minimum. The Reed-Solomon error correcting code can correct most errors caused by wear or damage to the disc.

Dynamic range of an analog recording is severely limited. The information is carried either by variations in the shape of the record groove, or by magnetic variations on a tape. Designing a reasonably-priced system which is sensitive both to very small changes and to very large changes is a significant challenge which has limited to quality of analog audio systems. In the digital system, the only difference between the smallest recorded sound (corresponding to the lowest quantization level) and the largest sound is the distribution of ones and zeros. The pits in the disc are just as deep in both cases, so the dynamic range is dependent only on the number of bits of quantization. With 14 bits of A/D, the largest value is 2^{14}, or about 16,000 times as large as the smallest value.

Problems

2.1 A 1-kHz sine wave is sampled 3,000 times each second and encoded using four-bit PCM. (You may choose three bits plus a sign or four-bit encoding).

(a) Sketch the PCM waveform for the first 3 msec, assuming that the NRZ-L format is used.

(b) Repeat part (a) for the NRZ-M differential format.

(c) Repeat part (a) for the biphase-L format.

2.2 You are given the NRZ-M signal shown in Fig. P2.2. Sketch the NRZ-L signal for the same binary information.

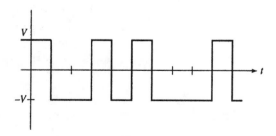

Figure P2.2

2.3 Modify the NRZ-M encoder and decoder to apply to NRZ-S. Do this by adding only one inverter to the system. Investigate whether there is more than one possible location for the inverter.

2.4 You are given the function $s(t)=2\sin\pi t$.

The signal is to be transmitted using four-bit PCM.

(a) Find the PCM waveform the first 5 sec.

(b) Find the mean square quantization noise power.

2.5 Repeat Problem 2.4 for a Gray code.

2.6 A PCM wave is as shown in Fig. P2.6, where voltages of +1 and -1 are used to send a one and a zero, respectively. Two-bit quantization has been used. Sketch a possible analog information signal $s(t)$ that could have resulted in this PCM wave.

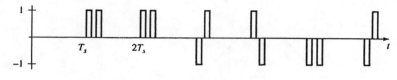

Figure P2.6

2.7 Two voice signals with 5-kHz maximum frequency are transmitted using PCM with eight levels of quantization. How many pulses per second are required to be transmitted?

2.8 The input to a delta modulator is $s(t) = 5t + 1$. The sampler operates at 20 samples per sec. The step size is 1 V. Sketch the output of the delta modulator for the first 2 sec.

2.9 Delta modulation is used to transmit a voice signal with a maximum frequency of 3 kHz. The sampling rate is set at 20 kHz. The maximum amplitude of the analog voice signal is 1 V. Design the modulator, and discuss your reasons for selecting a particular step size.

2.10 An adaptive delta modulator is used to transmit a signal $s(t) = 5t + 1.1$. The sampler operates at 10 samples/sec. The step size is either 0.5, 1, or 1.5 V, depending on the number of ones in a sequence of four bits. For zero or four ones, the maximum step size is chosen. For one or three ones, the step size is 1 V, and for two ones, the minimum step size is chosen. Sketch the staircase approximation to $s(t)$ and the transmitted digital signal for the first 3 sec.

2.11 Repeat Problem 2.10, using the Song algorithm.

2.12 Repeat Problem 2.10, using the space shuttle algorithm.

2.13 You are given the function $s(t) = 2\sin 2\pi t + 4\cos 4\pi t$.

(a) Find the binary sequence that results if this signal is to be transmitted using delta modulation with a sampling rate of 20 samples/sec. Choose an appropriate step size.

(b) Repeat part (a) for the Song algorithm.

2.14 Design a delta-PCM modulator for the signal $s(t)=\sin2\pi t+4\cos4\pi t$.

Samples are taken at the rate of 10 samples/sec, and the bit transmission rate is 20 bps. (i.e., the correction factor is coded into two-bit words.)

2.15 Find the weights for a predictor that operates on the three most recent samples in order to predict the next sample value. The autocorrelation of the signal is shown in Fig. P2.15. Evaluate the performance of the predictor, and compare the resulting error to that given in Example 2.3.

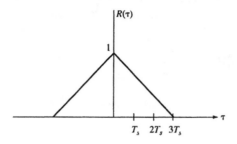

Figure P2.15

2.16 (a) Design a linear predictor to approximate the signal $s(t_o + T)$ as a linear combination of $s(t_o)$ and its derivative $s'(t_o)$. You may either derive a general result in terms of autocorrelations or use the autocorrelation given in Problem 2.15 to arrive at a numerical answer.
(b) Evaluate the mean square error.

2.17 Design a linear predictor to estimate the value of $s(nT+A)$ as a linear combination of $s(nT)$ for all n. Find the mean square error. Show that if the conditions of the sampling theorem are met, the mean square error goes to zero.

2.18 Design a PCM system that could be used to transmit the video portion of a black-and-white television signal. The video signal contains frequencies up to 4 MHz. A study of viewers has indicated that at least 70 gray-scale levels are required for a pleasing picture.

2.19 Prove that about 91% of the total power of a random NRZ-L signal is contained in frequencies below the bit rate. [Hint: Integrate the expression in Eq. (2.8)].

2.20 Prove that about 65% of the total power of a random biphase-L signal is contained in frequencies below the bit rate and that about 86% of the power is contained in frequencies below twice the bit rate. [Hint: Integrate the expression in Eq. (2.12)].

2.21 You are given the NRZ-S signal shown in Fig. P2.21.
(a) Sketch the NRZ-L signal for the same binary information.
(b) Sketch the bi-phase (Manchester) signal for the same binary information.

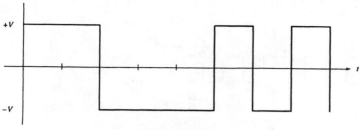

Figure P2.21

2.22 You are given the NRZ-L signal shown in Fig. P2.22.
(a) Sketch the NRZ-S signal for the same binary information.
(b) Sketch the bi-phase-M signal for the same binary information.

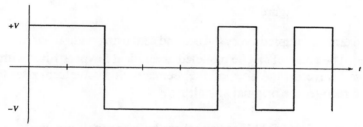

Figure P2.22 (same waveform as P2.21)

2.23 A signal is of the form $s(t)=\sin\pi t+3\sin3\pi t$. It is sampled at 1.5 times the Nyquist rate and transmitted using PAM. The pulse waveform, $s_c(t)$, is a periodic train of triangular pulses, as shown in Fig. P2.23. Find the Fourier transform of the modulated waveform.

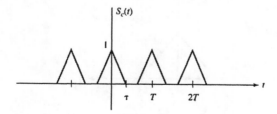

Figure P2.23

2.24 An information signal is of the form

$$s(t) = \frac{\sin\pi t}{\pi t}$$

Find the Fourier transform of the waveform that results if each of the two carrier waveforms shown in Fig. P2.24 is pulse amplitude modulated with this signal.

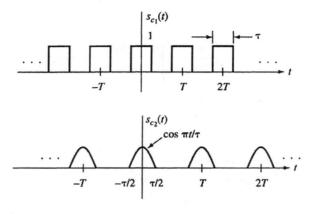

Figure P2.24

2.25 The signal $s(t)=\cos 2\pi t$ is sampled every 0.4 sec and sent using natural-sampled PAM with pulse widths of 0.1 sec. The channel can be modeled as an ideal lowpass filter with a cutoff at 10 Hz. Find the received waveform. Also, find the reconstructed waveform after the receiver uses a lowpass filter to recover the original signal, $s(t)$.

2.26 Consider a two-channel TDM PAM system where both channels are used to transmit the same signal $s(t)$ with Fourier transform $S(f)$, as shown in Fig. P2.26. The system samples $s(t)$ at the minimum rate. Find the Fourier transform of the TDM waveform, and compare it to the Fourier transform of a single-channel PAM system used to transmit $s(t)$.

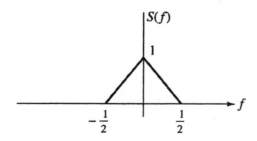

Figure P2.26

2.27 Three asynchronous sources transmit PAM waveforms to a buffer multiplexer. Each of the sources transmits at a pulse rate that is Gaussian distributed with a mean value of 1000 pulses/sec and a variance of 9. The channel is capable of transmitting 3000 pulses per second. How large must the buffer be such that the probability of overload is less than 1%?

2.28 Three information signals are to be sent using time-division multiplexed PAM. Suppose that the maximum frequency of each of the first two signals is 5 kHz and that the maximum frequency of the third signal is 10 kHz. Design the multiplex system and draw a block diagram.

2.29 Ten signals are to be time division multiplexed and transmitted using PAM. Four of the signals have a maximum frequency of 5 kHz, two have a maximum frequency of 10 kHz, two have a maximum frequency of 15 kHz, and two have a maximum frequency of 20 kHz.
(a) Design a multiplex system.
(b) How many pulses per second must be transmitted?

2.30 You wish to sample and time division multiplex 36 channels with the following maximum frequencies:

One channel has a maximum of 10 kHz.
Three channels have a maximum of 5 kHz.
Eight channels have a maximum of 2.5 kHz.
Eight channels have a maximum of 300 Hz.
Sixteen channels have a maximum of 150 Hz.
Design a system using sub- and supercommutation. Make any reasonable approximations.

2.31 A discrete-time analog signal is created by sampling a speech waveform at 10,000 samples per second. Each sampling pulse is 0.01 msec wide and is transmitted through a channel that can be approximated by a lowpass filter with cutoff frequency at 50 kHz. Evaluate the effects of channel distortion.

2.32 A PWM system multiplexes three signals derived from waveforms with the same maximum frequency f_m. Sample values are normalized (to lie between zero and unity), and the width, w, of each pulse is related to the normalized sample value $s_i(nTs)$ by

$$w(nT_s) = 1 + s_i(nT_s) \ \mu sec$$

(a) What is the maximum frequency f_m of the baseband signals that would permit this multiplexing to take place?
(b) What is the minimum bandwidth of the channel?

2.33 Show that a sample-and-hold circuit is an approximation to a lowpass filter, provided that the sampling is performed at the Nyquist rate or higher. [Hint: You may wish to find the step response of the sample-and-hold circuit and compare it to the step response of a lowpass filter.]

2.34 The system shown in Fig. P2.34 is similar to a sample-and-hold circuit, where the input can be considered to be an impulse-sampled version of the original waveform.
(a) Find the impulse response of this system.
(b) Find the transfer function, and compare it to the transfer function of a lowpass filter.

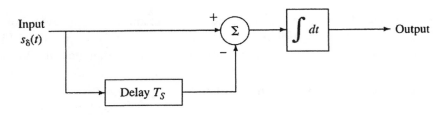

Figure P2.34

2.35 Two signals,

$$s_1(t) = \cos 2\pi t$$
$$s_2(t) = \cos \pi t + 2\cos 2\pi t$$

are sampled every 0.4 sec and are sent using multiplexed natural-sampled PAM. The channel can be modeled as an ideal lowpass filter with a cutoff at 10 Hz.
(a) Find the received waveform.
(b) Find the reconstructed waveforms after the receiver uses a lowpass filter and demultiplexer.
(c) Repeat part (b), with $s_1(t)$ changed to $\cos 1.9\pi t$.

2.36 You wish to investigate the use of a first-order lowpass filter (e.g., an RC circuit) to recover the original time signal from a PWM waveform. You can assume that each PWM pulse has its leading edge at the sample point. Analyze this system, and comment on how well it acts as a demodulator.

2.37 Consider the system shown in Fig. P2.37 We wish to compare output to input in order to evaluate the sample-and-hold circuit as a PAM demodulator. The comparison between $y(t)$ and $x(t)$ is performed by defining an error

$$e^2 = \frac{1}{T}\int_0^T [y(t) - x(t)]^2 dt$$

Find the value of this error term.

2.38 A sinusoidal signal, $s(t) = \sin 2\pi t$, is sampled every 0.4 sec and transmitted using PPM.
(a) Design the PPM system (i.e., choose a pulse width and a relationship between pulse position and sample value).
(b) White noise of power $N_o = 10^{-3}$ watt/Hz adds during transmission. Find the mean square timing error at the receiver.
(c) Find the approximate SNR after reconstruction at the receiver.

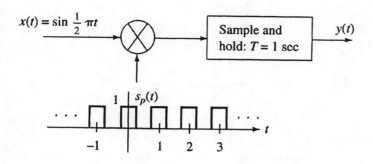

$$x(t) = \sin \frac{1}{2} \pi t$$

Sample and hold: $T = 1$ sec

$y(t)$

$s_p(t)$

Figure P2.37

2.39 A signal is given by

$$s(t) = 10\cos300\pi t + 20\cos900\pi t + 10\cos1200\pi t$$

(a) How many bits of quantization are required so that the signal to quantization noise ratio is at least 36 dB? Show all work.

(b) For the number of bits of quantization found in part (a), find the actual signal to noise ratio.

2.40 Binary ones and zeros are sent using the two baseband signals shown in Fig. P2.40. Noise of power 10^{-3} watts/Hz adds during transmission. Find the maximum number of bits per second that can be transmitted such that the error rate is less than 10^{-5}.

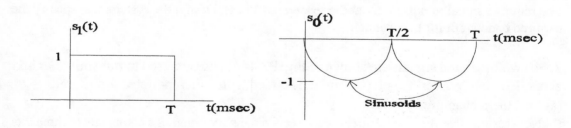

Figure P2.40

2.41 Binary information is sent using the baseband signals shown in Fig. P2.41.

(a) Design a detector.

(b) Suppose you are told that the bit error rate must be less than 10^{-5}. Find the maximum bit transmission rate if $N_o = 10^{-6}$.

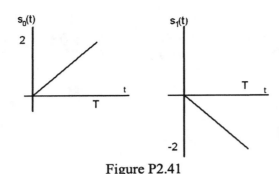

Figure P2.41

2.42 You are given the following signal:

$$s(t) = 5\cos(2\pi \times 200t) + 20\cos(2\pi \times 600t)$$

This signal is sent using PCM.

(a) Choose an appropriate sampling rate (justify your choice).

(b) How many bits of quantization are required to achieve a signal to noise ratio of at last 13 dB?

2.43 A PCM system is to be used to transmit a monochrome video signal. Assume that each frame contains 100,000 picture elements and that 32 levels of gray are desired. At least 50 frames per second are required to prevent flicker. Additive white Gaussian noise of power N_o = 4 ×10⁻⁶ watt/Hz is present in the channel, which can support voltages between -3V and +3 V. The required signal to noise ratio at the receiver is 30 dB. Design the system, and specify the required bandwidth for the channel.

2.44 Two baseband signals are shown in Fig. P2.44. These are used to transmit ones and zeros. Noise of power $N_o = 10^{-3}$ watt/Hz is added, and coherent detection is used.

(a) Find the probability of error.

(b) Now assume that the detector treats each received binary signal as a composite of three bits and processes the signal bit by bit. Majority logic is then used to decide which hypothesis is true. That is, if at least two received bits match a particular signal, the corresponding hypothesis is chosen. Find the probability of error, and compare it to your answer to part (a).

2.45 Suppose that in a bit synchronization scheme, a known message of 50 bits in length is sent in order to provide for symbol synchronization. The probability of correct reception is as shown in Fig. P2.45, where the abscissa is the amount of timing mismatch. That is, when the receiver clock is perfectly synchronized, there is a 0.99 probability of correct reception. With a mismatch of one-half of the bit period, the probability of correct detection drops to 0.5.

The system is considered synchronized if there are 45 or more agreements between the transmitted and detected waveforms.

Find the probability that the system locks to within 1 of correct synchronization.

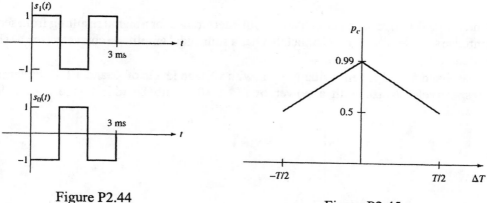

Figure P2.44

Figure P2.45

2.46 Assume that the early-late gate of Fig. 2.73 is in perfect lock with a baseband signal at the input. Sketch the waveform at the various points within the block diagram.

2.47 The DTTL is shown in Fig. 2.74. Repeat the waveform sketches of Fig. 2.75, assuming that the system is one-fourth of a bit period away from the locked position.

2.48 The system of Fig. 2.76 is used to recover clock information with a biphase-L coded input signal. Sketch the waveform at each point in this system if the input is a biphase-L signal representing the binary train 00101110.

2.49 Find the probability of false word synchronization using a 15-bit pseudonoise (PN) code as a preamble.

2.50 A frame synchronization scheme is to be designed with a search mode, a check mode, and a lock mode. Each frame contains 250 bits of information and a preamble of length N bits.
The frame must be transmitted within 1 msec. The bit error rate is given by

$$BER = 0.5 - 0.5e^{-f/2500}$$

where f is the number of bits per second being transmitted.
(a) If $N = 10$ and frame synchronization must be acquired within 10 frames (with 99% probability), find the threshold T_S, to be used in the search mode.
(b) Once the system is in lock, we wish it to remain in this condition with probability 99.9%. Find this threshold value, T_L.
(c) Repeat parts (a) and (b) if N is raised to 100 bits.

2.51 Calculate the correlation for the Barker code of length seven.

2.52 Calculate the correlation for the Neuman-Hofman code of length 24.

2.53 Repeat the design example of Section 2.4.3 for an environment that is no longer noisy. Assume that the bit error rate is 0.1% of the value given by Fig. 2.79.

2.54 Find the peak and mean square intersymbol interference for a signal resulting from sending ideal impulse samples through a channel that has a sinusoidal amplitude characteristic as shown in Fig. P2.54.

2.55 A baseband signal is sent using NRZ-L with voltage levels of -2 and +2 V for zeros and ones, respectively. Noise with a power of $N_o = 10^{-3}$ watt/Hz adds to the signal during

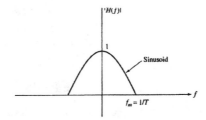

Figure P2.54

transmission. You may assume that the amplitude of the signal at the receiver is the same as that transmitted. Evaluate the performance of a single-sample detector.

2.56 (a) Design a matched filter detector to decide which of the two signals shown in Fig. P2.56 is being received in additive white Gaussian noise.
(b) Find the bit error rate if $N_o = 10^{-6}$ watts/Hz.

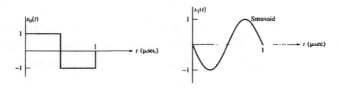

Figure P2.56

2.57 You are given the two baseband waveforms shown in Pig. P2.57. A matched filter detector is used to decide between the two possible transmitted signals. The additive white Gaussian noise has a power spectral density of $N_o/2 = 0.1$. Find the bit error rate.

2.58 Given the two signals shown in Fig. P2.58 design a matched filter detector, and evaluate its performance as a function of Δ.

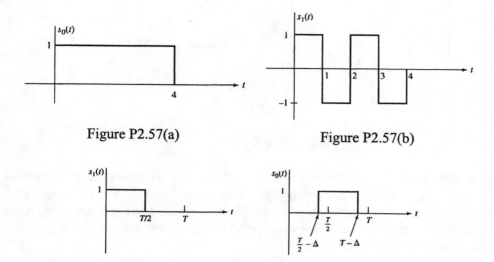

Figure P2.57(a)   Figure P2.57(b)

Figure P2.58

2.59 You are given the two signals shown in Fig. P2.59 where t is in msec. These are transmitted in the presence of additive Gaussian white noise.

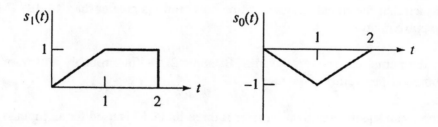

Figure P2.59

(a) Design a binary matched filter detector to decide which of the two signals is being sent. Your design must include the value of the threshold against which the output is compared.
(b) Find the maximum noise power per Hertz such that the bit error rate does not exceed 10^{-5}.

2.60 A biphase baseband communication system uses the two signals shown in Fig. P2.60 in order to transmit ones and zeros. The additive white noise has a power of $N_o = 10^{-4}$ watt/Hz.
(a) Design a detector to decide between ones and zeros.
(b) Find the maximum bit rate that can be used in order to achieve a bit error rate below 10^{-4}.

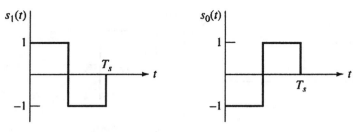

Figure P2.60

2.61 The signal $s(t)=\sin 2\pi t$ is sampled at a rate of 4 samples/sec. The samples are converted to a binary signal using a four-bit analog-to-digital converter. Compute the actual value of quantization noise at each sampling point, and find the mean square average noise. (You need do this only over one complete period of the waveform.) Compare your answer to that given by Eq. (2.57), and explain any discrepancies.

2.62 A signal $s(t)$ is uniformly distributed between -5V and +5 V. How many bits of uniform quantization are needed to achieve a signal-to-quantization noise ratio greater than 40 dB?

2.63 A signal is Rayleigh distributed with a mean value of one. How many bits of quantization are needed such that the signal-to-quantization noise ratio is greater than 40 dB? (Make any reasonable approximations.)

2.64 Verify the numerical answers given in Example 2.7. (You probably will want to use a software computer program.)

2.65 Find the mean square quantization error if three-bit PCM is used for an input signal that is Gaussian distributed. Assume that the quantization steps range over $\pm 3\sigma$, where σ^2 is the variance of the input.

2.66 A signal is Gaussian distributed with zero mean and variance of σ^2. The signal is to be transmitted using four-bit PCM, where the quantization steps are uniformly distributed over the range $\pm 3\sigma$. Find the mean square quantization error.

2.67 (a) A signal is derived by passing a 1-kHz square wave (zero average value) through an ideal lowpass filter with cutoff at 5.1 kHz. The resulting signal is quantized using six-bit PCM. Find the signal-to-quantization noise ratio.
(b) Repeat the analysis, assuming that μ-255 companding is used.

2.68 A signal is given by $s(t)=20\cos 100\pi t+17\cos 500\pi t$. How many bits of quantization are required so that the signal-to-quantization noise ratio is greater than 40 dB?

2.69 If μ-255 companding is used in a quantization system with the signal $s(t)=20\cos100\pi t+17\cos500\pi t$, how many bits of quantization are required to achieve a signal-to-quantization noise ratio of 50 dB? Compare your answer to that of Problem 2.68.

2.70 Find the mean square quantization error when a signal $s(t)$ with probability density as shown in Fig.P2.70 is:

(a) Uniformly quantized using five-bit quantization.

(b) Compressed according to the formula

$$F(s) = \sqrt{|s|}\,sgn(s)$$

and then uniformly quantized using five-bit quantization.

(c) Compressed according to the μ-255 rule and then uniformly quantized using five-bit quantization.

Repeat parts (a), (b), and (c) for four-bit quantization.

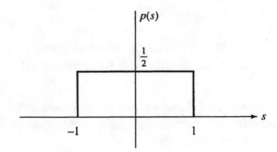

Figure P2.70

2.71 A compressor operates along a sinusoidal curve given by

$$F(s) = V_{max}\sin\left(\frac{\pi s}{2V_{max}}\right)$$

This curve is shown in Fig. P2.71. Find the improvement factor if:

(a) The signal is a sinusoid.

(b) The signal is triangularly distributed as shown in the figure.

(c) The signal is uniformly distributed as shown in the figure.

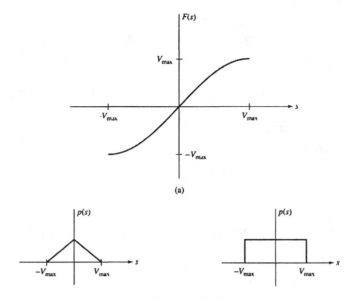

Figure P2.71

2.72 Repeat the analysis of Example 2.7, but using quantization regions that are selected according to Eq. (2.64). Compare your results with those obtained in the example.

2.73 A signal is derived by passing a 1-kHz square wave through a lowpass filter with cutoff at 5.1 kHz. Choose appropriate parameters for quantization by delta modulation, and find the signal-to-quantization noise ratio.

2.74 A signal to be transmitted is of the form $s(t)=10\cos1000\pi t+5\cos1500\pi t$. The signal is quantized using delta modulation.
(a) Choose an appropriate sampling rate and step size.
(b) Find the signal-to-quantization noise ratio for your design of part (a).

2.75 You are given a signal $s(t)$ with probability density $p(s)$, as shown in Fig. P2.75. The signal is to be converted from analog to digital form using five bits of uniform quantization.
(a) Design a serial analog-to-digital converter.
(b) Find the signal-to-quantization noise ratio for the converter of part (a).
(c) Suppose that the signal is now uniformly distributed between zero and two (instead of having the ramp distribution shown in the figure). How many decibels of improvement in signal-to-quantization noise ratio would result relative to the signal of part (a)?

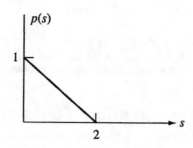

Figure P2.75

2.76 A signal has a probability density

$$p(x) = \begin{cases} Ke^{-2|x|} \\ 0 \end{cases}$$

(a) Find the signal-to-quantization noise ratio if three bits of uniform quantization are used.
(b) Find the signal-to-quantization noise ratio if three bits of quantization are used for μ-255 companding.

2.77 You wish to design a communication system in order to link two stations separated by 5,000 km. The system transmits 1 Mbps for 10 hours each day. Each bit error costs US$1, and each repeater station costs US$3 million (including maintenance costs over its life). A repeater station lasts five years. The bit error rate per link between two repeaters is given by $BER = 10^{-6}e^{-5000/X}$ where X is the spacing in km between repeaters. Develop formulas, and discuss how you would find the optimum number of repeaters.

CHAPTER 3
AMPLITUDE MODULATION

What we will cover and why you should care

Chapter 2 examined various forms of baseband communications where the information is transmitted using a signal with frequencies clustered around *dc*. This form of communication is adequate if the distances are short, if only one signal (or group of time-division multiplexed signals) is to be sent at any one time and if wires or cable are used. It is not, however, appropriate for transmission through the air, if distances are large, or the channel is to be shared by multiple users.

The current chapter introduces the concept of *modulation* as a technique for modifying a signal to match the channel properties. We consider amplitude modulation, followed in the next two chapters by frequency and phase modulation.

After introducing the concept of modulation and presenting the various forms of amplitude modulation, we turn our attention to the design of modulators and demodulators. We deal simultaneously with analog and digital communication. Once we know how systems are configured at the transmitter and receiver, we evaluate the performance of these systems. Knowing the performance and comparisons between various configurations, we are then in a position to explore the broad topic of design considerations. Trade-off decisions are required whenever the theory is applied to the real world.

The last section presents applications. While these are not specifically presented as *case studies*, they do serve as examples of how the theory has been applied to meet practical needs.

Necessary Background

To understand the concepts of modulation and the various amplitude modulation formats, you need familiarity with Fourier transform analysis and basic linear systems. A thorough understanding of some of the carrier recovery loops of Section 3.4.2 requires that you be familiar with basic feedback control systems, stability, and convergence. Evaluation of performance requires an understanding of probability theory and random processes.

Throughout this chapter, it is assumed that you have studied and understood the material in Chapter 2 of this book. Unless you have an appreciation of baseband communications, you cannot know why any other form is needed.

3.1 Concept of Modulation

Suppose you were given the job of transmitting either speech or baseband data through a channel. The first question you should ask yourself is whether the signals must be modified before injecting them into the channel. If the answer is "no", your job is very simple–you must only decide how to couple the signal into the channel (i.e., interface). In the following, we assume the signal must be modified.

Figure 3.1 shows the Fourier transform of a typical speech waveform and the power spectrum of NRZ-L random data (these are repeats of Figures 1.10 and 2.28).

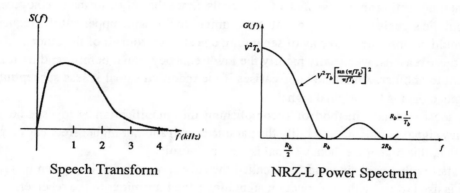

Speech Transform   NRZ-L Power Spectrum

Figure 3.1 - Frequency Content of Baseband Signals

In the case of short-range transmission, as in the local loop of a telephone circuit, in the path between the pre-amp and amplifier, or between the amplifier and speakers of a sound system, these baseband (low-frequency) signals are sent through wires. For longer distances, it is sometimes difficult to use wires since they require *rights of way*. Additionally, if transmission is *point-to-point*, one must specify the location of *every* terminal. Mobile communication by wire is almost impossible (we say *almost* since some receivers actually trail a wire behind which unwinds like a fishing line–but this is the exception. Imagine your car having to do this to provide you with a working mobile phone!). For these reasons, *broadcast* communication has been a popular form of transmission.

Suppose we take an audio signal and attempt to transmit it through the air. Let's choose a typical audio frequency of 1 kHz. The wavelength of a 1 kHz signal in air is approximately 300 km, (about 180 miles–do you remember the equation for wavelength from your Physics courses? The wavelength depends on the speed of light and the frequency. All you need to do is keep track of units. The speed of light is in meters per second, and frequency is number of cycles per second. You want to end up with meters per cycle, so there's no need to memorize equations). A quarter-wavelength antenna would then have to be 75 kilometers (45 miles) long, and erecting such antennas in backyards of homes would be a bit impractical! But even if you were willing to erect such antennas, you would still be left with two very serious problems. The first is related to the characteristics of air at audio frequencies. While propagation does occur at frequencies below 10 kHz, such signals are not efficiently transmitted through air. Even more serious is the second problem–that of interference. Suppose throughout the entire world we want to transmit more than one audio signal at a time (as is certainly true). That is, suppose that more than one of the dozens of local radio stations wished to transmit broadcasts simultaneously. Each station would have an antenna 75 km long on top of its studio (or on a mountain top) and the air would be polluted with many audio signals. The listener would erect an antenna 75 km high and receive a weighted sum (depending on relative distances from, and antenna patterns of, the different transmitting antennas) of all of the signals. Since the only information the receiver has about the signals is that they are all bandlimited to the same upper cutoff frequency, there would be absolutely no way of separating one station from all of the others.[1]

For these reasons, we usually modify the low frequency signal before sending it from one point to another. An added bonus arises if the modified signal is less susceptible to interference than is the original signal.

The most common method of accomplishing this modification is to use the low-frequency signal to modulate (modify the parameters of) another higher frequency signal. Most often, this higher frequency signal is a pure sinusoid.

We start with a pure sinusoid, $s_c(t)$, called the *carrier* waveform. It is given this name since it is used to *carry* the information signal from the transmitter to the receiver.

$$s_c(t) = A\cos(2\pi f_c t + \theta) \tag{3.1}$$

If the carrier frequency, f_c, is properly chosen, this carrier waveform can be efficiently transmitted. For example, suppose you were told that frequencies in the range between 1 MHz and 3 MHz propagate in a mode that allows them to be reliably sent over distances up to about 250 kilometers. If you chose the frequency, f_c, to be in this range, then the pure sinusoidal carrier would transmit efficiently. The wavelength of transmission in this range of frequencies is on the order of 100 meters, and reasonable-length antennas can be used.

We now ask the question whether this pure sinusoidal carrier waveform can somehow be altered in a way that (a) the altered waveform still propagates efficiently and (b) the information we wish to send is somehow superimposed on the new waveform in a way that it can be recovered at the receiver. We are asking whether there is some way that the sinusoid can *carry* the information along. The answer is "yes" as we now illustrate.

The expression in Eq. (3.1) contains three parameters which may be varied: the amplitude, A; the frequency, f_c; and the phase, θ. Using the information signal to vary A,

---

[1]Try this experiment: Walk into a crowded, noisy room, and try to distinguish one conversation from all of the others. Then record the sounds in the room, and try again to distinguish the sound, this time by listening to the recording. Ask yourself why there is a difference.

f_c, or θ leads to *amplitude modulation, frequency modulation*, or *phase modulation* respectively.

We will show that efficient transmission is achieved for each of these three cases. We will also show that if more than one signal is simultaneously propagated through the channel, separation of the signals at the receiver is possible. But it is critical to illustrate a third property that we have not yet mentioned. The information signal, *s(t)*, must be uniquely recoverable from the received modulated waveform. It wouldn't be of much use to modify a carrier waveform for efficient transmission and station separability if we could not reproduce *s(t)* accurately at the receiver.

3.2 Amplitude Modulation

Wwe separate our introduction to amplitude modulation into various categories depending on the bandwidth of the resulting waveform. As you will see shortly, we also categorize signals depending on whether a portion of the unmodulated carrier accompanies the modulated waveform.

3.2.1. Double Sideband Suppressed Carrier

If we modulate the amplitude of the carrier of Eq. (3.1), the modulated waveform of Eq. (3.2) results.

$$s_m(t) = A(t)\cos(2\pi f_c t + \theta) \tag{3.2}$$

The frequency, f_c, and the phase, θ, are constant. The amplitude, $A(t)$, varies in accordance with the baseband signal, *s(t)*. Recall that *s(t)* is the signal we want to be carried through the channel.

We simplify the expression by assuming that $\theta = 0$. This will not affect any of the basic results since the angle actually corresponds to a time shift of $\theta/2\pi f_c$. A time shift is not considered distortion in a communication system.

If somebody asked you how to vary *A(t)* in accordance with *s(t)*, the simplest approach you could suggest would be to make *A(t)* <u>equal to</u> *s(t)*. This would yield a modulated signal of the form,

$$s_m(t) = s(t)\cos 2\pi f_c t \tag{3.3}$$

This type of signal is given the name *double sideband suppressed carrier amplitude modulation* (DSBSC) for reasons that will soon become clear.

This simple guess for the amplitude, *A(t),* does indeed satisfy the criteria demanded of a communication system. The easiest way to illustrate this fact is to express $s_m(t)$ in the frequency domain, that is, to find its Fourier transform.

Suppose that we call the Fourier transform of *s(t)*, *S(f)*. We require nothing more of *S(f)* than it be the Fourier transform of a baseband signal. That is, *S(f)* must equal zero for

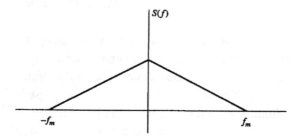

Figure 3.2 - Baseband $S(f)$

frequencies above some cutoff frequency, f_m (the "m" in the subscript stands for "maximum"). Figure 3.2 shows a representative sketch. We do not mean to imply that $S(f)$ must be of the shape shown. The sketch is meant only to indicate the transform of a general low-frequency bandlimited signal.

The modulation theorem (see Appendix A) is used to find $S_m(f)$.

$$S_m(t) = \mathcal{F}[s(t)\cos 2\pi f_c t] = \frac{1}{2}[S(f+f_c) + S(f-f_c)] \tag{3.4}$$

This transform is sketched as Fig. 3.3.

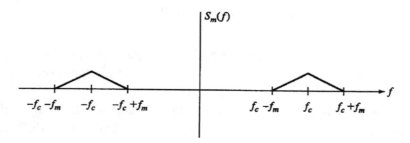

Figure 3.3 - $S_m(f)$, the Transform of $s_m(t)$

Note that modulation of a carrier with $s(t)$ has shifted the frequencies of $s(t)$ both up and down by the frequency of the carrier. This is analogous to the trigonometric result that multiplication of a sinusoid by another sinusoid results in sum and difference frequencies. That is,

$$\cos A \cos B = \frac{1}{2}\cos(A+B) + \frac{1}{2}\cos(A-B) \tag{3.5}$$

If $\cos A$ is replaced by $s(t)$, where $s(t)$ contains a continuum of frequencies between zero and f_m, the trigonometric identity can be applied term-by-term to get the same result.

Figure 3.3 indicates that the modulated waveform, $s_m(t)$, contains components with frequencies between f_c-f_m and f_c+f_m. As long as signals in this range of frequencies transmit efficiently and a reasonable length antenna can be constructed, we have solved the first of the two transmission challenges. Let's plug in some typical audio numbers. Let the maximum frequency of $s(t)$, f_m, be 15 kHz and the carrier frequency, f_c, be 1 MHz. Then the range of frequencies occupied by the modulated waveform is from 985,000 Hz to 1,015,000 Hz.

The second objective is that of channel separability. We see that if one information signal modulates a carrier of frequency f_{c1} and another information signal modulates a carrier of frequency f_{c2}, the Fourier transforms of the two modulated carriers do not overlap in frequency provided that f_{c1} and f_{c2} are separated by at least $2f_m$. This is illustrated in Fig. 3.4.

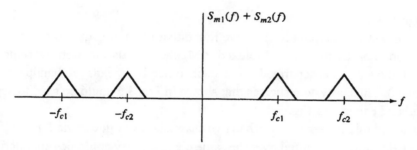

Figure 3.4 - Fourier Transform of Two AM Signals

Since the signals are "stacked" in frequency, we refer to this as *frequency division multiplexing* (FDM). It is the exact dual of time division multiplexing (TDM), which we introduced in Section 2.3.3.

If the frequencies of the two modulated waveforms are not too widely separated, both signals can even share the same antenna. That is, although the optimum antenna length is not the same for both channels, the total bandwidth can be made relatively small compared to the carrier frequency. In practice, an antenna is useable over a *range* of frequencies rather than just being effective at a single frequency. If this were not true, radio broadcasting would not exist.

As an example you don't have to readjust your car antenna length when you tune across the AM dial. The effectiveness of the antenna does not vary greatly from one frequency limit to the other. Instructions accompanying early car antennas (when receivers were less sensitive than they are today) suggested that their length be shortened to about 75 centimeters when changing from AM to FM (Don't try doing this if you have the type of antenna which is sandwiched within the windshield). Modern receivers have sufficient sensitivity that such tuning is no longer necessary, even with frequency changes of two

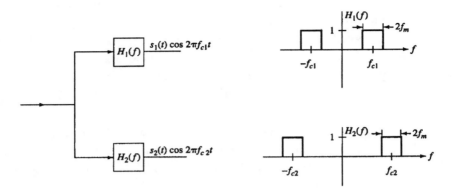

Figure 3.5 - Separating Two Non-Overlapping Channels

orders of magnitude.

If signals are non-overlapping in time, we found that gates or switches can be used to effect separation. For frequency-multiplexed AM, the signals are non-overlapping in frequency and they can be separated from each other by means of bandpass filters (frequency gates). Thus, a system such as that shown in Fig. 3.5 could be used to separate the two modulated carriers of Fig. 3.4.

The extension of this system to more than two channels is straightforward. Even if many modulated signals were transmitted over the same channel, they could be separated at the receiver using bandpass filters which accept only those frequencies present in the desired modulated signal. This is true provided that the separate carrier frequencies are wide enough apart to prevent overlapping of the Fourier transforms. We see from Fig. 3.4 that the minimum spacing is $2f_m$. In practice, a spacing larger than this is sometimes desirable for the following two reasons. First, even though we may view the information signal as limited to frequencies below f_m, no matter how sharply we lowpass filter, the signal still has some components above f_m. Second, if the minimum spacing is used, the bandpass filters which separate out the desired channel must be perfect with flat response in the passband and an infinite roll-off.

Example 3.1

An information signal is of the form

$$s(t) = \frac{\sin 2\pi t}{t}$$

This signal amplitude modulates a carrier of frequency 10 Hz. Sketch the AM waveform and its Fourier transform.

Solution: The AM waveform is given by the equation

$$s_m(t) = \frac{\sin 2\pi t}{t}\cos 20\pi t$$

This function is sketched in Fig. 3.6.

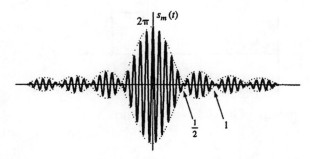

Figure 3.6 - AM Waveform
for Example 3.1

We note that when the carrier, $\cos 20\pi t$, is equal to 1, $s_m(t)=s(t)$, and when the carrier is equal to -1, $s_m(t)=-s(t)$. In sketching the AM waveform, we start by drawing $s(t)$ and its mirror image, $-s(t)$, as a guide. The AM waveform periodically touches each of these curves, and smoothly varies between these periodic points. In this manner, we develop the waveform sketch. In most practical situations, the carrier frequency is much higher than that illustrated in this example. In fact, it is so high that if you observed $s_m(t)$ on an oscilloscope, you would not be able to see the back and forth oscillations unless you greatly expanded the time axis. Instead, you would see the $s(t)$ and $-s(t)$ outlines, and what looks like shading between these two outlines.

The Fourier transform of the information signal, $s(t)$, is shown in Fig. 3.7. This is found by referencing the table in the Appendix.

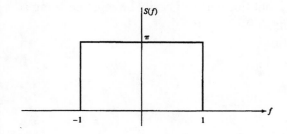

Figure 3.7 - Fourier Transform of $s(t)$

The transform of the modulated waveform is given by the following equation where we have applied the modulation theorem.

$$S_m(f) = \frac{S(f-10)+S(f+10)}{2}$$

This is shown in Fig. 3.8.

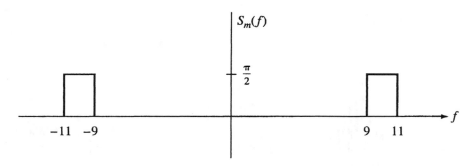

Figure 3.8 Fourier Transform of Modulated Waveform

We have indicated that an AM wave of the type discussed could be transmitted efficiently, and that more than one signal could share the channel. A critical property that must still be addressed is whether of not the information signal, $s(t)$, can be uniquely recovered from the AM waveform.

Since $S_m(f)$ was derived from $S(f)$ by shifting all of the frequency components of $s(t)$ by f_c, we should be able to recover $s(t)$ from $s_m(t)$ by shifting the frequencies again by the same amount, but this time in the opposite direction.

The *modulation theorem* (Appendix A) states that multiplication of a time function by a sinusoid shifts the Fourier transform of that time function *both up and down* in frequency. Thus, if we *remultiply* $s_m(t)$ by a sinusoid at the carrier frequency, the Fourier transform shifts *back down* to its low-frequency baseband position. This multiplication also shifts the transform up to a position centered about $2f_c$, but this part can easily be rejected using a lowpass filter. The process is illustrated in Fig. 3.9.

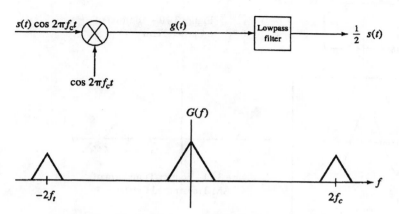

Figure 3.9 - Recovery of *s(t)* From $s_m(t)$

The recovery of *s(t)* is described by the following equations:

$$s_m(t)\cos2\pi f_c t = [s(t)\cos2\pi f_c t]\cos2\pi f_c t = s(t)\cos^2 2\pi f_c t$$

$$= \frac{s(t)+s(t)\cos4\pi f_c t}{2} \qquad (3.6)$$

We have used the trigonometric identity,

$$\cos^2(A) = \frac{1}{2} + \frac{1}{2}\cos(2A) \qquad (3.7)$$

The output of the lowpass filter is therefore *s(t)*/2, which is an undistorted version of *s(t)*.

This process of recovering *s(t)* from the modulated waveform is known as *demodulation*. We have taken this moment now to discuss demodulation rather than waiting until Section 3.4 where we will have more to say about it. Indeed, if *s(t)* could not be recovered from $s_m(t)$, there would be no reason to go on.

3.2.2. Double Sideband Transmitted Carrier

In the previous section, we studied double sideband suppressed carrier AM. We found that the waveform resulting from the multiplication of the information signal with a carrier sinusoid possesses desirable properties. In particular, the modulation process shifts frequencies from a band around *dc* to a band around the carrier frequency. This permits efficient transmission and also allows simultaneous transmission of more than one signal.

We now explore a modification of AM where we add a portion of the pure sinusoidal carrier to the modulated waveform. We will see in Section 3.4 that this addition greatly

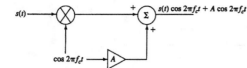

Figure 3.10 - Addition of
a Carrier Term

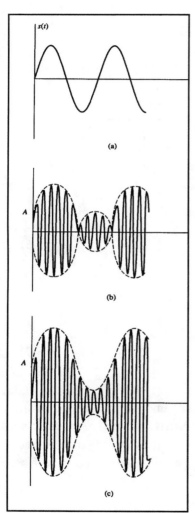

Figure 3.12 - AMTC
Waveform

Figure 3.11 - Fourier Transform of
AM Transmitted Carrier

simplifies the demodulation process.

Figure 3.10 shows the addition of a pure sinusoidal carrier to the double sideband suppressed carrier waveform. The resulting waveform is given by Eq. (3.8).

$$s_m(t) = s(t)\cos 2\pi f_c t + A\cos 2\pi f_c t \qquad (3.8)$$

This is known as *double sideband transmitted carrier* (DSBTC). The type of AM discussed in the previous section did not include an explicit carrier term. That is why it is labeled *suppressed carrier*. We begin by examining the time function and its Fourier transform.

The Fourier transform of transmitted carrier AM is the sum of the Fourier transform of suppressed carrier AM with the Fourier transform of the pure carrier. The transform of the carrier is a pair of impulses at $\pm f_c$ in frequency. The complete transform of the AM wave is therefore as shown in Fig. 3.11.

Now that we know the shape of the Fourier transform of DSBTC, let's turn our attention to the time function. The time function can be sketched if we first combine terms in Eq. (3.8).

The waveform can be rewritten as follows:

$$s_m(t) = [A + s(t)]\cos2\pi f_c t \tag{3.9}$$

This function is sketched in the same manner as that used to draw the suppressed carrier waveform. We first draw the outlines at $[A+s(t)]$ and $-[A+s(t)]$. The AM waveform periodically touches these two outline curves. We then fill in with a smooth oscillating waveform. This is illustrated for a sinusoidal $s(t)$ (i.e., someone whistling into a microphone) in Figure 3.12. Figure 3.12(a) shows the sinusoidal $s(t)$, Fig. 3.12(b) shows the AM waveform for a value of A less than the amplitude of $s(t)$, and Fig. 3.12(c) shows the waveform where A is greater than the amplitude of $s(t)$.

Efficiency

We ask you to accept for now that the addition of the carrier makes demodulation easier. The price we pay is efficiency. A portion of the transmitted power is being used to send a pure sinusoid which does not carry any useful signal information.

We see from Eq. (3.8) that the carrier power is the power of $A\cos2\pi f_c t$, or $A^2/2$ watts. The power of the signal portion is the power of $s(t)\cos2\pi f_c t$, which is the average of $s^2(t)$ divided by 2. The average of $s^2(t)$ is simply the power of $s(t)$, or P_s. Therefore, the signal power is $P_s/2$. The total transmitted power is the sum of these two terms. [2]

We define *efficiency*, η, as the ratio of signal power to total power. The efficiency is then given by

$$\eta = \frac{P_s/2}{(A^2+P_s)/2} = \frac{P_s}{A^2+P_s} \tag{3.10}$$

As an example of the application of Eq. (3.10), suppose we view the AM wave of Fig. 3.12(c) and set A equal to the amplitude of the sinusoid. P_s is then $A^2/2$ and the efficiency is

$$\eta = \frac{A^2/2}{A^2+A^2/2} = 33\% \tag{3.11}$$

The efficiency depends on the relative size of the modulated term when compared to the pure sinusoidal carrier term.

---

[2] The square of the sum contains a cross product, but if the average of $s(t)$ is zero, the average of this product is also zero.

MATLAB EXAMPLE

This example presents MATLAB code to plot an AM Transmitted Carrier waveform. You get to input the amplitude and frequency of the carrier and the amplitude and frequency of the modulating signal.

```
function ammodulator()
%EXAMPLE
%AM Modulation of a carrier signal  --> c(t) = cos(2*pi*fc*t)
%by an information signal s(t) = A + Bcos(2*pi*f*t).
%You will specify parameters A, B, fm, and fc of your choice.
clc; close all;
disp(' ')
disp('AM Modulation of a carrier signal  --> c(t) = cos(2*pi*fc*t)')
disp('by an information signal s(t) = A + Bcos(2*pi*f*t).');
disp(' ')
disp('You will specify parameters A, B, fm, and fc of your choice.')
disp(' ----------------------------------------------------------------')
disp(' ')
disp(' ')
A = input('Please input the modulating signal s(t) DC component A (e.g. 1): ');
B = input('the signal Amplitude B (e.g. 1): ');
fm = input('and its frequency fm (e.g. 100): ');
disp(' ')
fc = input('Please input the carrier frequency, fc (e.g. 1000): ');
t = 0:1/(10*(fc)):5/(fm); %linspace(0,5/fm,100*fc);          % time domain range
for x-axis
s = A+B*cos(2*pi*fm.*t);              % info signal
c = cos(2*pi*fc.*t);                  % carrier signal
p = s.*c;                                          % modulated signal
subplot(2,2,1);
plot(t,s);
title('The information signal')
xlabel('time, sec.'); ylabel('Amplitude');
```

```
subplot(2,2,2);
plot(t,c);
title('The carrier signal')
xlabel('time, sec.'); ylabel('Amplitude');
subplot(2,2,3);
plot(t,p);
title('The modulated signal')
xlabel('time, sec.'); ylabel('Amplitude');
```

We define a dimensionless quantity, m, as the ratio of the maximum amplitude of the modulated term divided by the amplitude of the carrier. That is,

$$m = \frac{\max |s(t)|}{A} \tag{3.12}$$

This dimensionless quantity, m, is known as the *index of modulation*. Viewing Fig. 3.12, we see that if the envelope of the waveform extends down to the zero axis, the index of modulation is one. As the index of modulation decreases, the efficiency also decreases. This index is sometimes expressed as a percentage by multiplying by 100.

MATLAB EXAMPLE

Use FFT to find the Fourier Transform of an AM Waveform. Try experimenting with different values of fm and of the modulation index, m.

```
% FFT of Transmitted Carrier AM
clc; close all;
disp(' ');
disp('Fast Fourier Transform of an AM waveform. Info signal sm(t) = cos(2*pi*fm*t) ')
disp('and carrier signal c(t) = cos(2*pi*fc*t). ')
disp(' ')
disp(' ')
disp('------------------------PRESS ANY KEY WHEN READY!---------------------------')
pause
disp(' ')
disp(' ')
```

```
fm = input('Please input the frequency fm (in hz) of the modulating signal, (e.g. 100): ');
m = input('The index of modulation - See Eq. (3.12) - Value between 0 and 1: ');
fc = 500;
N = max(fm,fc)+1024;                    % # sampling elements for fft computation
t = linspace(0,1,N);                    % Create N elements from 0 to 1 with increment
1/N
s = 1+m*cos(2*pi*fm*t);                 % information signal
c = cos(2*pi*fc*t);                     % carrier waveform
a = s.*c;                               % AM waveform
f1 = fft(a);                            % Compute the fft of AM
waveform
disp(' ')
disp(' ')
stem(t,abs(f1),'x');
title('The fft signal)')
xlabel('Normalized freq. \omega, rad/s'); ylabel('Amplitude');
```

Computer Simulation Example

The CD packaged with this text contains Tina software. When you run the program, you will find a file called *AM* . When you open this file, you screen should look like the figure at the top of the next page. This is a circuit consisting of only one item: a user-defined voltage source. If you double click on this source, and then click on the dots on the right part of the "signal" definition, you will be able to change parameters from those we have stored in the program. Then if you pull down the *Analysis* menu and click *Transient*, you are presented with the Transient dialogue box. Click *OK*, and the time function is presented. If you position the cursor on the waveform and click, the waveform turns red. Then right click the mouse, and you can select *Fourier Spectrum.* Clicking *OK* gives you the FFT of the AM waveform. You can now go back, repeat the process, but change the Fourier Spectrum frequency limits to be able to see the sidebands.

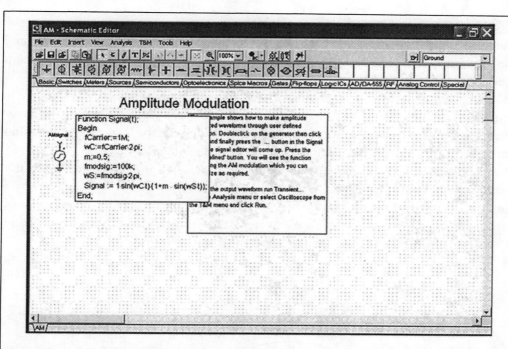

For example, the following FFT output resulted when we reset the frequency limits to range between 800kHz and 1200kHz.

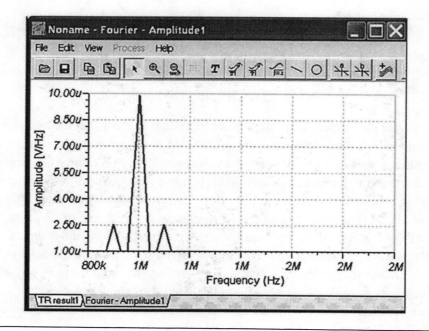

3.2.3. Single Sideband

In the AM systems we have studied, the range of frequencies required to transmit the signal is the band between f_c-f_m and f_c+f_m, where f_c is the carrier frequency and f_m is the maximum frequency of the baseband signal, $s(t)$. The total *bandwidth* is then $2f_m$.

The frequency spectrum is a natural resource. Conservation of this critical resource is of crucial importance. The more frequency bandwidth required for each channel, the fewer number of stations can communicate simultaneously. Wouldn't it be lovely if we could find a way to send this information using less than $2f_m$ of bandwidth?

Single sideband is a technique which allows transmission in half of the bandwidth required for AM double sideband.

In Fig. 3.13, we define that portion of $S_m(f)$ which lies in the band above the carrier as the *upper sideband*. The portion below the carrier is the *lower sideband*. A double sideband AM wave is composed of a lower and an upper sideband.

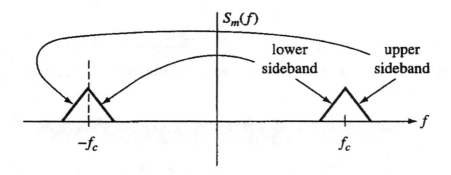

Figure 3.13 - Definition of Sidebands

We can use the properties of the Fourier transform to show that the two sidebands are dependent on each other. The transform of the AM wave is formed by shifting the signal transform, $S(f)$, up and down in frequency. The negative-f portion of $S(f)$ represents the lower sideband and the upper sideband is the positive-f portion of $S(f)$. We assume that the information signal, $s(t)$, is a real time function. Therefore the magnitude of $S(f)$ is even and the phase is odd (see properties of the Fourier transform in Appendix A). The negative-f portion of $S(f)$ can be derived from the positive-f portion by taking the complex conjugate. Similarly, the lower sideband of $s_m(t)$ can be derived from the upper sideband. Since the sidebands are not independent, it should be possible to transmit all essential information by sending only a single sideband. This is the essence of single sideband.

Figure 3.14 shows the Fourier transforms of the upper and lower sideband versions of the AM wave, denoted $s_{usb}(t)$ and $s_{lsb}(t)$ respectively.

The double sideband AM wave is the sum of the two sidebands,

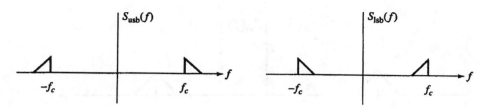

Figure 3.14 - Single Sideband Fourier Transforms

$$s_m(t) = s_{lsb}(t) + s_{usb}(t) \tag{3.13}$$

Since the single sideband waveform resides in a subset of the band of frequencies occupied by the double sideband waveform, it satisfies two of the requirements of a modulation system. That is, by proper choice of the carrier frequency, we can move the modulated waveform into a range of frequencies that transmits efficiently. We can also use different bands for different signals thereby allowing simultaneous transmission of multiple signals. The original signal can also be recovered from the SSB waveform, as we will see when we design demodulators later in this chapter.

3.2.4. Vestigial Sideband (VSB)

The advantage of SSB over DSB is the economy of frequency usage. That is, SSB uses half the corresponding bandwidth required for DSB transmission. The primary disadvantage of SSB is the difficulty in building a transmitter or an effective receiver. Getting ahead of the game, one problem with SSB is that when we attempt to build a sharp filter to remove one of the sidebands, the phase characteristic of that filter develops ripple. The closer one approaches the ideal filter amplitude characteristic, the worse becomes the phase characteristic. One area where frequency conservation becomes critical is television where bandwidths are orders of magnitude greater than those of voice. Phase distortion in a video signal causes offset of the resulting scanned image, and this is seen as ghost images on the screen. The eye is much more sensitive to such forms of distortion than is the ear to equivalent forms of voice distortion. We therefore have reason to explore a compromise between SSB and DSB.

Vestigial sideband (VSB) possesses a frequency bandwidth advantage approaching that of SSB without the disadvantage of difficulty in building a modulator. It is also easier to construct a demodulator.

As the name implies, VSB includes a *vestige*, or *trace*, of the second sideband. Thus, instead of completely eliminating the second sideband, as in the case of SSB, we eliminate most but not all of it.

Suppose we begin with DSB but filter out one of the sidebands. In contrast with the situation in SSB, we use a practical filter which does not closely approach the ideal infinite

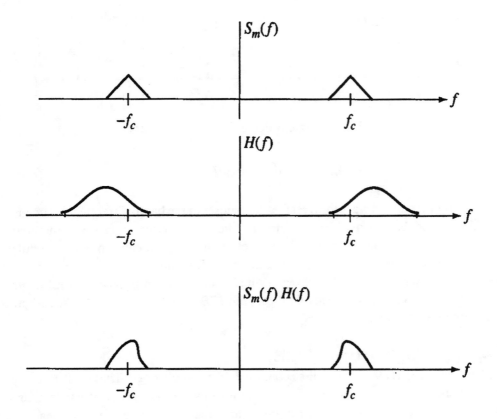

Figure 3.15 - Vestigial Sideband Generator

rolloff. The result might resemble that of Fig. 3.15, where we show the DSB transform, the filter characteristic, and the resulting output transform.

We explore modulators and demodulators in the following sections. An important application of VSB is in broadcast television.

3.2.5 ASK and MASK

Suppose now that the baseband signal is an NRZ-L digital signal. $s(t)$ then takes on only one of two values, A_0 or A_1. The modulated waveform, $s(t)\cos2\pi f_c t$, is then either $A_0\cos2\pi f_c t$ or $A_1\cos2\pi f_c t$ depending on whether a zero or a one is sent. A typical waveform might appear as in Fig. 3.16(a).

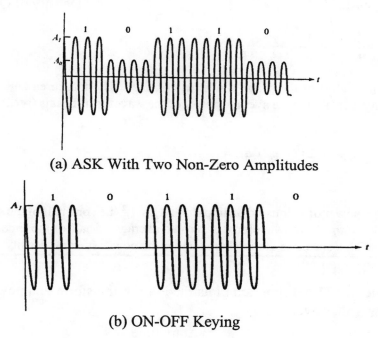

(a) ASK With Two Non-Zero Amplitudes

(b) ON-OFF Keying

Figure 3.16 - ASK Waveforms

In communicating a binary baseband digital signal using amplitude modulation, we see that the amplitude shifts between two possible levels. Even though this is simply a special case of AM, it is given a separate name: *amplitude shift keying* (ASK). The word "keying" is a carryover from the teletype days when a key was used to send dots and dashes.

Suppose that A_0 is set equal to zero. The amplitude is then shifting between A_1 and zero, and the waveform corresponding to Fig. 3.16(a) would appear as in Fig. 3.16(b). Note that we send a sinusoidal burst of amplitude A_1 to send a binary one, and turn off the generator to send a binary zero. This type of communication is given the descriptive name, *on-off keying* (OOK). When we look at system performance later in this chapter, we will find that setting A_0 to zero is the best choice for digital communication. In fact, there are few reasons for choosing any other value, so when people refer to ASK, you can automatically assume they mean OOK unless stated otherwise.

The Fourier transform of an amplitude modulated waveform consists of a shift of the baseband transform by the frequency of the carrier. The same is true for the power spectral density, so the power spectral density of a random OOK signal is as shown in Fig. 3.17.

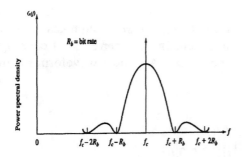

Figure 3.17 - Power Spectral
Density of OOK

This represents a shift of the shape shown in Fig. 2.28. The total area under the power spectral density is related to the average power of the waveform. That is (see Appendix C),

$$P_{ASK} = 2\int_0^\infty G_{ASK}(f)df = \frac{A_1^2}{4} \tag{3.16}$$

The average power of a sinusoid of amplitude A_1 is A_1^2 (i.e., one half of the square of the amplitude). Since the OOK waveform is zero an average of half of the time, its average power is $A_1^2/4$. This specifies the total area under the power spectral density.

MATLAB EXAMPLE

%This simple MATLAB program plots the Fourier transform and power spectral density of an ASK waveform.

```
Ac=1;                          %Amplitude
Tb=0.25;                         %bit period
R=1/Tb;                          %bit rate
c=Ac^2/8*R;            %constant
f=-10:.0013:10;        %freq range
fc=10;                      %center freq
x=pi*f*Tb;
psd= c*((sin(x)./ (x))).^2;
four=c*(sin(x)./ (x));
figure(1), plot(f,four,f, psd),
title('FFT and PSD of an ASK waveform'), ylabel('Magnitude, P(f)'),xlabel('f')
grid;
%figure(2), plot(f,psd),grid;
```

MASK

Figure 3.17 shows that the nominal bandwidth (out to the first zeros of the power spectral density) of binary ASK is twice the bit rate. Thus, for example, if you wanted to transmit 56 kbps, you would need a bandwidth of approximately 112 kHz. To put this in perspective, we note that non-conditioned telephone lines have a bandwidth of less than 4 kHz. So how do we get around this serious speed limitation? Of course, you could always slow the bit rate until the bandwidth matches that of the channel. In some applications, you would have the luxury of doing this. Suppose, for example, you were transmitting photographs from a remote location, and real time was not an issue. You could send the picture elements at very slow speed (perhaps taking several hours to send a single photo). Another approach would be to use data compression to reduce the number of bits you have to send. For now, we are going to concentrate on bandwidth reduction techniques.

As you study the derivation of the bandwidth of ASK, you will realize that it is inversely proportional to the length of the sinusoidal bursts used to send ones and zeros. In a binary system, each burst lasts T_b seconds, where T_b is the bit period. Now suppose that we were to combine each pair of bits into a *symbol*. The symbol would take on one of four possible values (0, 1, 2, or 3), and we would have a 4-ary system. If we now -transmit the symbols using sinusoidal bursts of four different amplitudes, we have an example of *MASK*, where $m=4$. Since each sinusoidal burst now lasts for $2T_b$, the bandwidth is exactly half that of binary ASK. Of course, we pay a price (yet another *trade-off* consideration). As we need more and more amplitude levels, the amplitudes get closer together and you would expect to make more errors. We examine performance later in this chapter.

Let's look at an example of quadrature (4-ary) ASK. In this case, the amplitude takes on one of four possible values: zero, A_1, A_2, or A_3. As an example, suppose we wish to transmit the bit sequence, 101101. We first combine the bits into pairs: 10 11 01, which

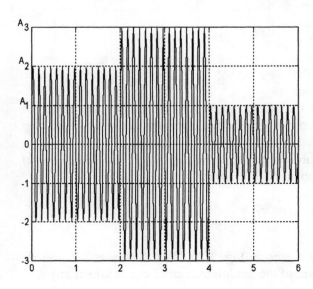

Figure 3.18 - MASK Signal

converts to 231 in a base-4 system. To transmit this using MASK, we sent a carrier of amplitude A_2 for the first two bit periods, then shift the amplitude to A_3 for the next two periods, and finally we set the amplitude to A_1 for the next two bit periods. This is illustrated in Fig. 3.18.

3.3 Modulators

Now that we know the basic format for both analog and digital amplitude modulated signals, it is time to look at some of the practical aspects. We need to see how to create an AM waveform. After we do that, we will have to explore ways to go backwards to generate the original signal from the modulated waveform. There are advantages to split our exploration into parts depending on the structure of the sidebands. We begin with double sideband, then study single sideband and vestigial sideband, and conclude with modulators ideally suited for the digital form of amplitude modulation.

3.3.1 Double Sideband

Figure 3.19 illustrates a simplified block diagram of an amplitude modulator. The system of Fig. 3.19(a) produces double sideband suppressed carrier AM, while the systems of Figs. 3.19(b) and 3.19(c) produce double sideband transmitted carrier AM.

You may ask why we devote an entire section of this text to modulators if Fig. 3.19 tells the whole story. Indeed, if we were simply interested in drawing system block diagrams there would be no need to go any further. However, if you ever intend to implement any system design, you must have some idea of the components that go into each block in the system diagram. The modulator represents a system that is *not* linear time invariant. This makes the implementation quite different from what you learned when you studied of linear filters.

Why is modulation not linear time invariant? Any linear time invariant system has an output whose Fourier transform is the product of the Fourier transform of the input with the system function, $H(f)$. If the Fourier transform of the input is zero over

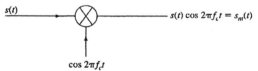

Figure 3.19 - Modulator

some range of frequencies, the output transform must also be zero for this range of frequencies. Therefore, a general property of linear time invariant systems is that they cannot generate any output frequency that does not appear in the input.

We now ask whether any linear time invariant system can have $s(t)$ as the input and $s_m(t)$ as the output. In other words, can we find an $H(f)$ for which

$$S_m(f) = S(f)H(f) \tag{3.17}$$

In the event that this is not yet clear, we present Fig. 3.20. This illustrates the Fourier transform of $s(t)$ and the Fourier transform of the modulated waveform. It asks if any $H(f)$

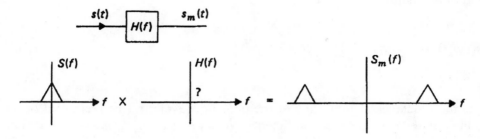

Figure 3.20 - Can a linear system create AM?

has the property that multiplication with $S(f)$ yields $S_m(f)$. Stare at the figure and see if you can find an $H(f)$ with the desired properties.

Hopefully you can see that this is impossible! For parts of the frequency axis where $S(f)$ is zero, there is no possible multiplying function which will produce the non-zero portions of $S_m(f)$. Modulation is a frequency-shifting process, and no linear system can accomplish this.

Synthesis of non-linear systems is generally quite difficult. Fortunately, simplifications are possible for the modulator. We begin our investigation with two classes of indirect amplitude modulator: The gated and square law.

The Gated Modulator

The *gated modulator* uses the fact that multiplication of $s(t)$ by *any* periodic function produces a series of AM waves at carrier frequencies that are multiples of the fundamental frequency of the periodic function. We illustrate this in Fig. 3.21 where we choose a pulse train for the periodic signal.

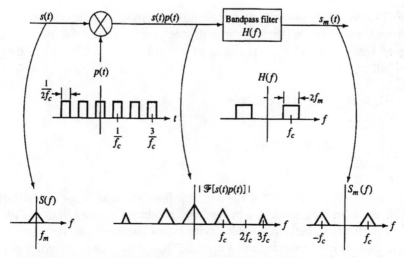

Figure 3.21 - Multiplication of $s(t)$ by Periodic Function

The output of the multiplier is given by

$$s(t)p(t) = s(t)\left[a_0 + \sum_{n=1}^{\infty} a_n\cos(2\pi nf_c t)\right] \tag{3.18}$$

where f_c is the fundamental frequency of the periodic waveform (the reciprocal of the period) and a_n are the Fourier series coefficients. We have assumed $p(t)$ is an even function simply to avoid having to write the sine terms in the series. The bandpass filter of Fig. 3.21 blocks all but one term in the series, with the result that the output is an AM wave. We have shown the filter as tuned to the fundamental frequency, but it could have been tuned to one of the harmonics thereby resulting in an AM wave at that higher carrier frequency. In practice, we often favor the lower harmonics since the Fourier coefficients generally decrease in magnitude with increasing n. At some point, the output AM waveform would be so small that it would be lost in the circuit noise.

What have we accomplished? If we cannot build a multiplier to take the product of $s(t)$ and a cosine waveform, what makes us think we can build the multiplier of Fig. 3.21? The answer lies in a specific choice of $p(t)$–a periodic pulse train gating function as shown in Fig. 3.22.

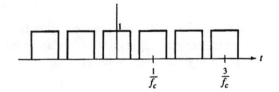

Figure 3.22 - Gating Function

Since $p(t)$ is always either zero or one, the multiplication can be viewed as a *gating* operation, where the input is switched on and off–the gate is opened and closed. The output of the bandpass filter is found by expanding $p(t)$ in a Fourier series and finding a_1. The modulator output is then

$$s_m(t) = a_1 s(t)\cos 2\pi f_c t$$

$$= \frac{2}{\pi} s(t)\cos 2\pi f_c t \tag{3.19}$$

The a_1 of Eq. (3.19) has been selected for a gating function which spends half of its time high and half at zero. In fact, an AM wave will be produced for any value of *duty cycle*.

The gating function can be implemented with either passive or active circuitry. Figure 3.23 shows two passive implementations. Figure 3.23(a) shows a switch which periodically shorts out the input. When the switch is open, the output equals the input. When the switch is closed, the output is zero. The resistance is the resistance of the source. The disadvantage

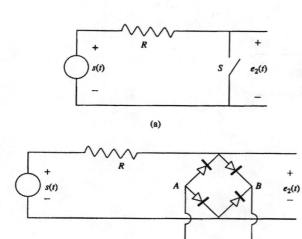

(a)

Figure 3.23 - Gating Circuits

(b)

of mechanical switching is that the switch must open and close at a rate equal to the carrier frequency (or a sub-multiple if we select a harmonic). If the carrier frequency is in the MHz range, mechanical switching is not practical.

Figure 3.23(b) presents a variation of the switch circuit where the switching is accomplished using a *diode bridge* circuit. When $\cos 2\pi ft$ is positive, the point labeled "B" is at a higher potential than the point labeled "A". In this condition, all four (ideal) diodes are open-circuited, and the circuit is equivalent to that of Fig. 3.23(a) with switch S open. On the other hand, when $\cos 2\pi ft$ is negative, point "A" is at a higher potential than point "B", and all four diodes are short circuits. This is equivalent to the switch being closed. The only limit to the rate of switching is imposed by the non-idealness of practical diodes (e.g., capacitance).

The gating can also be accomplished using active electronic devices such as transistors operating between cutoff and saturation. A transistor at cutoff is an open switch, while at saturation, it can be modeled as a closed switch.

The Ring Modulator

The *ring modulator* is a variation of the gated modulator. A ring modulator circuit is shown in Fig. 3.24.

The carrier, a square wave, is fed into the center taps of the two transformers. The output is a gated version of the input, and need only be filtered to produce AM. We illustrate sample waveforms in Fig. 3.24(b).

The Square Law Modulator

The principle of operation of the *square law modulator* is completely different from that of the gated or ring modulator. The square law modulator takes advantage of the fact that the square of a sum of two functions contains a cross product term which is the product of

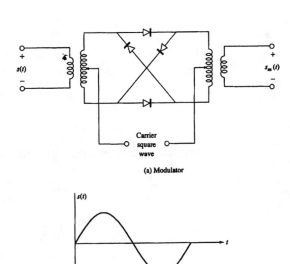

(a) Modulator

Figure 3.24 - The Ring Modulator

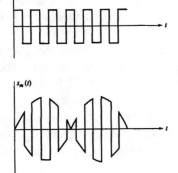

(b) Waveforms

the two functions. That is,

$$\left[s_1(t) + s_2(t)\right]^2 = s_1^2(t) + s_2^2(t) + 2s_1(t)s_2(t) \tag{3.20}$$

If the two functions are the baseband signal and the carrier, we have

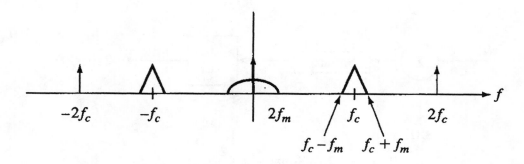

Figure 3.25 - Fourier Transform of Signal in Eq. (3.21)

$$\left[s(t)+cos2\pi f_c t\right]^2 \;=\; s^2(t)+cos^2 2\pi f_c t+2s(t)cos2\pi f_c t \tag{3.21}$$

The third term in Eq. (3.21) is the desired AM wave. We must find a way of separating it from the other two terms. Such separation is simple if the terms are non-overlapping either in time or frequency. Clearly they overlap in time, so our only hope is to look toward frequency.

Figure 3.25 shows the Fourier transform of the signal of Eq. (3.21). The impulses at the origin and at $2f_c$ result from expanding the squared cosine using the trigonometric identity,

$$cos^2 2\pi f_c t \;=\; \frac{1+cos4\pi f_c t}{2} \tag{3.22}$$

The continuous shape shown at low frequency represents the Fourier transform of $s^2(t)$. We do not know the exact shape of $s(t)$, but only that its Fourier transform is limited to frequencies below f_m. The Fourier transform of $s^2(t)$ is then limited to frequencies below $2f_m$. One way to show this is to observe that the transform of $s^2(t)$ is the convolution of $S(f)$ with itself. Graphical convolution easily shows that this transform goes to zero beyond $2f_m$. Another way is to consider $s(t)$ as a sum of individual sinusoids at frequencies below f_m. When this sum is squared, the result contains all possible cross products of terms. Trigonometric identities tell us that this leads to sums and differences of the various frequencies. If the original frequencies do not exceed f_m, none of these sums or differences can exceed $2f_m$.

Figure 3.25 indicates that as long as $f_c > 3f_m$, the terms do not overlap in frequency and the AM waveform can be separated using a bandpass filter. In most practical situations, $f_c >> f_m$, so the condition is easily met. The resulting square law modulator is shown in block diagram form in Figure 3.26.

This block diagram contains a summing device, a squarer and a bandpass filter. You

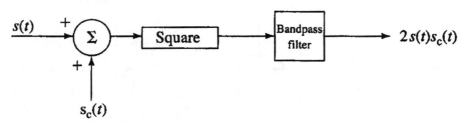

Figure 3.26 - Square-Law Modulator

know how to build a bandpass filter, so let's spend a moment examining the summing and squaring operations. Summing devices can be active or passive. Any resistive circuit with two sources produces weighted sums of these throughout the network (superposition). Alternatively, summing op-amp circuits can be used.

Square-law devices are not quite as simple. Any practical non-linear device has an output vs. input relationship that can be expanded in a power series. This assumes that no energy storage is taking place. That is, the output at any time depends only on the value of the input at that same time, and not on any past input values. With $y(t)$ as output and $x(t)$ as input, the non-linear device follows the relationship,

$$y(t) = a_0 + a_1 x(t) + a_2 x^2(t) + a_3 x^3(t) + ... \tag{3.23}$$

The term we are interested in is $a_2 x^2(t)$. If we could somehow find a way of separating this term from all of the others, the non-linear device could be used as a squarer.

Unfortunately, the various power terms overlap both in time and frequency (you should take the time to verify this–you already know enough trigonometry and Fourier transform theory to do so).

The non-linear device must essentially be a squarer–the a_n is Equation (3.23) must have the property that

$$a_n \ll a_2 \quad for \quad n>2 \tag{3.24}$$

There are several things to note about the non-linearity before we move on. The first is that if the $n=1$ and $n=2$ terms in the series predominate, the result is *transmitted carrier* AM. We can show this is true by first finding the output of the non-linear device. Neglecting terms beyond a_2, the output is

$$a_1 \left[s(t) + s_c(t) \right] + a_2 \left[s(t) + s_c(t) \right]^2 \tag{3.25}$$

$$= a_1 s(t) + a_1 \cos 2\pi f_c t + a_2 s^2(t) + a_2 \cos^2 2\pi f_c t + 2a_2 s(t) \cos 2\pi f_c t$$

The bandpass filter output is then equal to

$$[a_1 + 2a_2 s(t)]\cos 2\pi f_c t \qquad\qquad (3.26)$$

But suppose that the a_n are *not* insignificant for $n>2$. An AM signal will still be produced provided that $s(t)$ is made very small. Then $s^n(t) \ll s(t)$ for $n>1$, and the AMTC terms will predominate. This is not a desirable situation because of the small amplitudes that result.

Do squaring devices exist in real life? In fact, semiconductor diodes have terminal relationships that are good approximations to square-law devices over limited operating ranges.

A practical implementation of a square law modulator is shown in Fig. 3.27. This common-emitter transistor circuit uses the transistor non-linearity to produce the product of the signal with the carrier. The tuned circuit in the collector filters out the undesired harmonics.

Square law modulators are surprisingly easy to build in the real world. In fact, they often exist unintentionally. Modulation products appear in circuits when an electronic device is driven into non-linear operation. Significant effort is often expended to *prevent* a circuit from operating as a modulator.

We can relax the constraints on the non-linear device by constructing the *balanced modulator*. Figure 3.28 shows the block diagram and one possible implementation of such

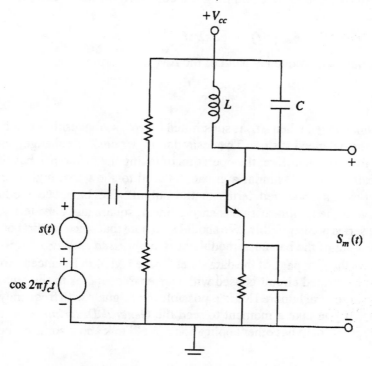

Figure 3.27 - Square-Law Modulator Implementation

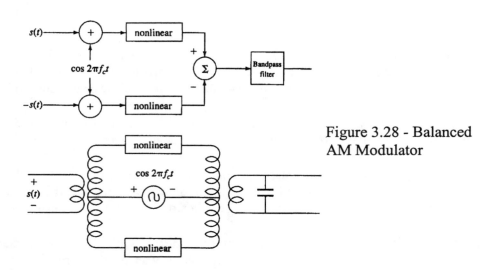

Figure 3.28 - Balanced
AM Modulator

a modulator. This system adds the carrier to $s(t)$ and places this through a non-linear device. The operation is repeated using $-s(t)$ as the baseband signal. The difference of the two outputs is taken, resulting in a cancellation of the terms due to odd powers in the expansion of Eq. (3.23). We illustrate this by examining the cube term in the equation. When we expand

$$[s(t) + \cos 2\pi f_c t]^3$$

the term which overlaps the frequency band of the AM waveform is

$$s^2(t)\cos 2\pi f_c t$$

This term remains unchanged when $-s(t)$ is substituted for $s(t)$, so the result is that it cancels from the output of the balanced system. The desired term, $s(t)\cos 2\pi f_c t$ changes sign when $-s(t)$ is substituted for $s(t)$. Therefore, the operation of taking the difference has the effect of doubling the desired term. A similar approach is used to show that higher order odd powers do not appear as undesired terms in the output. The balanced modulator is particularly effective if the nonlinearity has strong linear, square and cube terms, and all higher terms in the series are negligible. We should also note that since the first order term is eliminated, the output of the balanced modulator is *suppressed carrier* AM.

Figure 3.29 shows the first page of the data sheet for the LM1496 Balanced Modulator-Demodulator. This integrated circuit is used with carrier frequencies in the tens of MHz (although operation at several hundred MHz is possible) and signals with frequency content up to about 1 MHz. If you take a moment to read the *General Description* on that data sheet, you'll note that this chip has other applications. We'll talk about some of these later.

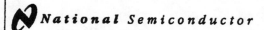

National *Semiconductor*

February 1995

LM1596/LM1496 Balanced Modulator-Demodulator

General Description

The LM1596/LM1496 are doubled balanced modulator-de-modulators which produce an output voltage proportional to the product of an input (signal) voltage and a switching (carrier) signal. Typical applications include suppressed carrier modulation, amplitude modulation, synchronous detection, FM or PM detection, broadband frequency doubling and chopping.

The LM1596 is specified for operation over the −55°C to +125°C military temperature range. The LM1496 is specified for operation over the 0°C to +70°C temperature range.

Features

- Excellent carrier suppression
  65 dB typical at 0.5 MHz
  50 dB typical at 10 MHz
- Adjustable gain and signal handling
- Fully balanced inputs and outputs
- Low offset and drift
- Wide frequency response up to 100 MHz

Schematic and Connection Diagrams

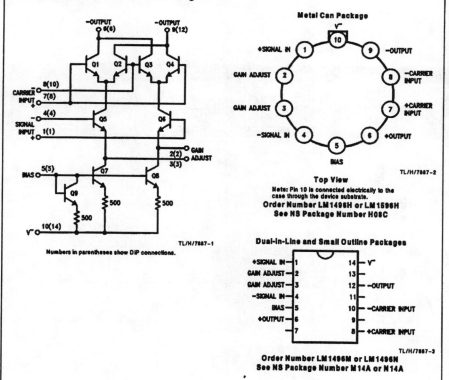

Figure 3.29 - LM1496 Balanced AM Modulator
(Courtesy of National Semiconductor)

The Waveshape Modulator

The *waveshape modulator* can be thought of as a brute force device. If you wanted to modulate a flow of water in a hose, you could hold your hand on the valve and keep turning it back and forth. The analogy to this simple system exists in electronics. You can envision building a power amplifier (or oscillator) that produces the carrier. Then simply vary the supply voltage to this amplifier in a manner that follows the baseband information signal. You can find details in most electronics textbooks.

Computer Simulation Example

The CD packaged with this text contains Tina software. When you run the program, you will find a file called *AM Modulator*. When you open this file, you screen should look like this:

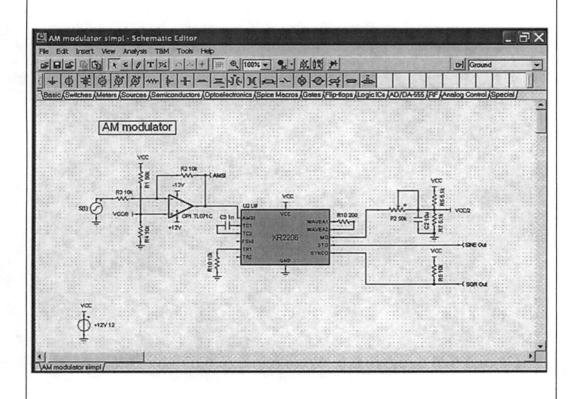

If you double click on the input source, you will see it is a triangular wave. After running the simulation, you can go back and change the frequency and/or amplitude of this source by clicking *Signal* in the dialogue box defining this source, and then clicking the "..." following the words, "triangular wave".

Note that the XR2206 is a function generator chip available from EXAR Corporation. You can get a data sheet at www.exar.com. The chip can be configured to produce sine, square triangular, ramp, and pulse waveforms, and any of these can be amplitude or frequency modulated.

Pull down the *Analysis* menu and select *Transient*. Run the simulation without changing any of the parameters. If the output curves print on top of each other, you can separate them by pulling down the *View* menu and selecting *Separate curves*. The output AM wave is labeled *SINE out*. Note that the carrier frequency is 100 kHz and the triangular wave modulating signal has a frequency of 10 kHz. You can get an approximation to the Fourier transform by clicking on the output waveform (it changes to red color). Then right click and select Fourier Spectrum. Change the frequency limits to something like 50k to 150k. The result will be a rough approximation since the waveform only contains several periods of the information signal. Now cancel the simulation and go back to the schematic. Double click on the input source, and then on *signal*. You will see it is a triangular wave. You can click on any of the other waveshapes and also vary the frequency and amplitude.

3.3.2. Single Sideband Modulators

Since the upper and lower sidebands of an AM wave are separated in frequency, filters can be used to select the desired sideband. We simply form a double sideband signal and chop off one of the sidebands. The top generator in Figure 3.30 shows a modulator for lower sideband single sideband while the bottom generator is for upper sideband.

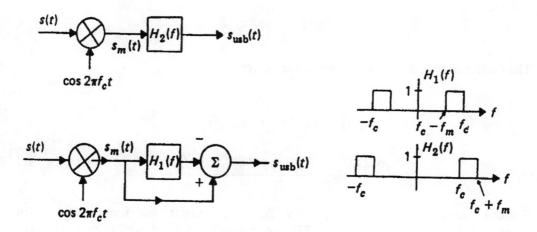

Figure 3.30 - Single Sideband Modulators

The bandpass filters of Fig. 3.30 must have very sharp roll off since there is no *guard band* between the two sidebands. For example, the filter of Fig. 3.30(a) must pass frequencies just above f_c while blocking those just below. There is some leeway in the filter design if the baseband signal, *s(t)*, has little low-frequency content. For example, if the energy below 25 Hz is negligible, there will be little energy within 25 Hz on either side of the carrier frequency. The demands on the filter can then be relaxed.

We now explore an alternate method of producing single sideband. This method substitutes a phase shifting and *quadrature* operation for the bandpass filter. The system is presented in Fig. 3.31 for lower sideband SSB.

To see how this generator operates, let us begin by restricting *s(t)* to be a pure sinusoid (i.e., whistle into a microphone). In this manner, the analysis requires only trigonometry.

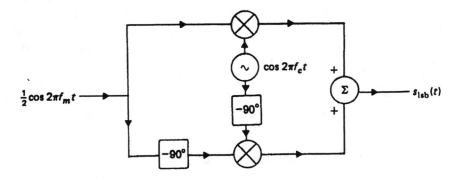

Figure 3.31 - Alternate Generator for LSB SSB

We let

$$s(t) = \cos 2\pi f_m t$$

The double-sideband AM waveform is then given by

$$s_m(t) = \cos 2\pi f_m t \cos 2\pi f_c t$$

$$= \frac{\cos 2\pi (f_c - f_m)t + \cos 2\pi (f_c + f_m)t}{2} \tag{3.27}$$

For this simple case, we can explicitly identify the upper- and lower-sideband time functions. The first term in Eq. (3.27) is the lower sideband while the second term represents the upper sideband. We now expand one of these, the lower sideband signal.

$$s_{lsb}(t) = \frac{\cos 2\pi (f_c - f_m)t}{2}$$

$$(3.28)$$

$$= \frac{\cos 2\pi f_c t \cos 2\pi f_m t + \sin 2\pi f_c t \sin 2\pi f_m t}{2}$$

We can now see why the system of Fig. 3.31 produces lower sideband. The first term in Eq. (3.28) is the AM waveform, and the second term results from shifting each cosine wave by 90 degrees. In fact, we put the amplitude of ½ on the input term in Fig. 3.31 so it would exactly match Eq. (3.28).

Now that we have verified this alternate approach for the special case of a sinusoidal modulating signal, let's examine the more general case. We could wave our hands like a magician and say that a general signal is composed of many sinusoids, so we need to just add the pieces together. But let's try to be more mathematically rigorous.

We begin by expressing the lower sideband Fourier transform as a product of the double sideband transform with a bandpass filter function. The filter only passes the lower sideband.

$$S_{lsb}(f) = S_m(f)H(f)$$

$$= \frac{S(f + f_c)U(f + f_c) - S(f - f_c)U(f - f_c) + S(f - f_c)}{2} \qquad (3.29)$$

We now express each unit step function of Eq. (3.29) in an alternate manner using the *sign* (sgn) function.

$$U(f + f_c) = \frac{1 + sgn(f + f_c)}{2}$$

$$(3.30)$$

$$1 - U(f - f_c) = \frac{1 - sgn(f - f_c)}{2}$$

We can now relate these expressions to the Hilbert transform (see Appendix A). The Fourier transform of the Hilbert transform is given by,

$$\hat{S}(f) = \frac{S(f)sgn(f)}{j} \qquad (3.31)$$

The Hilbert transform of a time function results from shifting all of the frequencies of the time function by -90 degrees. Substituting Eq. (3.31) and (3.30) into (3.29) yields,

$$S_{lsb}(f) = \frac{1}{2}\left[\frac{S(f + f_c) + S(f - f_c)}{2} + \frac{\hat{S}(f - f_c) - \hat{S}(f + f_c)}{2j}\right] \qquad (3.32)$$

Both ratios of Eq. (3.32) should look familiar. The first is in the form of the Fourier transform of an AM wave. The positive and negative frequency shifts represent multiplication of $s(t)$ by a cosine function. The second ratio corresponds to the time function resulting from multiplying $\hat{s}(t)$ by a sine wave. The result is that the lower sideband signal is given by

$$s_{lsb}(t) = \frac{1}{2}s(t)\cos 2\pi f_c t + \frac{1}{2}\hat{s}(t)\sin 2\pi f_c t \qquad (3.33)$$

This matches the operations in Fig. 3.31. The upper sideband can be derived from the lower sideband by observing that the sum of the two sidebands is the double sideband AM waveform. Therefore,

$$s_{usb}(t) = s_m(t) - s_{lsb}(t)$$

$$= \frac{1}{2}s(t)\cos 2\pi f_c t - \frac{1}{2}\hat{s}(t)\sin 2\pi f_c t \qquad (3.34)$$

You should take a moment to compare Eq. (3.34) to Eq. (3.28). If we had first done the general case, we could have derived Eq. (3.28) by simply substituting the cosine waveform for $s(t)$ in Eq. (3.34).

The double sideband AM wave is the sum of the two sidebands,

3.3.3. VESTIGIAL SIDEBAND (VSB)
The modulator for vestigial sideband is quite simple. We saw the vestigial sideband generator in Fig. 3.15. We repeat the figure here. We first generate double-sideband AM using any of the techniques described in this section. We then filter the double-sideband signal using a filter with characteristics as shown in Fig. 3.15. The result is to pass one of the sidebands and part of the other sideband.

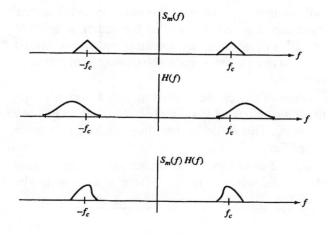

Figure 3.15 (repeat) -
Vestigial Sideband Generation

3.3.4 Digital ASK Modulators

The modulators discussed so far in this section can be applied to baseband signals that are both analog or digital. However, in the digital application, much simpler implementations are possible. When we speak of digital AM, we mean amplitude shift keying (ASK), and when we use the term ASK, we normally refer to on-off keying (OOK).

There are two approaches to generating the ASK waveform. One technique starts with the baseband signal and uses this to amplitude-modulate a sinusoidal carrier. Since the baseband signal consists of distinct waveform segments, the AM wave also consists of distinct modulated segments.

Another approach is to generate the AM wave directly without first forming the baseband signal. In the binary case, the generator would only have to be capable of formulating one of two distinct AM wave segments. For on-off keying, we need simply switch an oscillator on and off, as shown in Fig. 3.32. We've seen this system several times in our earlier work. It represents a natural-sampled PAM generator. It is also the "front end" of gated modulator. Implementation therefore follows the format presented earlier in this section.

Binary ASK modulators can be easily adapted to M-ary systems. For example, if we wish to generate 4-ary MASK using the indirect technique described above, we simply start

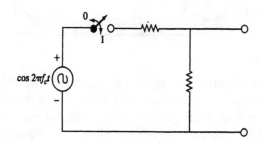

Figure 3.32 - Modulator for
OOK Binary ASK

with a baseband NRZ signal which would simply be a staircase-type function taking on one of four amplitude levels. This baseband signal is fed into any of the analog amplitude modulators. The result will be the MASK waveform. Since the amplitude (theoretical) changes instantaneously, the designer would have to pay attention to the transient response of the particular circuit used in the modulator.

Another technique is to feed the output of the carrier oscillator into a variable-gain amplifier. The gain must be capable of taking on one of M values (i.e., four values in 4-ary ASK, of which one would probably be zero). This could be done by switching resistance values at the input of an operational amplifier.

A modulator analogous to that of Fig. 3.32 would have four sinusoidal sources feeding a switch (or multiplexer IC). The switch would select which of the four sources would be connected to the output.

Computer Simulation Example

Run the Tina software included on the CD in the back of this book. Then open the file called *ASK Modulator*. Your screen should look like this:

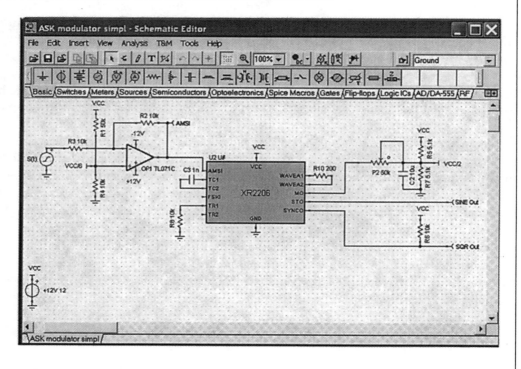

This circuit consists of an op amp (used as a buffer) and an integrated circuit. The XR2206 is the same function generator we used to generate analog AM. The ASK waveform is *SINE Out*. Once again, after running the transient response you can go back and change the frequency of the input waveform by double clicking on that source, then clicking on *Square wave* on the *Signal Line*, and then on the "...".

3.4 Demodulators

We divide demodulators into two broad classifications: coherent and incoherent. Coherent demodulators must be configured to take advantage of all received information including the amplitude, phase, and timing of the waveform. Incoherent demodulators do not need to establish absolute timing (phase) relationships.

3.4.1. Coherent Demodulation

We previously observed that $s(t)$ is recovered from $s_m(t)$ by *remodulating* $s_m(t)$ and then passing the result through a lowpass filter. This yields the demodulator system block diagram of Fig. 3.33. This is known as a *synchronous demodulator*. It gets its name from the observation that the oscillator is *synchronized* in both frequency and phase with the received carrier.

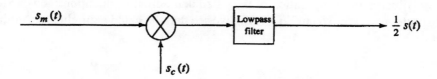

Figure 3.33 - AM Synchronous Demodulator

Since the multiplier in this figure looks no different from the multiplier used in the modulator, we might expect variations of the gated and square law modulators to be applicable to the process of demodulation.

Gated Demodulator

The gated demodulator is shown in Fig. 3.34.

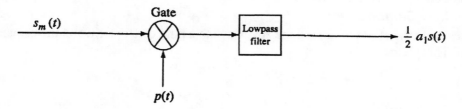

Figure 3.34 - Gated Demodulator

$p(t)$ is a gating function consisting of a periodic train of unit amplitude pulses (refer back to Fig. 3.21). It can be expressed in a Fourier series as

$$p(t) = a_0 + \sum_{n=1}^{\infty} a_n \cos 2\pi n f_c t \tag{3.36}$$

The input to the lowpass filter is then given by

$$s_m(t)p(t) = s(t)\cos 2\pi f_c t \left| a_0 + \sum_{n=1}^{\infty} a_n \cos 2\pi n f_c t \right|$$

$$= a_0 s(t)\cos 2\pi f_c t + \frac{s(t)}{2}\sum_{n=1}^{\infty} a_n[\cos(n-1)2\pi f_c t + \cos(n+1)2\pi f_c t]$$

$$(3.37)$$

The output of the lowpass filter is then given by

$$s_o(t) = \frac{1}{2}a_1 s(t) \tag{3.38}$$

and demodulation is accomplished.

We have illustrated the operation of the gated demodulator for suppressed carrier AM. If we substitute $A+s(t)$ for $s(t)$ in Eq. (3.37), we see that the gated demodulator produces an output of

$$s_o(t) = \frac{1}{2}a_1[A+s(t)] \tag{3.39}$$

This represents the original information signal shifted by an amplitude constant. If the system contains ac-coupled devices, the constant will not appear in the output. If all amplifiers in the system are dc-coupled, we may wish to remove the constant using a relatively large series capacitor which charges to the average value of the signal. We are assuming that the average value of the information, $s(t)$, is zero. If this were not true, removal of the constant would also remove some of the signal. Fortunately, most information signals have zero dc value.

Square Law Demodulator

We investigate the effect of adding the AM wave to a pure carrier term and then squaring the sum. This yields,

$$[s_m(t)+A\cos 2\pi f_c t]^2 \tag{3.40}$$

Let us first consider the suppressed carrier case. The expression of Eq. (3.40) then becomes

$$\{[s(t)+A]\cos 2\pi f_c t\}^2 = [s(t)+A]^2\cos^2 2\pi f_c t$$

$$= \frac{[s(t)+A]^2+[s(t)+A]^2\cos 4\pi f_c t}{2} \tag{3.41}$$

The second term in the numerator of Eq. (3.41) is an AM wave with a carrier frequency of

$2f_c$ Hz. It can therefore be easily rejected by a lowpass filter. We expand the first term as

$$s^2(t) + A^2 + 2As(t) \tag{3.42}$$

Unfortunately, the frequency content of $s^2(t)$ overlaps that of $s(t)$, and these cannot be separated. However, suppose we used a lowpass filter to separate the entire term

$$\frac{[s(t)+A]^2}{2} \tag{3.43}$$

from the expression of Eq. (3.41). Note that this lowpass filter must pass frequencies up to $2f_m$. We have then recovered the square of the sum of A with $s(t)$. We could then take the square root of this to get

$$0.707\,|s(t)+A| \tag{3.44}$$

Taking the magnitude of a signal represents a severe form of distortion. As a simple example, suppose the signal were a pure sinusoid. The magnitude would be a full-wave rectified sine wave with the fundamental frequency of twice the original frequency. The rectified sine wave no longer contains one single frequency, but includes harmonics. If we listened to this sound in a speaker, the original sinusoid would be a pure tone, while the full-wave rectified sine wave would be a raspy (harmonic content) tone one octave higher. If the original signal were composed of a mixture of many frequencies, the distortion effect would be far more severe. Indeed, full-wave rectified voice may not even be intelligible (try it in the lab!).

But suppose A is large enough such that $s(t)+A$ never goes negative. In that case, the magnitude of $s(t)+A$ is equal to $s(t)+A$ and we have accomplished demodulation. This means that the added carrier at the receiver must have an amplitude greater than or equal to the maximum negative excursion of $s(t)$. But don't get too excited. We will soon see that if A is large enough to meet this condition, there are much easier ways to demodulate.

Effects of Frequency Mismatch

The demodulators we have been discussing require that we generate a replica of the carrier at the receiver. The replica must be synchronized with the received carrier (frequency and phase matched). This may be very difficult to accomplish. So let's investigate the consequences of frequency and phase mismatches. We illustrate this for suppressed carrier. Suppose that the local oscillator of Fig. 3.33 is mismatched in frequency by Δf and in phase by $\Delta\theta$. The output of the multiplier is then

$$s_m(t)\cos[2\pi(f_c+\Delta f)t+\Delta\theta]$$

$$= s(t)\cos 2\pi f_c t\cos[2\pi(f_c+\Delta f)t+\Delta\theta]$$

$$\tag{3.45}$$

$$= s(t)\left[\frac{\cos[2\pi\Delta ft+\Delta\theta]}{2} + \frac{\cos[2\pi(2f_c+\Delta f)t+\Delta\theta]}{2}\right]$$

Since the expression in Equation (3.45) forms the input to the lowpass filter of the synchronous demodulator, the output of this filter is as given in Eq. (3.46).

$$s_o(t) = s(t)\frac{\cos(2\pi\Delta ft + \Delta\theta)}{2} \tag{3.46}$$

This is true since the second term of Eq. (3.45) has frequency content around $2f_c + \Delta f$, and is therefore rejected by the lowpass filter. The expression of Eq. (3.46) represents a signal, $s(t)$, multiplied by a sinusoid at Δf Hz. We can assume that Δf is small since we attempt to make it equal to zero. The modulation theorem then tells us $s_o(t)$ has a Fourier transform with frequencies ranging up to $f_m + \Delta f$. Even though the lowpass filter is designed to pass frequencies only up to f_m, it is reasonable to assume that this entire term passes through the filter (since $\Delta f \ll f_m$). Indeed, a practical filter would not have an infinitely sharp rolloff at f_m.

Note that if the phase and frequency are perfectly adjusted, Eq. (3.46) reduces simply to $s(t)/2$ as we already knew for the synchronous demodulator. It's nice to make simple reality checks from time to time. This gives us some confidence that we didn't make a careless mathematical error.

Suppose first that we are able to match the frequency precisely but that the phase is mismatched. Equation (3.46) then reduces to

$$s_o(t) = \frac{s(t)\cos\Delta\theta}{2} \tag{3.47}$$

This is an undistorted version of $s(t)$, so we would normally not be concerned. However, as the phase mismatch approaches $90°$, the output goes to zero. If noise is added to the signal, the attenuation presented by the $\cos\Delta\theta$ term could become a significant design consideration. That is, as $\Delta\theta$ deviates from zero, the signal to noise ratio decreases.

One method of making the receiver insensitive (robust) to phase variations is to use the *quadrature receiver* as shown in Fig. 3.35. We have indicated a phase shift of $\Delta\theta$ on both the sine and cosine multiplier signal. Equivalently, we could have indicated this phase shift on the input carrier.

The outputs of the two lowpass filters can be found using trigonometric identities to be

$$s_1(t) = \frac{1}{2}s(t)\cos\Delta\theta$$

$$\tag{3.48}$$

$$s_2(t) = -\frac{1}{2}s(t)\sin\Delta\theta$$

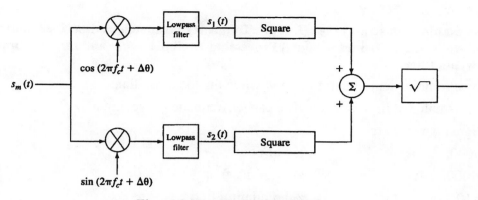

Figure 3.35 - Quadrature Receiver

After taking the square root of the sum of the squares, we find

$$s_o(t) = \frac{1}{2}\sqrt{s^2(t)} = \frac{1}{2}|s(t)| \qquad (3.49)$$

As in the case of the square-law demodulator, we find that undistorted demodulation is possible only if $s(t) \geq 0$. This means that the quadrature demodulator only works for transmitted carrier AM. Indeed, for AMTC we will find much simpler ways to demodulate. We present the quadrature demodulator only to develop this important building block for later application.

Let us now return to Eq. (3.46) and switch roles. Assume that the phase has been perfectly matched but that the frequency is mismatched. The output of the synchronous demodulator is then given by

$$s_o(t) = \frac{s(t)\cos 2\pi\Delta ft}{2} \qquad (3.50)$$

The frequency mismatch is usually small (we try to make it zero), so the result will be a slowly varying amplitude (beating) of $s(t)$. If, for example, $s(t)$ were an audio signal and the frequency mismatch is 1 Hz, the effect would be to multiply $s(t)$ by a 1 Hz sinusoid. This is like taking the volume control of your radio and smoothly varying it from zero to maximum twice each second! Clearly this is totally unacceptable. With a carrier frequency of 1 MHz, the 1 Hz mismatch represents only one part in 10^6. But suppose you were an expert at frequency matching, and your mismatch was only 10^{-3} Hz. Then your volume goes from maximum to zero once every 500 seconds. Unless we can derive the exact carrier from the incoming wave, or both the transmitter and receiver carriers are derived from the same source, synchronous demodulation is doomed to highly limited use.

MATLAB EXAMPLE

This example illustrates the effect of a frequency mismatch as described above. It uses two functions from the Signal Processing Toolbox: *butter* and *filter* to simulate a lowpass filter.

```
%Synchronous demodulator for AM when the local oscillator
%is mismatched in frequency. The carrier frequency is set at 9kHz.
%Signal parameters
n=9000;                    %# of points
fs=9000;                   %sampling freq
fm=10;                     %modulating freq
fc=900;                    %carrier freq
mu= 0.8;                   %modulation index
df = input('Please input the frequency mismatch df : (example .5)');
fmis=fc+df;            %freq mismatched
t=0:(1/fs):((2/fs)*(n));           %time normalized
%signals
info=cos(2*pi*fm*t);       %info signal
carr=cos(2*pi*fc*t);       %carrier signal
lo=cos(2*pi*(fc+f) *t);    %local oscillator
%signal operations
AMsig=(1+mu*info).*carr;   %AM signal
AMdem=AMsig.*carr;                   %Signal demodulation
AMmis=AMsig.*lo;                     %Signal demodulated with freq mismatched
%signal filtering and detection
[a,b]=butter(3,0.1,100, [fm, 50+fm]*2/fs); %lowpass filter
sf1=filter(a,b,AMdem);               %perfect signal demodulation
sf2=filter(a,b,AMmis);                %mismatched in freq signal demodulation
subplot(2,1,1), plot(t,AMsig), title('Modulated signal')
subplot(2,1,2), plot(t,sf1,'b',t,sf2,'r.'), title('Demodulated signal, matched in blue, mismatched in  red')
```

Single Sideband Demodulation

The synchronous demodulator can be used to demodulate single sideband. This can be shown either pictorially in the frequency domain or mathematically in the time domain. Looking first at frequencies, we know that multiplication by a sinusoid shifts the Fourier transform both up and down. Figure 3.36(a) shows the Fourier transform that results when $s_{usb}(t)$ is multiplied by a sinusoid at a frequency of f_c, and Fig. 3.36(b) shows a similar result for the lower sideband signal. In both cases, a lowpass filter would recover a replica of the original information signal.

We can illustrate this in the time domain by multiplying the single sideband waveform by a cosine at the same frequency as the carrier. We derived a time expression for single sideband in Section 3.3, and came up with the results on Eqs. (3.33) and (3.34). We repeat these results as Eqs. (3.51) and (3.52). The lower sideband time function is

$$s_{lsb}(t) = \frac{1}{2}s(t)\cos 2\pi f_c t + \frac{1}{2}\hat{s}(t)\sin 2\pi f_c t \qquad (3.51)$$

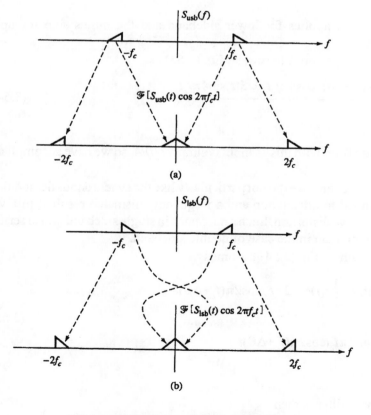

(a)

(b)

Figure 3.36 - Demodulation of SSB

The upper sideband function is given by

$$s_{usb}(t) = s_m(t) - s_{lsb}(t)$$

$$= \frac{1}{2}s(t)\cos2\pi f_c t - \frac{1}{2}\hat{s}(t)\sin2\pi f_c t \qquad (3.52)$$

Now that we have time functions for the single sideband waveforms, we can return to the analysis of the synchronous demodulator. We use the time domain expression of Eqs. (3.51) and (3.52) for the single sideband waveforms.

$$s_{ssb}(t)\cos2\pi f_c t = \frac{s(t)\cos^2 2\pi f_c t \pm \hat{s}(t)\sin2\pi f_c t\cos2\pi f_c t}{2} \qquad (3.53)$$

The plus sign in Eq. (3.53) applies for lower sideband and the minus sign for upper sideband.

We use trigonometric expansions to rewrite Eq. (3.53) as

$$s_{ssb}(t)\cos2\pi f_c t = \frac{s(t) + s(t)\cos4\pi f_c t \pm \hat{s}(t)\sin4\pi f_c t}{4} \qquad (3.54)$$

The output of the lowpass filter with this as input is simply $s(t)/4$, so we have accomplished demodulation.

Recall that in double sideband, we did not particularly like the synchronous demodulator since a phase mismatch led to attenuation and a frequency mismatch resulted in a very serious form of multiplicative distortion. Such a mismatch in single sideband is also serious, but slightly more forgiving than in the case of double sideband.

If the phase is mismatched, Eq. (3.54) becomes

$$s_{ssb}(t)\cos2\pi(f_c+\Delta\theta)t = \frac{1}{2}s(t)\cos2\pi f_c t\cos2\pi(f_c+\Delta\theta)t$$

$$\pm \frac{1}{2}\hat{s}(t)\sin2\pi f_c t\cos2\pi(f_c+\Delta\theta)t \qquad (3.55)$$

The output of the lowpass filter is then

$$\frac{s(t)\cos\Delta\theta}{4} \qquad (3.56)$$

As in the case of double sideband AM, the phase mismatch causes an attenuation, but no distortion.

If the frequency is now mismatched, Eq. (3.54) becomes

$$s_{ssb}(t)\cos 2\pi (f_c + \Delta f)t =$$

$$\frac{s(t)\cos 2\pi \Delta ft + s(t)\cos 4\pi (f_c + \Delta f)t}{4} + \frac{\hat{s}(t)\sin 2\pi \Delta ft + \hat{s}(t)\sin 4\pi (f_c + \Delta f)}{4} \qquad (3.57)$$

$$\frac{s(t)\cos 2\pi \Delta ft}{4} + \frac{s(t)\cos 4\pi (f_c + \Delta f)t}{4} + \frac{\hat{s}(t)\sin 2\pi \Delta ft}{4} + \frac{\hat{s}(t)\sin 4\pi (f_c + \angle}{4}$$

Only the first and third term in Eq. (3.57) go through the lowpass filter. If only the first term appeared in the output, the effect would be the same as that experienced in double sideband. However, the addition of the Hilbert transform term changes the output. The specific form of that change can be seen if we consider the special case of a sinusoidal modulating signal. That is, let $s(t) = \cos 2\pi f_m t$. The output of the lowpass filter is then proportional to $\cos 2\pi (f_m \pm \Delta f)t$, with the plus sign obtaining for lower sideband and the minus sign for upper sideband. (You can prove this either from Eq. (3.57) or by writing the single sideband waveform as a single sinusoid at a frequency of either $f_c - f_m$ or $f_c + f_m$, depending on whether lower or upper sideband is being considered). The effect of the frequency mismatch is therefore a frequency offset in the demodulated wave. If the information signal is a sum of sinusoids, the demodulated signal will be a sum of shifted sinusoids. Thus, in the general case, we see an overall shifting of frequencies of $s(t)$ by the amount of the frequency mismatch.

You might think this results in a change in pitch of the sound, but such is not the case. For example, suppose you hummed into a microphone. The resulting waveform is periodic with a fundamental frequency equal to the frequency at which you are humming. Harmonics occur at multiples of this frequency. If each frequency component is shifted by the same number of Hz, the harmonic relationships are destroyed and the resulting sound changes. The effect on music is generally considered to be unacceptable. The effect on voice is sometimes acceptable since the resulting sound is usually intelligible (for small Δf relative to the frequencies present). Some have described this as a "Donald Duck" effect. Therefore, while mismatches are to be avoided, the effects may be considered less devastating than in the case of double sideband.

Vestigial Sideband Demodulation

Vestigial sideband results from passing double sideband AM through a bandpass filter which admits essentially all of one sideband and part of the other. The vestigial sideband Fourier transform is therefore given by

$$S_v(f) = S_m(f)H(f) = \frac{S(f + f_c) + S(f - f_c)}{2} H(f) \qquad (3.58)$$

$H(f)$ is the transfer function of the filter which removes most of one sideband. Suppose now that this signal forms the input to a synchronous (coherent) demodulator. When the VSB signal is multiplied by a cosine at the carrier frequency, the Fourier transform shifts both up and down by the carrier frequency. The part that shifts down passes through the lowpass filter. The part that shifts up resides around $2f_c$, and is rejected by the filter. The filter output then has a transform given by

$$S_o(f) = \frac{S(f)[H(f+f_c)+H(f-f_c)]}{4} \tag{3.59}$$

Equation (3.59) can be used to set the conditions on the filter. The bracketed sum is shown in Fig. 3.37 for a typical $H(f)$.

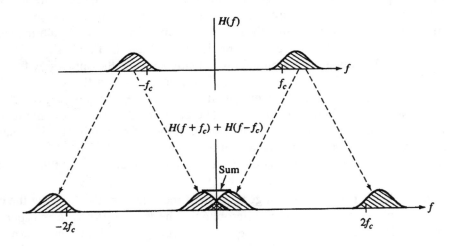

Figure 3.37 - Bandpass Filter Constraint for VSB

The filter transfer function, $H(f)$, must display odd symmetry for frequencies around the carrier such that the sum of the two terms approximates a constant characteristics. The tail of the filter characteristic must be asymmetric about $f=f_c$. That is, the output half of the tail must fold over and fill in any difference between the inner half values and the value for an ideal filter.

Simulation Example:

The Tina software included on the disk at the back of this text includes a simulation example called **AM Demodulator**. Open this file, and your screen should look like:

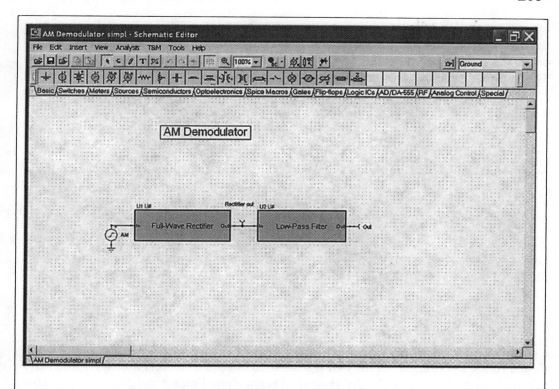

We are simulating the rectifier detector using two macro blocks. Run the Transient analysis by pulling down the *Analysis* menu and selecting *Transient*. For now, use the default settings. You should get two plots, the output of the rectifier and the output of the filter. If they appear on top of each other, select *View* and then *Separate curves*.

Note that you are seeing a transient response as the output dc level adjusts upward. If you increase the simulation run from 500 to 2000 μsec, you will see the transient almost die out (it may take a while for you computer to run the simulation because of the large number of data points). You can change the display axes by double clicking on the appropriate axis and changing the settings.

ASK Demodulation

We now turn our attention to the coherent detector for digital AM, that is ASK. If we were to use the synchronous demodulator for ASK (OOK), the output would be a piecewise constant function. Suppose, for example, that the input is a sinusoidal burst of amplitude A when a one is being sent, and that the input is zero when a zero is being sent. The output of synchronous demodulator would be a unipolar baseband signal ranging in amplitude between zero and A volts.

Matched filter detector

In our discussion of baseband digital communication in the previous chapter, we concluded that the matched filter (or correlator) detector was the "best" receiver for binary digital communication. We can now apply that receiver to the ASK signal. The receiver consists of two legs matched to the two possible received signals. The difference is taken between the two filter outputs, and that difference is compared to a threshold. However, for OOK, one of the signals is zero so one of the legs of the matched filter detector can be eliminated. The resulting detector is shown in Fig. 3.38.

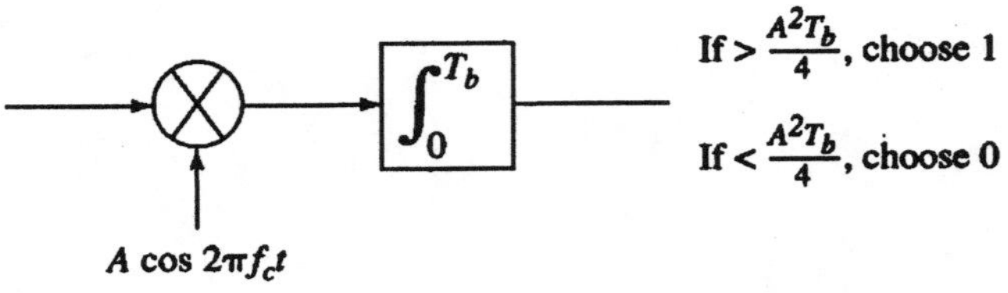

If $> \frac{A^2 T_b}{4}$, choose 1

If $< \frac{A^2 T_b}{4}$, choose 0

Figure 3.38 - Matched Filter Detector for OOK

$A \cos 2\pi f_c t$

Our only design task is to decide what threshold the output must be compared to. The threshold is given by Eq. (3.60) [This is a repeat of Eq. (2.35)].

$$y_o = \frac{E_1 - E_0}{2} \tag{3.60}$$

E_1 and E_0 are the energies of the two signals used to send a one and a zero respectively. When a binary one is transmitted, a sinusoidal burst of amplitude A and duration T_b (the bit period) is sent. Thus, the energy is given by the power multiplied by time, or

$$E_1 = \frac{A^2 T_b}{2} = \frac{A^2}{2R_b} \tag{3.61}$$

R_b is the bit rate in bits/sec, and it is the reciprocal of T_b. Since the signal used to send a binary zero is zero, the energy is zero (i.e., $E_0 = 0$). The threshold is then given by

$$y_o = \frac{A^2 T_b}{4} = \frac{A^2}{4R_b} \tag{3.62}$$

If the detector output exceeds this value, we decide that a one is being sent. If the output is less than this value, we decide that a zero is being sent.

3.4.2 Carrier Recovery in AMTC

We have seen that synchronous demodulation requires *perfect* matching of the frequency, and a phase mismatch which is not close to 90°. Frequency matching is possible if the AM waveform contains a periodic component at the carrier frequency. That is, the Fourier transform of the received AM waveform must contain an impulse at the carrier frequency. This is the case with AMTC.

We assume that the received signal is of the form $s(t)cos2\pi f_c t + A cos2\pi f_c t$. One way to extract the carrier is with a very narrow bandpass filter tuned to the carrier frequency. In the steady state, all of the carrier term will pass through this filter while only a portion of the modulated carrier will go through. The Fourier transform of the filter output is given by

$$S_o(t) = \frac{S(f-f_c)+S(f+f_c)+A\delta(f+f_c)+A\delta(f-f_c)}{2} \qquad (3.63)$$

for

$$f_c - \frac{B}{2} < f < f_c + \frac{B}{2}$$

In Eq. (3.63), B is the filter bandwidth. The inverse transform is then given by

$$s_o(t) = A cos2\pi f_c t + \int_{f_c-B/2}^{f_c+B/2} S(f-f_c)cos2\pi ft df \qquad (3.64)$$

The integral in Eq. (3.64) is bounded by

$$\frac{1}{2\pi t}S_{max}(f)\,B \qquad (3.65)$$

The smaller the bandwidth of the filter, the closer is the output to the pure carrier term.

An alternative to the narrow filter is a *phase-lock loop*. This is illustrated in Fig. 3.39. The phase-lock loop is discussed in detail in the next chapter. For now, we indicate that, if properly designed, the loop will lock on to the periodic component in the input to produce

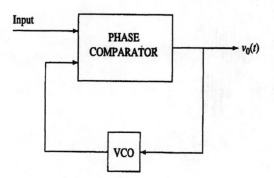

Figure 3.39 - The Phase-Lock Loop

a sinusoid at the carrier frequency.

This discussion leads to the detectors for AMTC shown in Fig. 3.40. Figure 3.40(a) shows the bandpass filter used for carrier recovery, and Fig. 3.40(b) shows the phase-lock loop.

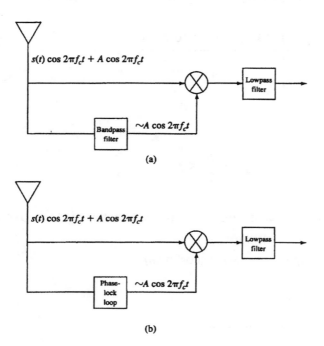

Figure 3.40 - Carrier Recovery in AMTC

3.4.3. Incoherent Demodulation

Coherent demodulators (detectors) require reproduction of the carrier at the receiver. Since the exact carrier frequency and phase must be matched at the detector, accurate timing information is needed. If the carrier term is sufficiently large in AMTC, it is possible to use incoherent detectors which do not have to reproduce the carrier or know timing information. Let us suppose that the amplitude of the carrier is sufficient such that $A+s(t) \geq 0$. We sketch a typical AM waveform in Fig. 3.41.

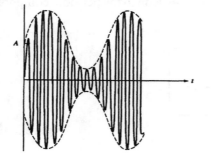

Figure 3.41 - AMTC
$A+s(t)>0$

Square Law Detector

We noted earlier that the square-law demodulator is effective for this AMTC case. We repeat that demodulator as Fig. 3.42.

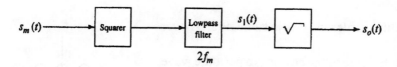

Figure 3.42 - Square-Law Detector for AMTC

The output of the squarer is

$$[A+s(t)]^2\cos^2 2\pi f_c t \ = \ \frac{[A+s(t)]^2+[A+s(t)]^2\cos 4\pi f_c t}{2} \qquad (3.66)$$

The output of the lowpass filter (which passes frequencies up to $2f_m$) is

$$s_1(t) \ = \ \frac{[A+s(t)]^2}{2} \qquad (3.67)$$

If we now assume that A is large enough such that $A+s(t)$ never goes negative, the output of the square rooter is

$$s_o(t) \ = \ 0.707[A \ + \ s(t)] \qquad (3.68)$$

and demodulation is accomplished.

Rectifier Detector

The squarer can be replaced by other forms of nonlinearity. In particular, consider the rectifier detector shown in Fig. 3.43.

The rectifier can be either half wave or full wave. We will consider the full wave here, and ask you to examine the half-wave as part of the problems at the back of this chapter.

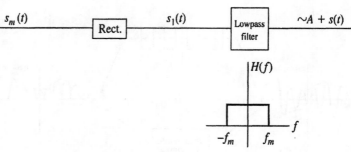

Figure 3.43 - Rectifier Detector

Full-wave rectification is equivalent to the mathematical operation of taking the absolute value. The output of the rectifier is then given by

$$s_1(t) = |A+s(t)| \, |\cos 2\pi f_c t| = [A+s(t)] \, |\cos 2\pi f_c t| \tag{3.69}$$

The last equality is true since we assume $A+s(t)$ never goes negative.

The absolute value of the cosine is a periodic wave as shown in Fig. 3.44. Its fundamental frequency is $2f_c$. We rewrite $s_1(t)$ by expanding the rectified cosine in a Fourier series.

Figure 3.44 - Rectified Sine Wave

$$s_1(t) = [A+s(t)][a_0 + a_1 \cos 4\pi f_c t + 2_2 \cos 8\pi f_c t + ...] \tag{3.70}$$

The output of the lowpass filter is then given by Eq. (3.71) and demodulation is achieved.

$$s_o(t) = a_0[A + s(t)] \tag{3.71}$$

Before leaving the rectifier detector, we will point out the mechanism by which this detector reconstructs the carrier waveform. This should help give you a feel for the underlying principle of the detector operation. Figure 3.45 shows that full-wave rectification of the AM wave is equivalent to multiplying the waveform by a square wave at the carrier frequency. That is, the process of taking the absolute value flips around the negative portion of the carrier. Flipping is equivalent to multiplication by -1. Therefore, the rectifier, which does not need to know the exact carrier frequency, is performing an operation which is equivalent to multiplication by a square wave carrier at the exact frequency and phase of the received carrier.

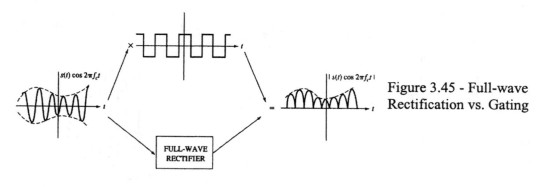

Figure 3.45 - Full-wave Rectification vs. Gating

Envelope Detector

The final detector we examine is by far the simplest. Let us observe the AMTC waveform of Fig. 3.46. If $A+s(t)$ never goes negative, the upper outline, or *envelope* of the AM wave is exactly equal to $A+s(t)$. If we can build a circuit which follows this outline, we have built a demodulator.

It may be helpful for some of you to borrow an example from mechanics. Suppose that instead of representing a voltage waveform as a function of time, the curve represents the shape of a wire. One can visualize a cam moving along the top surface. If the cam is attached by means of a shock absorber, or viscous damper device, it will approximately follow the upper outline of the curve. This is shown in Fig. 3.47. This is much the same as the behavior of an automobile suspension system. You want the car to follow the outline of the road, but you do not wish it to track every ripple and bump in the road surface.

The higher the carrier frequency, the more smoothly the cam will describe the upper outline provided that it can respond fast enough to follow the shape of the outline (you would probably not want your car to fly through the air between peaks of the road surface....unless you are filming a San Francisco chase scene for an action movie). The outline, or envelope of the waveform, has a maximum frequency of f_m, while the ripples (carrier) have a frequency of f_c. As long as $f_c \gg f_m$, intuition tells us we can design the mechanical system.

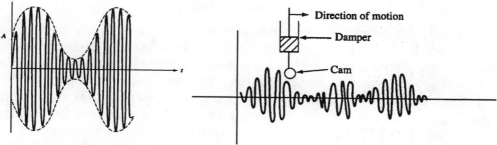

Figure 3.46 - AMTC Waveform   Figure 3.47 - Mechanical Outline Follower

We now construct the electrical analogy to this mechanical system. The mass of the cam is represented by a capacitor ($F=ma=mv'$ is replaced by $i=Cv'$). The viscous friction provides a force proportional to velocity, much as a resistor provides a current proportional to voltage. The cam is not attached to the wire–the wire (road) can push on the cam but cannot

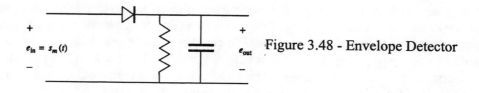

Figure 3.48 - Envelope Detector

pull on it. This is a mechanical diode. The equivalent circuit is then as shown in Fig. 3.48.

The circuit of Fig. 3.48 is known as an *envelope detector*. When properly designed, it serves as a demodulator and is clearly far simpler to build than the demodulators we have discussed earlier in this chapter. We will look closely at the operation of this simple circuit, but before we do so, you should observe that the overall effect of the diode is to create an *RC* circuit with a relatively fast capacitor charging rate and a relatively slow discharge rate. This proves significant in the analysis.

We first examine the operation of this circuit and then explore the appropriate choice of parameter values. Let us begin by removing the resistor, as shown in Fig. 3.49(a). This circuit is known as a *peak detector*.

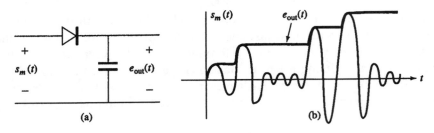

Figure 3.49 - The Peak Detector

The analysis of the peak detector requires only two observations: (a) The input can never be greater than the output (for an ideal diode) and (b) The output can never decrease with time. The first observation is true since if the input did exceed the output, the ideal diode would be supporting a positive forward voltage. The second observation follows from the fact that the capacitor has no discharge path. Figure 3.49(b) shows an AMTC waveform (the carrier frequency has been drawn much lower than it would be in practice for illustration) and the output of the peak detector. The output is always equal to the maximum past value of the input. [Can you think of some practical applications of the peak detector?]

If a discharging resistor is now added to the circuit, the output decays along an exponential between peaks of the AM wave. This is shown in Fig. 3.50. If the time constant of the *RC* circuit is appropriately chosen, the output approximately follows the outline and the circuit acts as a demodulator. The output contains ripple at the carrier frequency, (residual *rf*) but this does not cause a problem since we are only interested in frequencies below f_m.

The *RC* time constant must be short enough such that the envelope can track the changes in peak values of the AM waveform. A guideline that is often used is to set the time constant so that the initial slope of an exponential with that time constant approximately matches the maximum negative slope of the envelope. We can consider the case where $s(t)$ is a pure sinusoid at a frequency of f_m. Using the maximum frequency in our analysis provides the fastest possible change in peak values. At this frequency, the maximum slope is found from the derivative of the sinusoid. Using a maximum amplitude of A (i.e., index of modulation of unity or 100% modulation), we find the maximum slope is $2\pi f_m A$. The initial (maximum)

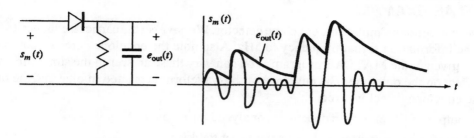

Figure 3.50 - Addition of a Discharging Resistor

slope of an exponential with starting amplitude A and time constant RC is A/RC. Therefore, if we set the RC time constant to $1/2\pi f_m$ we meet the condition. If the modulation index, m, is less than unity, we would scale accordingly by setting the time constant to $1/2\pi m f_m$. For example, with f_m of 5 kHz, the time constant would be set to 31.8 µsec. This rule of thumb represents a first cut at envelope detector design. In reality, a sinusoid has its maximum slope in the middle and minimum slope at the extremes. Therefore, choosing the time constant based upon extremes means our detector will not track all carrier peaks in between these extremes. We are saved by noting that typical information signals spend a small fraction of their time at the highest frequency. Also, the large difference between carrier and envelope frequency allows a lot of leeway in choosing the time constant. If the information signal, $s(t)$, is such that we expect significant periods near the highest frequency (e.g., a soprano singing a particularly high-pitched selection), we could safely choose a time constant considerably smaller than that indicated above.

Example 3.2

Design an envelope detector for use in demodulating an AMTC waveform. The carrier frequency is 1 MHz and the information signal is a voice waveform.

Solution: Once the diode is selected, all that is required in the design of the envelope detector is to choose the value of R and C in the circuit. The highest frequency of the envelope of the AM waveform is f_m, which we will assume is 5 kHz for voice. The envelope detector must be capable of responding to the fastest possible changes in the signal. Using our guideline, the time constant is set of 31.8 µsec. Even though the envelope is very rarely at the maximum frequency for any sustained period of time, we can certainly afford to play it safe and design for a rather short time constant. The fact that a system is designed for audio does not mean that all transmitted signals will be voicelike—only that they will occupy audio frequencies.

Once the time constant is chosen, we must specify the diode type, the resistance and capacitance, and the power rating of the components. The resistance is normally chosen with a view toward input and output impedance matching. For purposes of the input to the envelope detector, we would like R to be as large as possible to avoid loading of the previous circuitry. Let us choose $R = 1$ kΩ. To achieve a time constant of 31.8 µsec, the capacitor value is approximately 0.03 µF.

MATLAB EXAMPLE

These instructions simulate the envelope detector. We have set the carrier frequency to 9kHz and the information signal frequency to 5Hz. Also note the envelope detector component values given in the MATLAB program. You can vary these and rerun the simulation to see the effect on the outcome. The envelope of the waveform is printed in blue and the output of the envelope detector is red.

```
%the output of the envelope detector is analyzed based on a
%AM-modulated wave form. The carrier is set to 9kHz.
clear all
%Signal parameters
n=9000;                         %# of points
fs=9000;                        %sampling freq
fm=5;                       %modulating freq
fc=1e5;                         %carrier freq
mu=.8;                      %modulation index
%Diode parameters
Rx=20;
R=1000;
C=1.0e-7;
t=0:(1/fs):((1/fs)*(n));            %time normalized
%signal operations
info=cos(2*pi*fm*t);            %info signal
carr=cos(2*pi*fc*t);            %carrier signal
AM=(1+mu*info).*carr;           %AM signal
env=(1+mu*cos(2*pi*fm*t));
V=zeros(1,n+1);
for j=2:(n+1)  %time constant RC=1/2*pi*fm*mu
  if AM(j-1)>V(j-1), k=1;
  else, k=0;
  end
 V(j)=V(j-1)+(3/2*pi*fm*mu)*((1/Rx)*k*(AM(j-1)-V(j-1))-1/R*V(j-1));
end
subplot(2,1,1)
plot(t,AM),title('Modulated Signal'),xlabel('time'),ylabel('Volts')
subplot(2,1,2)
plot(t,V,'r',t,env,'b'),title('Envelope'),xlabel('time'),ylabel('Volts')
```

Single Sideband

In the case of double sideband, we made demodulation simpler by adding a carrier. We can also add a carrier to the single sideband waveform, recognizing that the carrier would be at the edge of the band of frequencies occupied by the waveform. The carrier could be extracted using a filter at the receiver, or a phase lock loop could be used.

Can we use an incoherent detector such as the envelope detector? To answer this question, let us examine transmitted carrier lower sideband.

$$s_{lsb}(t) + A\cos 2\pi f_c t = \left[A + \frac{s(t)}{2}\right]\cos 2\pi f_c t + \frac{\hat{s}(t)\sin 2\pi f_c t}{2} \tag{3.72}$$

The envelope of this waveform is found by combining the cosine and sine into a single sinusoid with a time varying amplitude and phase. The envelope is given by

$$\sqrt{\left[A + \frac{1}{2}s(t)\right]^2 + \left[\frac{1}{2}\hat{s}(t)\right]^2} \tag{3.73}$$

In general, this does not look anything like $s(t)$. However, if A is very large, the first term predominates and the expression is approximately equal to $A + s(t)/2$. Of course large values of A mean inefficient operation with most of the transmitted energy going into the carrier. Therefore, a system of this type finds limited application.

Vestigial Sideband

Suppose that we add a carrier term to a VSB signal. The VSB transmitted carrier waveform is then of the form,

$$s_v(t) + A\cos 2\pi f_c t \tag{3.74}$$

This carrier term can be extracted at the receiver using either a very narrow bandpass filter or a phase lock loop. If the carrier term is large enough, an incoherent detector (e.g., envelope detector) can be used. We said this in SSB, where the carrier had to be much larger than the signal. In DSB, the carrier need only be of the same order of magnitude as the signal. The required carrier size for VSB is between these two extremes. While addition of a strong carrier significantly decreases efficiency, the ease of construction of an envelope detector makes this the system of choice in television, which we discuss in the applications section at the end of this chapter.

ASK Incoherent Demodulator

Suppose that we now examine the incoherent detector for ASK. We illustrate the conceptual system in Fig. 3.51

The nonlinear device followed by the lowpass filter could represent a rectifier detector, a square law detector, or an envelope detector. Suppose that it is an envelope detector. The input to the detector is a burst of a sinusoid of amplitude A when a binary one is being sent,

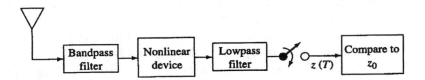

Figure 3.51 - Incoherent Detector for OOK

and the input is zero if a binary zero is being sent (assuming OOK). The output of the envelope detector, $z(t)$, is therefore a unipolar baseband signal with amplitude A or zero depending on whether a binary one or zero is being sent. We can then use baseband techniques to decide which bit is being sent. Figure 3.51 illustrates the approach of sampling the baseband output and comparing that sample to z_o. This threshold would be the midpoint between the two values, $A/2$.

Simulation Example

The CD Tina software contains a circuit called *ASK Demodulator*. Open this file and your screen should look like this:

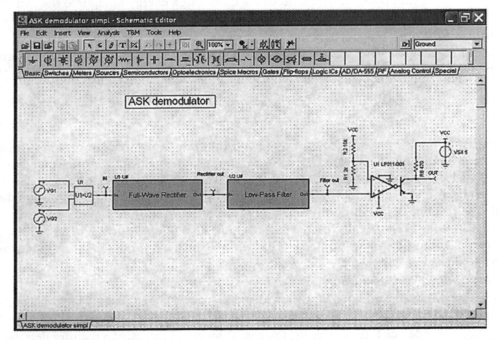

The ASK waveform is developed by multiplying a sine wave by a pulse train. This forms the ASK signal labeled "IN". It is full-wave rectified and lowpass filtered. This signal is then amplified and clipped to form the output.

Run the simulation by pulling down the *Analysis* menu and selecting *Transient*. If the curves are displayed on top of each other, you can separate them by selecting *View* and then *Separate Curves*. After running the simulation, you will see plots of the input, the rectifier output, the lowpass filter output, and the system output. If you want to add any other curves, click the right hand icon to add a curve, and select VP_14, which you will see is the node number assigned to the multiplier output.

3.4.4. IC Modulators/Demodulators

Several integrated circuit manufacturers have produced balanced modulators and demodulators. Among these are the Philips MC1496/MC1596 and the Analog Devices AD630. These ICs contain differential amplifiers that are either driven into saturation or simulate an electronic commutator (a device that alternately multiplies by positive and negative values). We refer you to the literature for details of the electronics. We will concentrate upon applications.

Figure 3.52 shows the MC1496 used as an AM Modulator. The same circuit can be used to generate suppressed carrier or transmitted carrier AM by choosing different resistor values in the carrier adjust circuitry. This same chip is used for demodulation of AMTC. The circuit is shown in Fig. 3.53.

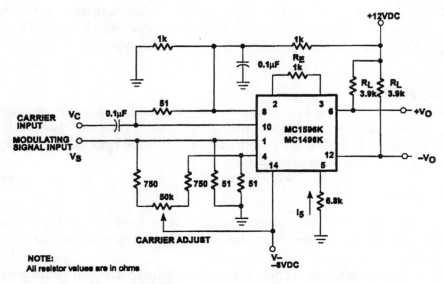

Figure 3.52 - AM Modulator (Courtesy of Philips Semiconductor)

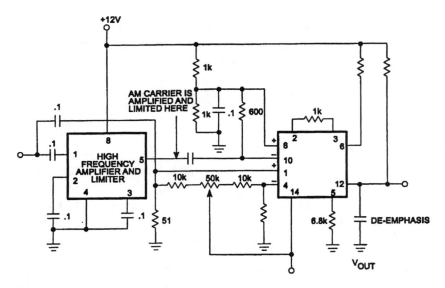

Figure 3.53 - AM Demodulator (Courtesy of Philips Semiconductor)

The carrier for this is derived by driving the high-frequency amplifier into saturation thereby providing an amplified and limited output which resembles a square wave at the carrier frequency. This carrier feeds into one of the MC1496 inputs along with the AM wave into the other. The output must be lowpass filtered to recover the information signal.

3.5 Performance

In the process of communicating a signal, noise arises in various ways. The information signal, $s(t)$, is corrupted by some noise before it even reaches the modulator in the transmitter. This noise is generated by electronic devices in the modulator. Thus, the signal at the output of the modulator is of the form

$$[s(t) + n_e(t)]\cos2\pi f_c t \tag{3.75}$$

where $n_e(t)$ is the additive noise from the electronics. Additional noise exists because the carrier sinusoid is not a pure cosine wave (due to nonlinear distortion), but contains harmonic distortion.

The modulated signal experiences multiplicative noise in the process of being transmitted from transmitter to receiver. This type of noise is due to turbulence in the air and reflection of the signal. The turbulence causes the characteristics of the transmission medium to change with time. The reflections, when recombined with the primary path signal, either enhance the signal strength or subtract from it. Taking the multiplicative noise into account, the signal at the receiving antenna is of the form

$$A[1 + n_m(t)]s_m(t) \tag{3.76}$$

where $n_m(t)$ is the multiplicative factor and we assume that the transmitted signal is $s_m(t)$ (i.e., we have neglected the modulator noise).

The AM signal experiences additive noise during transmission. This noise is generated by a multitude of sources including passing automobiles, static electricity, lightning, power transmission lines, and sunspots. If one could listen to this noise, it would sound like static as heard on radio transmission with occasional crackling sounds added. Assuming that the transmitted signal is $s_m(t)$ (i.e., neglecting the types of noises discussed earlier in this section), the received signal is of the form

$$As_m(t) + n_c(t) \tag{3.77}$$

where $n_c(t)$ is the additive noise in the channel.

Additional noise occurs in the receiver. Electronic devices and components are present, thus generating thermal and shot noise. In addition, the wires in the receiver act as small antennas, thereby picking up some transmission noise. For purposes of analysis, this receiver noise can be treated as additive noise and included in $n_c(t)$, as long as it occurs prior to detection in the receiver.

Of the various types of noise introduced, additive transmission noise is generally the most annoying type. It normally contains the most power of the various types discussed above. This is not to imply that other types of noise are not critical. Multiplicative transmission noise (turbulence) becomes significant as frequencies approach those of light. To a certain extent, uhf television signals experience such noise. Very low-frequency multiplicative noise causes fading in microwave systems.

3.5.1. Coherent Detection

We'll first examine the various coherent detectors presented in the previous section. These include the synchronous demodulator for analog AM, and the matched filter detector for ASK.

Double Sideband

We start with double sideband suppressed carrier transmission using synchronous demodulation. Assume that we have been able to exactly match the carrier frequency and phase.

The received waveform at the input to the receiver is

$$r(t) = Ks(t)\cos 2\pi f_c t + n(t) \tag{3.78}$$

where K is a constant which accounts for antenna gain and attenuation during transmission and $n(t)$ is the additive noise. We assume that $n(t)$ is white Gaussian noise with two-sided power spectral density, $N_o/2$. That is, the noise power is N_o watts/Hz.

Let us begin by finding the signal to noise ratio at the *input* to the synchronous demodulator of Fig. 3.54.

Why have we added a bandpass filter that did not appear in our earlier discussions? This is known as a *predetection filter*. It would pass the carrier and the two sidebands. In a theoretical analysis, this filter is redundant since any frequencies rejected by it would also be rejected by the final lowpass filter. It is included both for practical reasons and to

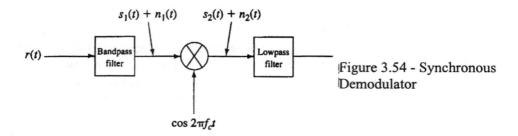

Figure 3.54 - Synchronous Demodulator

simplify the analysis. The electronic device which performs the multiplication could get overloaded and driven into saturation if the input contains too much energy outside of the band of interest. Additionally, the white noise at the input to the system contains (theoretically) infinite power. This would complicate the analysis.

The signal power at the input to the detector is the average power of $Ks(t)\cos2\pi f_c t$. This is one half of the average of the square of the cosine amplitude, or

$$\frac{\overline{K^2 s^2(t)}}{2} = \frac{K^2 P_s}{2} \tag{3.79}$$

where P_s is the average power of the information signal, $s(t)$. The average power of the filtered noise is N_o times the bandwidth of the filter, or $2f_m N_o$. The input signal to noise ratio, SNR_i, is then

$$SNR_i = \frac{K^2 P_s}{4N_o f_m} \tag{3.80}$$

We now wish to derive the signal to noise ratio at the output so we have a measure of performance. The signal at the output of the synchronous demodulator is $Ks(t)/2$, so its average power is $K^2 P_s/4$. To find the noise at the output of the detector, we must turn our attention to the time domain. We cannot analyze noise in the frequency domain since the demodulator performs the non-linear operation of multiplication.

The bandlimited noise at the detector input can be expanded into its quadrature components (see Appendix C). Thus,

$$n_1(t) = x(t)\cos2\pi f_c t - y(t)\sin2\pi f_c t \tag{3.81}$$

The power spectral densities of $x(t)$ and $y(t)$ are calculated as in the example in the Appendix. These are shown in Fig. 3.55.

The noise at the input to the lowpass filter is

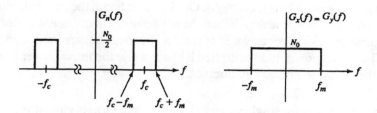

Figure 3.55 - Quadrature Expansion of Noise

$$n_2(t) = [x(t)\cos2\pi f_c t - y(t)\sin2\pi f_c t]\ \cos2\pi f_c t$$

$$= \frac{x(t)}{2} + \frac{x(t)\cos4\pi f_c t - y(t)\sin4\pi f_c t}{2} \qquad (3.82)$$

We have used the trigonometric identities,

$$\cos^2 2\pi f_c t = \frac{1}{2} + \frac{1}{2}\cos4\pi f_c t$$

$$\sin2\pi f_c t\cos2\pi f_c t = \frac{1}{2}\sin4\pi f_c t$$

The only term in Eq. (3.82) which passes through the lowpass filter is the first term, so the noise at the output of the detector is $x(t)/2$. The power of this is the power of $x(t)$ divided by 4. The power of $x(t)$ is found by integrating $G_x(f)$, so the total noise power at the detector output is $N_o f_m/2$. Now that we have the signal power and the noise power, the output signal to noise ratio, SNR_o, is found.

$$SNR_o = \frac{K^2 P_s}{2f_m N_o} \qquad (3.83)$$

Comparing this to Eq. (3.80), we find that

$$SNR_o = 2SNR_i \qquad (3.84)$$

The demodulation process has doubled the signal to noise ratio. Let's try to give an intuitive justification of this result. In double sideband, the two sidebands are related to each other. Knowing one of the sidebands, you can completely derive the second. The synchronous demodulator essentially realigns the two sidebands to add to each other, so the effect on the signal is similar to adding a signal to itself. This *coherent* addition doubles the amplitude, and therefore multiplies the power by a *factor of 4*. On the other hand, the noises in the two sidebands are unrelated (i.e., independent). When these two noise sources are added, it is like adding two independent noise processes. In that case, the mean square values add, and the power *doubles*. The signal power has multiplied by four while the noise power has only doubled. Therefore, the signal to noise ratio doubles.

Single Sideband

We now repeat the above analysis for single sideband. Let's begin by applying intuition. We would not expect the doubling of SNR that we found in double sideband transmission. The received signal is of the form

$$r(t) = \frac{Ks(t)\cos 2\pi f_c t \pm K\hat{s}(t)\sin 2\pi f_c t}{2} + n(t) \qquad (3.85)$$

Once again K is a constant which accounts for attenuation during transmission, and $n(t)$ is the noise at the output of the predetection filter of Fig. 3.54.

The signal power at the input to the detector is the average of the square of the signal portion of $r(t)$. This is given by

$$\frac{K^2[\overline{s^2(t)\cos^2\pi f_c t} + \overline{\hat{s}^2(t)\sin^2 2\pi f_c t}]}{4}$$

$$\pm \frac{K^2\overline{2s(t)\hat{s}(t)\cos 2\pi f_c t \sin 2\pi f_c t}}{4} \qquad (3.86)$$

In Eq. (3.86), the bar over the function represents time average. The last term in Eq. (3.86) is equal to zero since the average of cosine multiplied by sine is zero (take the integral over one period). The squares of the sinusoids have an average value of 1/2, so the input signal power becomes

$$P = \frac{K^2(P_s + P_{\hat{s}})}{8} \qquad (3.87)$$

P_s is the power of $s(t)$ and $P_{\hat{s}}$ is the power of the Hilbert transform, $\hat{s}(t)$. The Hilbert transform results from putting $s(t)$ through a filter with $H(f) = -j\,\text{sgn}(f)$, as shown in Fig. 3.56.

The output of the filter has a power spectral density given by

$$G_{\hat{s}}(f) = |H(f)|^2 G_s(f) \qquad (3.88)$$

The square magnitude of the filter characteristic is unity, so the power of the Hilbert transform is the same as the power of the original signal. Therefore, the input signal power is found from Eq. (3.89) to be, $K^2 P_s/4$. The input noise power is the product of N_o with the

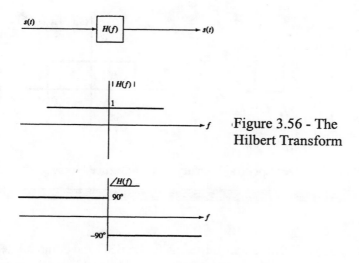

Figure 3.56 - The
Hilbert Transform

bandwidth of the bandpass filter, or $N_o f_m$. The signal to noise ratio at the detector input is then

$$SNR_i = \frac{K^2 P_s}{4 N_o f_m} \tag{3.89}$$

We now turn our attention to the detector output. The signal at the output is given by $Ks(t)/4$. If we expand the detector input noise in a quadrature expansion as in Eq. (3.80) for double sideband, the output noise is again given by $x(t)/2$. However, the power spectral density of $x(t)$ is not the same as that given in Fig. 3.55. Figure 3.57 shows the power spectral density of $x(t)$. The output signal power is $K^2 P_s/16$ and the output noise power is $N_o f_m/4$. The output signal to noise ratio is then given by

$$SNR_o = \frac{K^2 P_s}{4 N_o f_m} \tag{3.90}$$

which *is the same* as the signal to noise ratio at the detector input. That is,

$$SNR_o = SNR_i \tag{3.91}$$

This result distinguishes single sideband from double sideband. Indeed, you would hope to get some benefit from using twice the bandwidth.

$|G_n(f)$
$\frac{N_0}{2}$
f
$-f_c$ $-f_m$ $-f_c$
f_c f_c+f_m
$|G_x(f) = G_y(f)$
$\frac{N_0}{2}$
f
$-f_m$
f_m

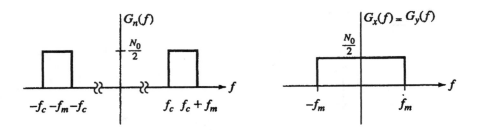

Figure 3.57 - Power Spectral Density of Quadrature Noise

Example 3.3

A baseband signal, $s(t) = 5\cos 2000\pi t$ is transmitted using DSBSC and demodulated using a synchronous demodulator. Noise with a power of 10^{-4} watts/Hz is added to the signal prior to reception. Find the signal to noise ratio at the output of the receiver.

Solution: We do not have any information about the attenuation due to the channel and also due to the antenna patterns. We shall make the unrealistic assumption that the received signal is identical to that transmitted. We can scale the signal to noise ratio according to the square of any attenuation factor.

The signal to noise ratio is found directly from Eq. (3.83).

$$SNR_o = \frac{K^2 P_s}{2N_o f_m} = \frac{2\times 25/2}{2\times 10^{-4}\times 1000} = 125 \rightarrow 21\ dB$$

Amplitude Shift Keying

We studied the matched filter detector for binary ASK. Now we examine the performance of that system. The detector is shown in Fig. 3.58 [a repeat of Fig. 3.38].

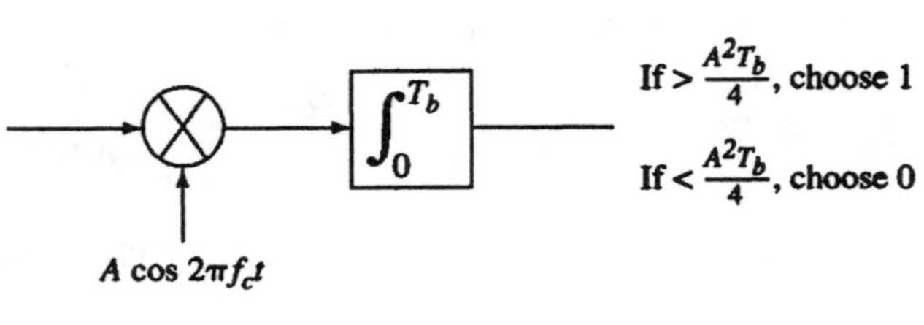

$\int_0^{T_b}$
If > $\frac{A^2T_b}{4}$, choose 1
If < $\frac{A^2T_b}{4}$, choose 0
$A\cos 2\pi f_c t$

Figure 3.58 - Matched Filter Detector for ASK

We borrow the results of the previous chapter to analyze the performance of the matched filter detector. We need simply find the average energy per bit and the correlation [see Eq. (2.87)].

If T_b is the bit transmission period (the reciprocal of the bit rate–i.e., $1/R_b$), the average energy per bit is

$$E = \frac{E_1 + E_0}{2} = \frac{(A^2 T_b/2) + 0}{2} = \frac{A^2 T_b}{4} \tag{3.92}$$

The correlation coefficient is given by

$$\rho = \frac{\int_0^{T_b} s_1(t) s_0(t) dt}{E} = 0 \tag{3.93}$$

This quantity is zero since $s_0(t) = 0$. The bit error rate is found from Eq. (2.85).

$$P_e = \frac{1}{2} erfc\left(\sqrt{\frac{E(1-\rho)}{2N_o}}\right) = Q\left(\sqrt{\frac{E(1-\rho)}{N_o}}\right)$$

$$= \frac{1}{2} erfc\left(\sqrt{\frac{A^2 T_b}{8N_o}}\right) = Q\left(\sqrt{\frac{A^2 T_b}{4N_o}}\right) \tag{3.94}$$

We shall examine this expression further after we evaluate the performance of the incoherent detector.

Example 3.4

Show that OOK is the "best" choice for amplitude shift keying. That is, show that the smallest possible error rate occurs when one of the two amplitudes is set equal to zero.

Solution: We need to take the practical constraints into account to solve this problem. If we don't, we will find that the best performance results when one of the amplitudes approaches infinity. So let's assume that the system limits either amplitude or average power. The amplitude limitation would apply in cases where the electronics are being driven to the limits of linear operation or the channel limits amplitude (e.g., broadcast systems in the United States must obey FCC rules). The average power limitation would apply in remote portable applications (e.g., a satellite which must draw all of its energy from solar cells), or if a regulatory body limits average power.

We shall assume that the amplitude is limited, and that the sinusoidal burst used to transmit a binary one has amplitude A_{max}. The amplitude of the burst to transmit a binary zero is kA_{max}, where k is a constant between zero and one. Our task is to prove that the "best" value of k is zero.

The performance of the coherent detector depends on two quantities: the average energy and the correlation coefficient. The energies of the two sinusoidal bursts are given by

$$E_0 = \frac{(kA_{max})^2 T_b}{2}$$

$$E_1 = \frac{(A_{max})^2 T_b}{2}$$

The average energy is then

$$E = \frac{(k^2+1)(A_{max})^2 T_b}{4}$$

The correlation coefficient is

$$\rho = \frac{\int_0^{T_b} s_1(t)s_0(t)dt}{E}$$

$$= \frac{k(A_{max})^2 \int_0^{T_b} \cos^2 2\pi f_c t\, dt}{\dfrac{(k^2+1)(A_{max})^2 T_b}{4}} = 2\frac{k}{k^2+1}$$

The probability of error is given by

$$P_e = \frac{1}{2}erfc\left(\sqrt{\frac{E(1-\rho)}{2N_o}}\right)$$

This quantity decreases as $E(1-\rho)$ increases. In order to minimize the probability of bit error, we want to maximize

$$E(1-\rho) = \frac{(k^2+1)(A_{max})^2T_b}{2} \quad \frac{k^2+1-2k}{k^2+1}$$

$$= \frac{(A_{max})^2T_b}{2}\left(k^2 - 2k + 1\right)$$

We assume that A_{max} and T_b are fixed. This quantity is then proportional to

$$k^2 - 2k + 1 = (k-1)^2$$

We wish to maximize this for k between zero and one. As k goes from zero to one, $(k-1)^2$ goes from one to zero. So the best choice for k is zero, which leads to OOK.

MASK

M-ary ASK (MASK) is simply an amplitude modulated version of M-ary baseband. Suppose, for example, we combine bits in pairs to form 4-ary ASK. Each symbol represents two bits, and there are four possible symbols. The waveforms used to transmit these symbols are pulse sinusoids of different amplitudes. We can express them as:

$$\begin{align}
s_0(t) &= A_0\cos 2\pi f_c t \\
s_1(t) &= A_1\cos 2\pi f_c t \\
s_2(t) &= A_2\cos 2\pi f_c t \\
s_3(t) &= A_3\cos 2\pi f_c t
\end{align} \tag{3.95}$$

The four amplitudes are often chosen to be symmetrical around the origin. For example, they may be chosen as -3, -1, +1, and +3. We will not do so in this text since we always assume that *amplitudes* in ASK or MASK are *non-negative numbers*. We assume this because a negative amplitude really means a phase shift of 180 degrees. We would therefore consider the amplitude choice of -3, -1, +1, and +3 to represent a combination of phase shift keying and amplitude shift keying. You should be prepared to find texts that call this MASK. Life would be very boring if we all agreed on everything.

The MASK detector can be coherent or incoherent. If we used negative amplitudes as discussed in the previous paragraph, we would not be able to use the incoherent detector. The two detectors are shown in Fig. 3.58.

Analysis of the performance of the coherent detector parallels the approach used in M-ary baseband in Section 2.5. The output of the integrator of Fig. 3.58(a) is a random variable with conditional probability density function similar to that of Fig. 2.112 which we repeat below.

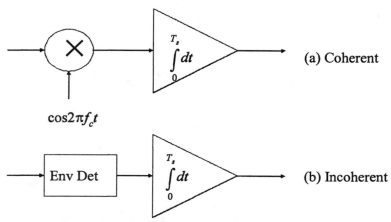

Figure 3.58 - Detector for MASK

We are showing this for 4-ary ASK. The four mean values we found for baseband must now be replaced with the new mean values. If the input to the detector is $A_i cos2\pi f_c t$ (where i is either 0, 1, 2, or 3), the output is $A_i T_s/2$. The variance of each Gaussian density in Fig. 2.112

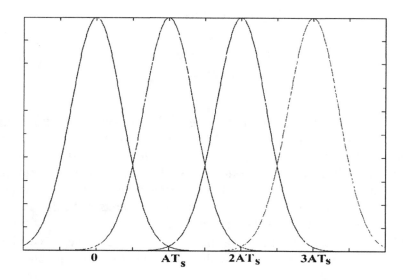

Figure 2.112 - Repeat of Figure for M-ary baseband

is found by feeding white Gaussian noise into the system of Fig. 3.58(a) and finding the average of the square of the output. This is given by $N_o T_s/2$. The probability of mistaking a transmitted symbol as the adjacent symbol value (e.g., send amplitude A_1 and call it A_2 at the receiver) is found by taking the integral under the tail of the Gaussian density. To simplify the expression, let's define d as the distance from one amplitude value to the midpoint between two adjacent values. That is,

$$d = \frac{A_i - A_{i-1}}{2} \tag{3.96}$$

The probability of adjacent symbol error is then given by

$$P = \frac{1}{\sqrt{2\pi}\sigma} \int\limits_{dT_s}^{\infty} \exp\left(-\frac{x^2}{2\sigma^2}\right) dx = \frac{1}{2}erfc\left(\sqrt{\frac{d^2 T_s}{2N_o}}\right) = Q\left(\sqrt{\frac{d^2 T_s}{N_o}}\right) \tag{3.97}$$

We have a little more work to do to find the probability of symbol error. The expression in Eq. (3.99) is the probability of mistaking one symbol for the adjacent symbol. Suppose we transmit $s_1(t)$. We can either mistake this for $s_0(t)$ at the receiver, or for $s_2(t)$. The probability of symbol error for this transmission is therefore twice that of Eq. (3.97). This is true for both of the "inner" signal amplitudes. But for the lowest and highest amplitudes, A_0 and A_3, we can only make an error in one direction, so Eq. (3.99) applies. Assuming the four symbols are equally probable, the average symbol error probability is 1.5 times that of Eq. (3.97), or

$$P_m = 0.75erfc\left(\sqrt{\frac{d^2 T_s}{2N_o}}\right) = 1.5 Q\left(\sqrt{\frac{d^2 T_s}{N_o}}\right) \tag{3.98}$$

We could derive a similar result for M-ary ASK where M is not equal to 4. We could then plot the symbol error probability as a function of signal to noise ratio. The signal to noise ratio could be referenced to either the signal energy per bit or signal energy per symbol. The energy per bit is simply the signal energy per symbol divided by $\log_2 M$. Thus, for example, for 4-ary ASK, the bit signal energy is one-half of the symbol signal energy. Similarly, we can transform from symbol error rate to bit error rate. If Gray coding is used, and we can assume the probability of mistaking a symbol for a *non-adjacent* symbol is much smaller than that for an adjacent symbol, we can assume each symbol error translates to one bit error. Thus the bit error rate is equal to the symbol error rate divided by $\log_2 M$. For example, in the 4-ary example, the bit error rate is one-half of the symbol error rate.

Figure 3.59 shows the symbol error rate as a function of bit signal to noise ratio. Note that as M increases, the error rate increases. For a given bit signal to noise ratio, as M increases, the levels must get closer together thereby increasing the symbol error rate.

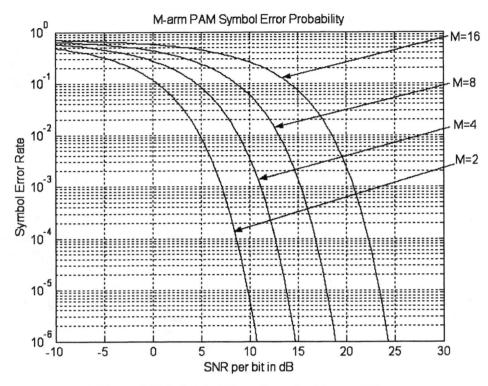

Figure 3.59 - Symbol Error Rate for M-ary ASK

3.5.2 Incoherent Detection

Now that the performance of coherent detectors has been examined, we turn our attention to the incoherent detectors. Since incoherent detectors do not use all of the information present in the received signal (i.e., they ignore phase information), your intuition should tell you they will not perform as well as the coherent receivers. The more information we have and the more work we are willing to do, the better the results will be.

AMTC

We consider the case of transmitted carrier, where the carrier amplitude is large enough to permit incoherent demodulation. The demodulator we examine is the envelope detector as shown in Fig. 3.60. We assume the bandpass filter is designed to pass the AM signal. So it's centered at a frequency of f_c and has a bandwidth of $2f_m$.

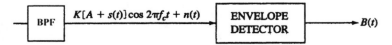

Figure 3.60 - Envelope Detector for AMTC

The received waveform is of the form

$$r(t) = K[A + s(t)]\cos2\pi f_c t + n(t) \tag{3.99}$$

The signal power at the input to the detector is $K^2 P_s/2$. The constant, A, is not considered to be part of the signal since it carries no information. The power of the noise at the detector input is $2N_o f_m$ and the input signal to noise ratio is

$$SNR_i = \frac{K^2 P_s}{4N_o f_m} \tag{3.100}$$

To find the output of the detector, we expand the input noise into quadrature form and then combine terms into a single sinusoid.

$$[KA + Ks(t) + x(t)]\cos2\pi f_c t - y(t)\sin2\pi f_c t$$

$$= B(t)\cos[2\pi f_c t + \theta(t)] \tag{3.101}$$

where

$$B(t) = \sqrt{[KA + Ks(t) + x(t)]^2 + y^2(t)} \tag{3.102}$$

and

$$\theta(t) = -\tan^{-1}\left(\frac{y(t)}{KA + Ks(t) + x(t)}\right) \tag{3.103}$$

The output of the envelope detector is $B(t)$. Unfortunately, the expression for $B(t)$ contains non-linear operations, which lead to higher order noise components, and cross products between signal and noise. We cannot obtain any general results for the output signal to noise ratio and would have to assume a specific form for $s(t)$ to carry this further.

We shall try to gain some insight into the situation by considering the limiting cases of input signal to noise ratio. That is, we will consider the case where the signal is much larger than the noise, and we will also consider the opposite situation.

We will find it helpful to view Eq. (3.101) in phasor form. Figure 3.61 illustrates this where we are aligning the abscissa with the cosine.

If $A+s(t)$ is much larger than the noise, $B(t)$ can be approximated by

$$B(t) \approx K[A+s(t)] + x(t) \tag{3.104}$$

This can be seen either from Eq. (3.102) or from the phasor diagram of Fig. 3.61, where we

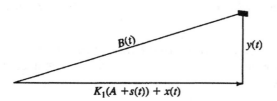

Figure 3.61 - Phasor Diagram of
Envelope Detector Input

assume $y(t) << K[A+s(t)]+x(t)$. We now need to find the signal to noise ratio at the output of the detector. We can save time by noting that the output of Eq. (3.104) is exactly twice the output of the synchronous demodulator for double sideband suppressed carrier. Since both the signal and noise are doubled, the signal to noise ratio remains unchanged at that given in Eq. (3.85).

$$SNR_o = \frac{K^2 P_s}{2f_m N_o} \tag{3.105}$$

The ratio of output SNR to input SNR is therefore the same as for double sideband coherent demodulation. That is,

$$SNR_o = 2SNR_i \tag{3.106}$$

This identical result can be deceiving. You might be tempted to say that incoherent detectors perform as well as coherent detectors. But bear in mind that the price we are paying is lower efficiency. In comparing the various systems, it is important to do so under equivalent conditions. We can get an approximate result by assuming $s(t)$ is a pure sinusoid, $\cos 2\pi f_m t$. Since this sinusoid has a maximum negative excursion of -1, the minimum value of A is 1. Using this value, $A+s(t)$ is given by $1+\cos 2\pi f_m t$. The signal power at the output is 1/2 watt and the power of the dc term in the output is 1 watt. Therefore the true signal to noise ratio at the output is 1/3 of that found in Eq. (3.107). In comparing systems, we therefore often use the following expression for incoherent detection.

$$SNR_o = \frac{2 SNR_i}{3} \tag{3.107}$$

We now consider the other extreme in incoherent detection, that is, a very low signal to noise ratio. To analyze this situation, we will redraw the phasor diagram of Fig. 3.61, but this time referenced to the noise signal. That is, we add the signal vector to the larger noise vector. This realigned diagram is shown in Fig. 3.62.

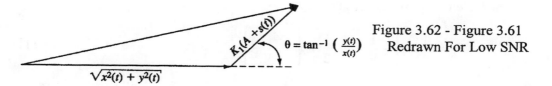

Figure 3.62 - Figure 3.61
Redrawn For Low SNR

Note that the angle between the signal and noise is given by

$$\theta(t) = \tan^{-1}\left(\frac{y(t)}{x(t)}\right) \qquad (3.108)$$

The resultant vector is approximately given by

$$\sqrt{x^2(t) + y^2(t)} + K[A + s(t)]\cos\theta(t) \qquad (3.109)$$

The only place where the signal appears is in the last term, and it is multiplied by a random noise term, $\cos\theta(t)$. Thus we have *both additive and multiplicative noise*. It can be shown that $\theta(t)$ is uniformly distributed between 0 and 360°. It should therefore not be surprising to note that the signal *cannot* be recovered from the envelope detector output.

As the signal to noise ratio decreases from a high value, a threshold is reached. For signal to noise ratios above this threshold, the output signal to noise ratio is linearly related to the detector input signal to noise ratio [as, for example, in Eq. (3.109)]. For signal to noise ratios below this threshold, the dependence approaches a quadratic relationship. That is, for each decrease in input SNR by a factor of two, the output SNR decreases approximately by a factor of four.

ASK

The analysis of the incoherent digital detector is complicated by the fact that the output of the envelope detector is not Gaussian distributed. This is true since the detector performs a non-linear operation. If a binary zero is being transmitted, the input to the envelope detector is bandpass noise, and the output is

$$z(t) = \sqrt{x^2(t) + y^2(t)} \qquad (3.110)$$

where $x(t)$ and $y(t)$ are the low-frequency Gaussian processes in the narrowband noise expansion. The envelope, $z(t)$, then follows a *Rayleigh probability density*. In the case of a one being transmitted, the analysis is more complex, and the output follows a *Ricean density*. Therefore, in contrast to the case represented by Fig. 2.94 (i.e., two bell-shaped conditional probability curves) the two conditional densities have different shapes and the symmetry we have previously used to simplify the analysis no longer exists. However, using realistic numbers, we operate far out on the tails of the density functions, and the two densities exhibit similar characteristics. Performing the analysis yields the simple result that the bit error rate is approximately given by

$$P_e = \frac{1}{2}\exp\left(\frac{-E}{2N_o}\right) = \frac{1}{2}\exp\left(\frac{-A^2T_b}{8N_o}\right) \tag{3.111}$$

Figure 3.63 shows the bit error rate for coherent detection [Eq. (3.95)] and for incoherent detection [Eq. (3.108)] as a function of the signal to noise ratio in dB. The signal to noise ratio is E/N_o. [Caution: You might be tempted to define SNR as $A^2T_b/2N_o$. But in OOK, the average energy is $A^2T_b/4$, not $A^2T_b/2$.]

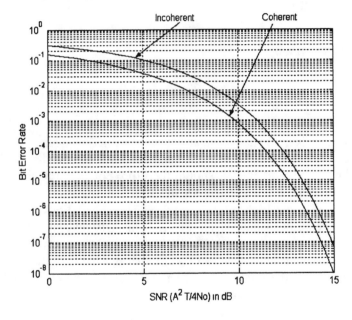

Figure 3.63 - Performance of BASK Detectors

As expected, the incoherent detector does not perform as well as the coherent detector. In system design, you would have to weigh the performance penalty against the advantages gained through ease of detector construction.

MATLAB EXAMPLE

This is the MATLAB program that was used to generate the curves of Fig. 3.63.

```
%Performance of BASK detectors (Coherent and Incoherent)
clear all
SNRdb=-20:.5:20;          %SNR in dBs
SNR=10.^(SNRdb./10);      %linear SNR
```

```
z=sqrt(SNR/2);
z2=(SNR/2);
pe1=(1/2)*erfc(z);
pe2=(1/2)*exp(-z2);
semilogy(SNRdb,pe1,'b',SNRdb, pe2,'r')
title('Coherent and Incoherent BER for a BASK detector'),
ylabel('BER'), xlabel('SNR in dB'), grid
```

Example 3.5

Binary information is transmitted at 10 kbps using OOK. The carrier frequency is 10 MHz, and received carrier amplitude is 10^{-2} volts. The additive noise power is 5×10^{-10} watts/Hz.
(a) Design a coherent detector and find the bit error rate.
(b) Design an incoherent detector and find the bit error rate.

Solution: (a) The coherent detector is shown in Fig. 3.64(a). It is simply a matched filter detector which has been simplified since one of the signals is zero.

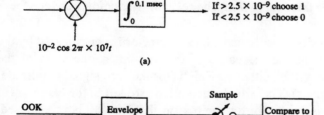

(a)

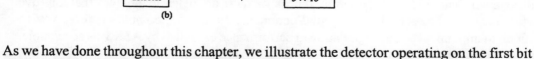

(b)

Figure 3.64 - Detectors for Example 3.5

As we have done throughout this chapter, we illustrate the detector operating on the first bit interval from 0 to 0.1 msec. After making that decision, the detector would then shift its attention to the second interval, from 0.1 to 0.2 msec.

To find the probability of bit error, we either need to find E and ρ, or we could plug numbers into Eq. (3.97) since we are sure the derivation of that equation applies in this case. The bit error rate is then given by

$$P_e = \frac{1}{2}erfc\left(\sqrt{\frac{10^{-4} \times 10^{-4}}{4 \times 10^{-9}}}\right) = \frac{1}{2}erfc(1.58) = 0.013$$

(b) The incoherent (envelope) detector is shown in Fig. 3.64(b). The performance is given by Eq. (3.111).

$$P_e = \frac{1}{2}\exp(-2.5) = 0.041$$

Two things should be noted about this result. First, the incoherent detector does not perform as well as the coherent detector, as would be expected. In fact, the incoherent detector's error rate is approximately three times as large. Second, this error rate for either detector is extremely high. For the coherent detector, we would expect an average of about 13 errors out of every 1000 transmitted bits. In fact, the rate is so high that the approximations leading to Eq. (3.111) are of questionable validity. The derivation of that equation requires that we operate far out on the tails of the probability density [far enough so that the Ricean density resembles a Rayleigh density]. Therefore, although the error result for the coherent detector is accurate, the result for the incoherent detector should be considered to be a first approximation.

MASK

The incoherent detector for MASK isan extension of the ASK detector shown in Figure 3.64(b). The only difference is that the threshold comparison needs to be expanded to cover multiple levels. If a noise-free MASK signal forms the input to an envelope detector, the output would be M-ary baseband. The various amplitude levels would be the amplitude levels of the sinusoids. As an example, in 4-ary ASK, suppose the amplitude levels are 0, A, $2A$ and $3A$. Then these four amplitudes would be the (noise-free) levels expected at the output of the envelope detector. The thresholds would be set at the midpoints between these intervals.

Evaluation of the performance of the incoherent MASK detector is complicated by the fact that the output noise is not Gaussian. Because of the non-linear operation of the envelope detector, the output noise will be either Rayleigh (for the zero signal amplitude) or Ricean (for the non-zero amplitudes). The actual calculations are left to the exercises at the end of this chapter. We do this primarily for the reason that MASK is rarely used. Indeed, even basic ASK is not often used because there are better techniques that improve performance. In the early days of digital communication, more complex receivers were difficult to implement so the degradation in performance exhibited by ASK was acceptable. This is no longer the case today.

3.6 Design Considerations

How do you apply all the theory you have learned to solve a practical problem? The following discussion is intended to help you make the required trade-off decisions. Some readers may be hoping for a recipe for system design–a type of "step-by-step" procedure. We shall resist this approach for several reasons. First, it downgrades the subject from *educating* to *training*. Second, there are no easy answers. The reason that communications is such an exciting field is that it is challenging. The *trade-off* considerations are numerous and the inputs to the design process keep changing with time. In the past, hardware implementation considerations often provided the most important trade-off input. With advances in electronics, in particular, VLSI and VHSIC, and with the seamless movement between software and hardware, implementation has become less of a significant consideration in system design. Many systems which were previously only of theoretical interest to provide an upper bound for system performance (i.e., they were viewed as being too difficult to implement) are now in use. As such, we can only hope to give the *tools* for communication system design, and relegate detailed decisions to the ingenuity of the engineer. Many designs are *open-ended*, and there is more than one "correct" answer. Indeed, this may well be the driving force for job security for communication engineers. After all, we would not want things to be too easy (well, maybe sometimes).

A number of the topics presented in this text are meant to stand alone. In contrast, the design engineer must assimilate a vast array of information and make trade-off decisions.

Design in any engineering discipline can be broken into five major steps. These are:

- **Define the problem**: This involves carefully stating the purpose of the design. The source of this information might be the customer, market survey specialists, an RFP (request for proposal), orders from your supervisor, or an assignment from your professor.

- **Subdivide the problem**: All but the simplest designs lend themselves to breaking up into smaller parts. These may include signal selection, transmitter design, receiver design, signal processing, and packaging. In major designs, individual teams are assigned to the various smaller pieces of the overall design problem.

- **Document your work**: This step pervades all of the other steps. You must communicate your work to others, whether they are members of the team working on the project, or the customer who is requesting the design. One of the most serious sins an engineer can commit is to fail to document work. If we had to constantly "reinvent the wheel", society would never have progressed technologically.

- **Build a prototype**: When an author writes a book, he or she would like to think that the printed page is the "whole truth and nothing but the truth". One of the most sobering experiences for the new engineer is to construct something according to the theory and find that it does not perform as expected. This does not necessarily mean that the theory is wrong–the model and assumptions used may not be matching the real world situation. The prototype is sometimes in the form of a computer simulation, but you must be cautious to assure that the simulation is closely modeling the real world (of course if the final design is implemented by computer, the simulation *may well be* the real world).

■ **Finalize the design**: This includes *fine-tuning* the design based on experience with the prototype. It also includes the important step of final documentation. This is when you take all of the earlier documentation and combine it into a comprehensive final report.

3.6.1. Analog design tradeoffs

So far in this text, we have considered baseband and amplitude modulated analog waveforms. We summarize the important results as they apply to bandwidth, performance, and system complexity.

Bandwidth

Single sideband AM has a bandwidth equal to the highest frequency of the baseband signal. This is the minimum bandwidth of all of the analog AM systems discussed in this text.

Vestigial sideband AM has a bandwidth greater than the maximum baseband frequency, but less than twice this quantity. Typically, the bandwidth is on the order of $1.25f_m$.

Double sideband AM (transmitted and suppressed carrier) has a bandwidth equal to twice the highest frequency of the baseband signal.

The bandwidth of *pulse modulated waveforms* depends on the width of the pulses. For example, in PAM, the maximum pulse width is limited by the sampling period. As the pulse width approaches that period, the bandwidth of the signal (nominal out to the first zeros of the transform) approaches $2f_m$. If time division multiplexing is being used, the pulse width is a small fraction of the sampling period, and the bandwidth is greater than $2f_m$.

Performance

In *single sideband AM* with coherent detection, the signal to noise ratio at the output of the receiver is the same as that at the input to the receiver (within the band of the signal).

Double sideband AM with coherent detection achieves an output signal to noise ratio that is twice that at the input to the receiver. The doubling is realized because of the redundancy between the upper and lower sideband.

The output signal to noise ratio for *transmitted carrier AM* with incoherent detection approaches twice that of the receiver input as the carrier to noise ratio gets larger. However, in such cases the efficiency decreases.

System complexity

Double sideband suppressed carrier AM has a bandwidth of $2f_m$. It has an efficiency of 100% (i.e., no power is wasted in sending a pure carrier). Modulation is performed by a multiplier while demodulation requires coherent circuits with the accompanying difficulty of reconstructing the carrier at the receiver.

Double sideband transmitted carrier AM has a bandwidth of $2f_m$. It has an efficiency of less than 50% because power is being wasted in sending a pure carrier. It enjoys the easiest implementation of a demodulator (envelope detector) of all schemes of modulation. It cannot support a signal with a non-zero *dc* level since that information would be lost in demodulation.

Single sideband AM has the smallest bandwidth of all of the systems we studied, f_m. It is 100% efficient since no power is wasted in sending a pure carrier. Complexity of implementation of modulators and demodulators is high due to the filtering required in the transmitter and the carrier recovery and coherent detection required in the receiver.

Vestigial sideband suppressed carrier AM has a bandwidth larger than f_m but less than $2f_m$. The modulator is easier to construct than that of SSB, but the demodulator requires carrier recovery and also requires a carefully controlled filter shape to properly combine the sidebands.

Vestigial sideband transmitted carrier AM has a bandwidth larger than f_m but less than $2f_m$. The modulator is easier to construct than that of SSB, and if the carrier is large enough, an envelope detector can be used thus making demodulation extremely simple.

3.6.2 Digital design tradeoffs

Digital systems have design rules similar to those of analog systems. However, since the digital systems attempt to reconstruct symbols rather than waveforms, performance is more often measured by error rates than by signal to noise ratio. We concentrate our study on system performance and bandwidth.

Performance comparisons

We have not yet examined enough different forms of digital modulation to make performance comparisons meaningful. Nonetheless, we introduce the concept in this chapter, and expand on it in later chapters. We characterize performance by the *bit error rate*, and this has been plotted as a function of signal to noise ratio several times in Chapters 2 and 3.

Figure 3.65 is a repeat of Figure 3.63. You can see that the incoherent detector does not perform as well as the coherent detector.

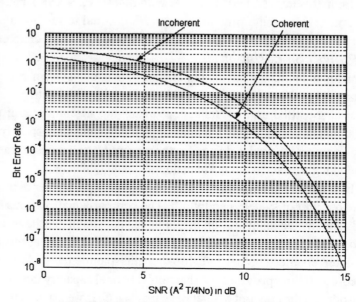

Figure 3.65 - Bit Error Rate Comparisons

Bandwidth comparisons

We define *nominal bandwidth* of a pulsed waveform as the distance out to the first zero of the power spectral density. Thus, we include the main lobe. Some applications of NRZ-L signaling define bandwidth out to half of this amount, and call this the *minimum bandwidth*. It can be shown that 91% of the total power is contained in the *nominal* bandwidth, while 78% is contained in the *minimum* bandwidth.

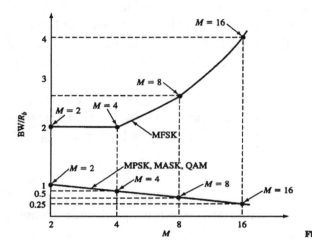

Figure 3.66 - Bandwidth vs. Bit Rate

Figure 3.66 shows the ratio of minimum bandwidth to the bit rate for various forms of signaling. Note that of the various curves shown, we have only discussed the MASK curve so far in this text. Note that MASK with M=2 (i.e., BASK), has a bandwidth to bit rate ratio of unity. Thus, the minimum bandwidth is equal to the bit rate, R_b. We discuss the other curves in this figure as more systems are added in later chapters.

BPS/Hz comparisons

Until this point, we have been concentrating our system comparative evaluations on bit error rate and on system bandwidth. In one class of practical design situation, the bit error rate is specified and we must "optimize" the design for that given error rate. Thus, for example, a particular data communication system may specify a bit error rate of 10^{-7}, and the engineer is asked to design for the maximum transmission rate within a given bandwidth and at a given signal to noise ratio.

In situations such as those described above, a parameter of interest is the bit transmission rate per unit of system bandwidth. The comparison of various transmission systems using this parameter is of major importance. Although we have only considered amplitude modulation so far in this text, we introduce the concept at this time, and will expand on it in later chapters. When we look at this parameter in later chapters, we will use the theoretical limit provided by Shannon's channel capacity theorem as a system goal. Shannon introduced the concept of channel capacity, C. If we assume that the additive noise is white and Gaussian, the channel capacity is given by

$$C = BW\log_2\left(1 + \frac{S}{N}\right)$$ (3.112)

where BW is the channel bandwidth and S/N is the signal to noise ratio. The units of the capacity are bits per second. Shannon showed that as long as the bit rate, R_b, is less than C, an arbitrarily small error rate can be achieved. Doing so may require complex error control coding, but the Shannon limit provides a useful goal against which to measure system performance.

3.6.3 Digital Communication Design Requirements

In performing a digital system design, one begins with a set of requirements and constraints for the system. These requirements and constraints might include the following:
- The required bit transmission rate.
- The maximum allowable bit error rate.
- The maximum system bandwidth.
- The maximum transmitted signal power (and resulting maximum signal to noise ratio).
- The maximum construction cost (i.e., complexity of detector).
- The maximum power utilization of the detector.
- The maximum acquisition time of the detector.

We now expand upon each of these requirements and constraints.

Required bit transmission rate

The required bit transmission rate is a function of how the system is to be used. For example, if a corporation wishes to transmit financial data to its data processing division, it knows how much data must be sent and how soon that data must reach the receiver.

The required bit transmission rate should normally be reduced as much as possible prior to beginning the communication system design. Even this reduction contains trade-off considerations. There are costs involved in data compression. First, the actual hardware is not without costs–production, real estate, and power. Special purpose chips for compression are readily available, and while the cost is not high, this must be considered in the overall design. Such circuitry takes up space on a circuit board, and while this might seem negligible, it must be considered. Compression can also lead to increased errors. Finally, any extra steps which must be performed at the transmitter require power and time. Some remote transmitters (e.g., space-based) have very limited power budgets, and the power used by compression devices must be considered. Of course, any reduction in bit rate due to compression results in a reduction of the energy needed to transmit the encoded data.

The required bit transmission rate for a digitally encoded analog signal (e.g., speech) depends on the frequency content of the analog waveform, the acceptable levels of signal to quantization noise ratio, and the properties of the signal. The frequency content determines the minimum sampling rate. The signal to quantization noise ratio determines the number of bits of quantization needed to achieve the required specifications. The properties of the signal determine whether more sophisticated compression techniques (e.g., ADM and DPCM) can be used to reduce the number of bits that must be transmitted.

We therefore see that the required bit transmission rate is a function of the information that must be transmitted and the amount of work we are willing to do prior to transmission. Although the latter of these two considerations often interacts with other decisions in the design process, we normally consider the transmission rate to be a specified quantity at the beginning of a design process.

Maximum bit error rate

The sensitivity of the data normally determines the maximum allowable bit error rate. In a speech or video signal, we know (from experience and human tests) the minimum signal to noise ratio for acceptable reception. For data, the requirements of the system usually provide a maximum bit error rate. For example, military and financial applications would normally require a smaller bit error rate than would entertainment applications. If we find that this maximum rate cannot be achieved within the constraints to be discussed next, we can resort to forward error correction in the form of coding.

Maximum system bandwidth

The channel through which the signal is transmitted normally determines the maximum system bandwidth. This is set both by the transmission characteristics of the medium (e.g., the capacitance and inductance of a coaxial cable), and by regulations (e.g., the FCC allocates bandwidth in terrestrial transmissions within the United States). This constraint, when combined with the required bit transmission rate, helps in the choice of modulation scheme. For example, we will see in the next two chapters that frequency modulation requires more bandwidth than does amplitude modulation.

We must also look at the signaling format and its relationship to system bandwidth. As an example, we saw that biphase baseband formats have frequency spectra that go to zero as the frequency approaches zero. This is in contrast to NRZ systems, where the frequency spectrum illustrates energy content down to *dc*. Some baseband channels do not transmit *dc* (i.e., they are *ac*-coupled). Similarly, some carrier systems have problems with signals that have significant energy near the carrier frequency. In such cases, biphase formats provide advantages over NRZ formats.

Maximum transmitted signal power

The maximum transmitted signal power affects the received signal to noise ratio, and therefore the bit error rate. This maximum transmitted power has constraints placed on it arising from a number of sources. The electronics could dictate a maximum amplitude for the output signal. If the signal is being broadcast, the regulatory agencies would place a maximum transmitted power constraint. The actual budget for the project might limit the amount of power than could be purchased. And if the power is derived from alternative energy sources (e.g., solar), the total amount of energy in a time period is limited.

Noise reduction is sometimes possible as a means of raising signal to noise ratio. If we have control over the transmission path, it is sometimes possible to shield the system from additive noise. If fading is causing selective reduction of signal to noise ratio, frequency or space diversity techniques might be used. That is, if one *signal path* is subject to strong noise or signal fading, we could change the transmission path. If one *frequency range* is subject to such problems, we can vary the frequency range.

Construction cost

Incoherent detectors are inherently simpler than coherent detectors.

Non-adaptive signal encoders are simpler than adaptive encoders.

In the above observations, we could almost universally substitute the statement "systems with poorer performance are less expensive than systems with better performance". Alas, the more work we are willing to perform, the better job we can do in communicating information. Hardware implementation costs are decreasing rapidly, so it will be a rare design that will be dictated by such costs. Nonetheless, when designing a small number of systems, off-the-shelf components should be used. Major integrated electronics manufacturers can supply data sheets describing the systems which have been implemented in LSI and VLSI chips.

Maximum power utilization of detector

We already discussed maximum transmitted signal power. We now turn our attention to the detector. The more complex the detector, the more power it needs to operate. Phase lock loops require more power than do passive filters (a "safe" statement for me to make since passive filters require *no* power). Quadrature detectors must perform more operations than binary detectors. Some detectors are in remote locations. This is particularly true in data acquisition. Some of these detectors rely on solar power for their operation. In such cases, a careful power budget analysis may place constraints on the system, and may require one to live with a higher bit error rate in order to meet the power requirements.

Maximum acquisition time of detector

Coherent detectors require time to acquire the carrier signal. In the case of transmitted carrier, the phase lock loop in the detector takes time to lock on to the carrier term in the received signal. In the case of suppressed carrier, squaring or Costas loops require even more time to lock on to the carrier. This acquisition time may not pose a serious problem in transmissions that are not time sensitive. That is, we can afford to devote a number of bit transmissions at the beginning of the message to give the detector time to adjust. Other transmissions may either be very short, or may require transfer of data to begin very soon after transmission begins. In such cases, our options are to do a significant amount of work to provide coherence through external operations (e.g., derive local carriers from an external source locked to the transmitter), or to use incoherent detectors. The incoherent detectors require a higher signal to noise ratio to provide the same level of performance as do coherent detectors. They therefore represent a viable alternative if the signal energy can be readily increased (through increase in signal power or decrease in transmission rate) or if the bit error rate is below specifications by a sufficient amount.

We hope that this brief discussion of design decisions has whet your appetite. If nothing else, it should point out the importance of paying close attention to detail. You can solve typical end-of-chapter problems by referencing particular equations or tiny subsets of the theory. To do competent design, you must be familiar with the broad spectrum of concepts in order to make the necessary trade-off decisions.

3.7 Applications

It is dangerous for a textbook to include applications since they change so rapidly. However, we'll take the risk and include several examples here so that you can see how the theory has been applied to the real world in the past. The information in this section is current as of the writing of this text. Updated specifications can be obtained from handbooks and from the web.

3.7.1 Broadcast AM

The electromagnetic spectrum has been described as a natural resource. The dictionary definition of *resource* is "something that lies ready to use or that can be drawn upon for aid or to take care of a need." *Natural resource* is defined as "those actual and potential forms of wealth supplied by nature." It is interesting that this term is appropriate for something that is not really an entity. The electromagnetic spectrum does indeed lie ready for use to take care of a need—the need to communicate. It is a potential form of wealth, and while it is not expendable in the sense of oil or natural gas, it does get used up in the sense that only a limited number of users can have access to it at any one time.

For these reasons, regulation of this valuable resource is necessary. It's a tribute to the ingenuity of lawyers that it has been legally possible to regulate use of the air. Not only transmission, but also reception, has been regulated. Interesting legal issues surround reception of police radio or scrambled satellite entertainment channels. In the early days of satellite entertainment, a legendary individual who was challenged for erecting a backyard antenna to receive restricted satellite entertainment channels countersued the broadcaster for injecting signals onto his private property without his permission. Thus we provide job security for lawyers.

When U.S. regulatory bodies were debating technical solutions to eavesdropping in cellular telephone, they opted instead to simply make it illegal to intercept such signals (Were this not intended to be a typically dry and boring college textbook, I'd interject some editorial comments at this point.)

Nonetheless, the need for regulation is universally acknowledged. The *International Telecommunications Union* (ITU) based in Geneva, Switzerland, is charged with worldwide regulation and includes membership from nations throughout the world. It has designated broad bands of frequency and issued regulations within each band. The bands are:

VLF	3-30	kHz
LF	30-300	kHz

MF	300-3000	kHz
HF	3-30	MHz
VHF	30-300	MHz
UHF	300-3000	MHz
SHF	3-30	GHz
EHF	30-300	GHz

VLF transmissions are in the form of surface waves, and they can travel over long distances. Radio-frequency energy in this frequency band is capable of penetrating oceans, so this band is used in submarine communications. The VHF and UHF bands are used for television. Additionally, cellular telephone uses portions of the UHF band.

The regulatory body within the United States that set rules (within the ITU general guidelines) is the *Federal Communications Commission* (FCC). The commission's regulations fill many volumes, and are available in technical libraries.

In the United States the AM broadcast band traditionally extended from 535 to 1605 kHz. [Other parts of the world have a slightly higher upper frequency cutoff (1606.5 kHz).] In the late 1990's, the FCC approved an *Expanded Band Allocation* covering frequencies between 1605 kHz and 1705 kHz. Within this band, stations are licensed by the FCC to operate commercial broadcast services. Transmission is by AMTC, and the maximum information frequency, f_m, is specified as 5 kHz. Assigned carrier frequencies must therefore be separated by 10 kHz, and they range from 540 kHz to 1700 kHz. With this spacing, the entire band can support up to 117 carrier assignments. Licensing of stations depends on many factors, the most important being (1) location of the antenna, (2) radiated power, (3) antenna pattern, and (4) times of broadcast. With the expanded band, another consideration is harmonic distortion. For example, the third harmonic of the station at 540 kHz is at a frequency of 1620 kHz. Until the band was expanded, this frequency was not part of the AM broadcast range, and interference with the corresponding AM station was not a factor to be considered.

The *antenna location* is important since a low-power station in a rural area may have fewer constraints than a high-power station in a large metropolitan area where many broadcasters are competing for the limited frequency space. If two high-power stations were assigned adjacent carrier frequencies, the filtering demands on the receiver would be excessive. Antenna height and altitude also affect the range of transmission.

The *radiated power* affects the range of transmission. The FCC must be careful not to assign identical carrier frequencies to two separated stations that will be received at the same point.

The *antenna pattern* also affects the range as a function of bearing from the antenna. A 5-kW station that broadcasts omnidirectionally creates far less interference with a station 500 km away than a 5-kW station that beams all of its power in that particular direction.

Time of broadcast affects range. Transmission characteristics of air at medium frequencies depend on temperature and humidity. It is not uncommon for a station that is heard at distances up to 150 km during the day to be received at 500 km after dark. Therefore, some stations are licensed to operate only during daylight hours.

With this as an introduction, we are now in a position to understand the operation of the standard broadcast AM receiver. We probably will not be able to repair it (a technician with a familiarity of electronics is needed for this) but we certainly can understand its block diagram.

Several basic operations can be identified in any broadcast receiver. The first is *station separation*. We must pick out the one desired signal and reject all the other signals. The second operation is that of *amplification*. The signal intercepted by the radio antenna is far too weak (on the order of 1 μV) to drive the electronics in the receiver without first being amplified. The third operation is that of *demodulation*. The incoming signal is amplitude modulated and contains frequencies centered about the carrier frequency.

Separating one channel from the others requires a very accurate bandpass filter with a sharp frequency cutoff characteristic. Suppose for example, that we wish to listen to a station with carrier at 1.01 MHz (1010 on the AM dial). That station occupies 1.005 to 1.015 MHz in frequency. The adjacent stations occupy 0.995 to 1.005 and 1.015 to 1.025 MHz, so the bandpass filter must be very close to ideal. The FCC attempts to help this situation by avoiding licensing of powerful stations in adjacent frequency slots. Nonetheless, the receiver must assume such local stations do exist. Assuming that the listener would want the capability to tune the receiver to any station in the band, the filter would have to be adjustable. That is, the band of frequencies that it passes must be adjustable.

To make a bandpass filter approach the ideal characteristics, we need multiple stages of filtering. A single RLC circuit does not have a high enough Q to accomplish the station separation. The tuning of multiple-section filters is no easy task. For example, you might have to vary three unequal capacitors in a particular manner to achieve adjustment of the center frequency of a third-order Butterworth filter. Early AM receiver contained such filters and typically had three tuning dials (variable capacitor controls) that had to be simultaneously adjusted. The family gathered around the *wireless*, and the head of the household (in those days, probably a man) would twiddle the three knobs. It was a major cause for celebration when a station (complete with cracking sounds) appeared at the speaker.

Fortunately, Edwin Armstrong changed all of this when he invented the *superheterodyne receiver* in 1918. This simple concept eliminated the need for complex adjustment of the filter, and ushered in the radio era.

Recall that multiplication of a signal by a sinusoid shifts all frequencies up and down by the frequency of the sinusoid. Because of this, station selection can be accomplished by building a *fixed* bandpass filter and shifting the input frequencies so that the station of interest falls in the passband of the filter. Looked at another way, we construct a viewing window on the frequency axis. Instead of moving this window around to view a particular portion of the axis, we keep the window stationary and shift the entire axis. The shifting process is known as *heterodyning*, and the resulting receiver is the *superheterodyne* receiver.

The receiver block diagram is shown in Fig. 3.67.

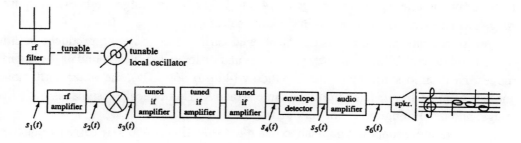

Figure 3.67 - AM Broadcast Receiver

The antenna receives a signal that is a weighted sum of all signals in the air. After some filtering, which we will examine in a moment, the incoming signal is amplified in an *rf* (radio frequency) amplifier. The signal, $s_2(t)$, is then shifted up and down in frequency by multiplying by a sinusoidal generator, called the *local oscillator*. This operation is known as *heterodyning* or *mixing*.

The output of the heterodyner is applied to the sharp bandpass filter consisting of multiple filtering stages. This filtering is normally combined with amplifying. The fixed bandpass filter is set to 455 kHz, called the *if* (intermediate frequency), and has a bandwidth of 10 kHz matching that of each station. This frequency of the bandpass filter is not within the AM broadcast band, and is specified by the FCC. If stations were authorized to broadcast at this frequency, some of that signal would enter the *if* portion of the receiver (since every piece of wire acts as an antenna) and would be heard on top of the desired station. In most receivers, the *if* filter is made up of three tuned circuits that are aligned so as to generate a Butterworth filter characteristic (poles around a semicircle in the *s* plane). At $s_4(t)$ we have a modulated signal whose carrier frequency has been shifted to 455 kHz.

The receiver then envelope detects $s_4(t)$ and amplifies (usually power amplifiers–push-pull) before applying the signal to a loudspeaker. After detection, the signal is at *af* (audio frequency). Some filtering may be done to provide treble and base controls.

Looking back at the local oscillator, how should we determine its frequency? Suppose you wish to listen to a station at the lower end of the dial, say 540 kHz carrier frequency. To shift this to 455 kHz, you would have to multiply by a sinusoid of either 85 or 995 kHz. Now suppose you wish to listen to the station at the top of the dial, with carrier at 1700 kHz. The local oscillator setting must be either 1245 or 2155 kHz. To tune any station in the band, the oscillator must then be tunable over either the range from 85 to 1245 kHz or from 995 to 2155 kHz.

The second (higher) of these two ranges is selected for practical reasons. The local oscillator is set at the *sum* of 455 kHz and the desired carrier frequency. The resulting oscillator must tune over a range where the highest frequency is a little more than two times the lowest. If the lower range had been selected, the highest frequency would be 14.6 times the lowest. It is much easier to construct variable oscillators for ranges that vary over a factor of two to one than over a range of 14.6 to 1. The higher range might even require a *range switch* much as early radios had a band selector switch.

One significant problem was not mentioned earlier. Heterodyning produces both an upshift and downshift in frequency (i.e., sums and differences). While one of these shifts moves the desired station into the *if* window (450 to 460 kHz), the other shift moves another band of frequencies into this same window. This undesired signal is called an *image*, and its elimination is not very difficult.

As an example, suppose you wish to listen to the station at 600 kHz carrier frequency. The local oscillator is set to 600+455=1055 kHz. Multiplication by this sinusoid places the desired 600 kHz station right into the *if* filter passband. But there is another station at a carrier of 1055+455=1510 kHz which will multiply by the local oscillator to produce a component at 455 kHz. This *image station* would be heard right on top of the desired station.

The separation between the image and the desired station is twice the *if* frequency, or 910 kHz. This places the image 91 frequency slots away from the desired station. The 89 stations between these two are eliminated by the *if* filter.

A bandpass filter with a bandwidth of less than 1820 kHz would accomplish the separation. This filter must pass the desired station while rejecting the station 910 kHz away. The filter must be tunable, but it need not be a sharp bandpass filer. We do not care what it does to the 89 stations between the image and desired signal. A single tuned stage is therefore sufficient. This is shown in Fig. 3.68.

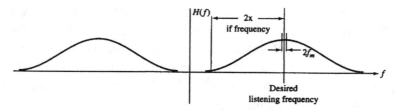

Figure 3.68 - Image Rejection Process

In practice, when a station is selected, this sloppy *rf* rejection filter is tuned at the same time that the frequency of the local oscillator is changed. Before electronic tuning, the shaft of the tuning dial connected to two separate variable capacitor sections. One of these formed part of the image rejection filter, and the other formed part of the tuning circuit of the local oscillator. This is indicated on the block diagram of Figure 3.66 as a dashed line connecting the two functions.

3.7.2 Television

Closed-circuit television was discussed in Section 2.6.2. In that section, we examined the ways in which a signal could be developed to convey video. We derived the block diagram of a monochrome monitor (see Fig. 2.110). We now expand that discussion to include the modulation techniques.

Modulation Techniques

The video signal in standard TV (NTSC standard) has a maximum frequency of about 4 MHz. The FCC allocates 6 MHz of bandwidth to each television channel, as specified by the following table. This 6 MHz must contain both the video and audio sections of the transmitted signal.

Channel	Frequency Range (MHz)	Comments
2	54-60	
3	60-66	
4	66-72	
5	76-82	Note 4 MHz gap between channels 4 and 5.
6	82-88	
7	174-180	Between 6 and 7 is FM radio, aircraft, government, railroad and police
8	180-186	
9	186-192	
10	192-198	
11	198-204	
12	204-210	
13	210-216	
14-83	470-890	

The use of double-sideband AM must be rejected since this would require over 8 MHz of bandwidth for each channel.

Single-sideband transmission is also rejected. Its generation requires very sharp filtering of the double-sideband signal to remove one of the sidebands. However, it is difficult to control the phase characteristic of a filter that has a very sharp amplitude characteristic. In designing practical filters, we can approach either the ideal phase or ideal amplitude characteristics as closely as desired, but to achieve both simultaneously is extremely difficult. In audio applications, phase deviations are not very serious. They represent varying delays of the frequency components of the message, and the human ear is not sensitive to such variations. In a video signal, these varying delays would be manifest as shifts in position on the screen. These are commonly referred to as *ghost images* and are highly undesirable.

For the reasons given above, the video portion of the TV signal is sent using vestigial sideband. The entire upper sideband and a portion of the lower sideband are sent. A large carrier is added so an envelope detector can be used for demodulation.

Figure 3.69 shows the frequency composition of a TV signal. Note that the audio and video are frequency multiplexed, and their carriers are separated by 4.5 MHz. The audio is sent using FM, which we describe in Chapter 4.

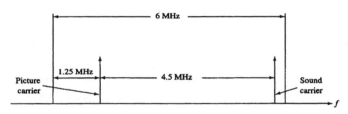

Figure 3.69 - Frequency composition of TV signal

Block Diagram of TV Receiver

We are now in a position to add blocks to the closed-circuit TV diagram of the previous chapter. The expanded diagram is shown in Fig. 3.70.

We examined the section of this diagram to the right of the video detector (an envelope detector) when we studied video monitors in Section 2.6.2. The portion to the left of the detector should look familiar. It is really no different from the AM radio receiver of Section 3.7.1. The only things that change are the various frequencies and the filter bandwidths.

The *rf* amplifier and the *if* amplifier now has a bandwidth of 6 MHz. The *if* frequency is set at 40 MHz. Since the video carriers range from 55.25 MHz to 211.25 MHz (for Channels 2 through 13), the local oscillator must tune from 95.25 MHz to 251.25 MHz.

High-Definition TV

At the time that standards were developed for television, few people dreamed of its evolution into a type of universal communication terminal. While these traditional standards are acceptable for entertainment video, they are not sufficient for many evolving applications, such as videotext. For those, we require a high-resolution standard. *High-definition TV* (HDTV) is a term applied to a broad class of systems. These systems have

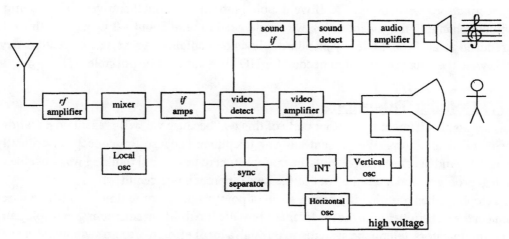

Figure 3.70 - Block diagram of monochrome TV receiver

received worldwide attention.

Of course, if we wanted to start from scratch, we could set the bandwidth of each channel to a number greater than 6 MHz, thereby achieving higher resolution. Since we don't have that luxury, a new approach had to be devised. Standards have evolved from the Grand Alliance (members are AT&T, General Instrument Corporation, Massachusetts Institute of Technology, Philips Electronics North America Corporation, David Sarnoff Research Center, Thomson Consumer Electronics, and Zenith Electronics Corporation). These standards specify a digital system with the video coding based on the *Moving Pictures Expert Group* (MPEG) standard.

When HDTV is sent through terrestrial broadcast, the FCC has mandated that the channel has a 6 MHz bandwidth (just as in the case of conventional TV). HDTV broadcasts are made on *taboo* channels–those which are not used due to interference with other channels. For example, in large metropolitan areas in the United States, commercial broadcasts occur on channels 2, 4, 5, 7, 9, 11 and 13. The taboo channels in these areas are 3, 6, 8, 10 and 12.

The HDTV aspect ratio (ratio of width to height) is 16×9, which matches that of film. Standard NTSC has an aspect ratio of 4×3. The resolution (both horizontal and vertical) of HDTV is approximately twice that of NTSC television.

Significant data compression occurs to reduce the transmission rate to less than 20 Mbps. As an example of how extreme this compression is, let's take a moment to calculate the bit rate for standard television using PCM. Recall that each frame in standard television contains 211,000 pixels. If we assume that three signals are needed to transmit full color (i.e., the primary colors), we need to send 633,000 numbers each 1/60 sec. (we are using progressive scanning), we need a total of almost 38 million numbers per second. Now suppose we code each of these into 8 bits (256 levels). This translates into 304 million bits per second *for conventional TV*. If we double both the horizontal and vertical resolution (without changing the aspect ratio), we would need to send about 1.2 Gbps. Yet this rate is reduced to 20 Mbps! Clearly standard television contains a great deal of redundancy. Otherwise the data compression needed for HDTV would not be possible.

3.7.3 Cellular Telephone

This section is different from other parts of this text because we will present no equations nor detailed diagrams. We present cellular telephone because the concept is critically important, and because it represents an example where traditional thinking was not able to solve a problem, and a unique "out of the box" approach was required.

Telephone has traditionally been a form of point-to-point communication where wires connect the transmitter with the receiver. To provide flexibility in connecting multiple pairs of communicators, telephone systems go through a lot of effort to design switching systems that can rapidly and efficiently make the necessary connections.

Wireless telephone started with "radio telephones" which were relatively high power devices with large batteries. In fact, in 1921 the Detroit police department used mobile radio at a carrier frequency of about 2 MHz to communicate between the base station and the police cars. This represents one of the earliest recorded uses of two-way mobile communication. The challenge was to transmit over greater and greater distances. But the problem with increasing range was greater demand for frequency assignments. That is, as telephones can transmit over greater and greater distances, sharing of frequency assignments became more difficult. Nobody within the large reception area of a particular telephone could use the same frequency range. When transmission occurs through the air (terrestrial communications) as opposed to through cables, frequency is a very scarce resource.

The breakthrough occurred when engineers decided to substitute spacial multiplexing for frequency or time multiplexing. Instead of trying to transmit over larger and larger distances, the cellular concept is to limit the distance, typically to only a few kilometers. The concept was reported upon in 1979 in the *Bell System Technical Journal*. If a particular telephone can transmit over only a short distance, the same frequency band can be utilized by another transmitter spaced at least twice that distance from the first transmitter. Of course the receiver must be somewhere within the transmission range of the transmitter. Multiple receiving antennas are needed, and the concept of transmission cells arose.

You can visualize dividing the entire area into approximately hexagonal cells, each about 10 square miles. This is illustrated in Fig. 3.71. A base station, consisting of a tower and small equipment building, is located near the center of each cell. The mobile cell phones establish communication with the nearest (i.e., strongest signal) base station, which is usually the closest one. Mobile cell phones typically transmit at one of two strengths: 0.6 watts or 3 watts. The control circuitry and signaling provide for monitoring received signal strength and instructing the mobile set to go to the higher power only when necessary.

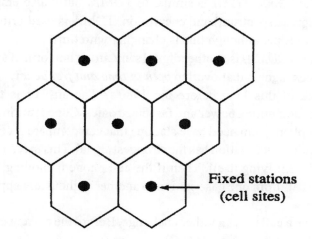

Figure 3.71 - Cell Structure

Fixed stations
(cell sites)

Adjacent base stations do not simultaneously use the same frequency bands. However, two base stations separated by at least one cell from each other can use the same frequency band. As demand increases (and we run out of frequencies), cells can be made smaller and smaller. The concepts of micro-cells and even nano- or pico-cells has been introduced in the literature. Indeed, we may some day see small base stations mounted on ever corner or on every light pole.

Now that you have a general idea, let's look at some of the design specifics. The analog system that has been in broad general use since about 1983 is known as *Advanced Mobile Phone Service (AMPS)*. This uses FM (which you will study in the next chapter) for transmitting the voice signals, while digital signals are used for call setup. Assigned frequencies range from 824 MHz to 894 MHz. The carrier (i.e., telephone service provider) is given 832 voice frequencies each with a bandwidth of 30 kHz. The provider uses two different frequencies for duplex voice communication (you can simultaneous talk and listen). These two frequencies are separated by 45 MHz. If you do the arithmetic, you will find that there is only enough room for two providers in each geographic area. The AMPS standard wanted to provide for competition, but not encourage a proliferation of competing antenna towers scattered across the landscape.

If you examine the cell structure in Fig. 3.71, you see that each cell has six other cells adjacent to it. Thus, with each provider having 416 duplex voice channels, each base station can only use 1/7th of these, or about 59 channels. If more than 59 people within a cell want to make a call, some of these will be blocked.

Time Division Multiple Access (TDMA) is one of the digital standards. The digitized voice signal is sent in bursts so that each user is only using a portion of the time. Each channel is divided into three time slots, so three simultaneous conversations can be transmitted on one channel. Control circuitry makes sure the signals do not overlap.

Global System for Mobile Telephone (GSM) is similar to TDMA, but it also employs frequency hopping (a form of spread spectrum) and encryption. TDMA is used primarily in North America and GSM in Europe, although this is changing with time.

Code Division Multiple Access (CDMA) is another digital standard. It is a form of spread spectrum that is used to multiplex signals that overlap *both in time and frequency*.

Why didn't somebody think of this long before the 1980's? The answer is that a workable system requires a lot of computing power and fast electronics. Concentrating first on originating a call from a cell phone, you need base stations that can compare reception from a single cell phone and decide which station has the strongest signal. This means every cell phone must have a way of identifying itself. Then if the cell phone is moving, there must be a way to "hand off" the call from one base station to another without any apparent interruption in the conversation.

Finding a cell phone to receive a call is yet another challenge. Before the current digital systems, the *Mobile Telephone Switching Office (MTSO)* would receive the call. This office provides the overall coordination and the interface between the Public Switched Telephone Network and the mobile system. The MTSO would then send a paging signal which would be broadcast by all of the base station transmitters throughout the region. If you left your local region, additional procedures would be necessary (in the "old" days, you actually had to call in to let the system know where you were traveling).

Roaming has improved considerably, although those concerned with privacy may not consider this progress. When you first turn on a cell phone, the phone listens for a System ID (SID). This is a control signal sent throughout the region. If your pre-programmed SID in the phone does not match this signal, the phone knows it is roaming and a light typically lights. The phone transmits a registration request which allows the system to keep track of which cell the phone is in. When the phone detects a stronger signal from another cell, it re-registers itself. The computer knows where you are within the resolution of a cell. Newer systems can locate you with much higher resolution. The design of these systems was driven by the need to locate people in an emergency (911 service in the U.S.).

The cellular concept was truly a revolution in communications. The concept can be extended to wireless communication within any limited area, such as an office building or college campus. One can even set up adaptable cells. For example, during major events, a mobile base station can be set up to allow communication within any defined area.

If all of these developments excite you, then you have chosen the right major or profession. If they scare you, we can only warn that "you ain't seen nothin' yet".

Problems

3.1 Given an information signal $r(t)$, with

$$R(f) = A(f)e^{j\theta(f)}$$

(i.e., $R(f)$ is complex), find the Fourier transform of $r(t)\cos 2\pi f_c t$. Also find the Fourier transform of

$$r(t)\cos\left(2\pi f_c t + \frac{\pi}{4}\right)$$

3.2 The signal shown in Fig. P3.2 amplitude modulates a carrier of frequency 100 Hz.
(a) If the modulation is DSBSC, sketch the modulated waveform.
(b) The modulated wave of part (a) forms the input to an envelope detector. Sketch the output of the detector.
(c) A carrier term is now added to the DSBSC waveform. What is the minimum amplitude of the carrier such that envelope detection can be used?
(d) For the modulated signal of part (c), sketch the output of an envelope detector.
(e) Draw a block diagram of a synchronous detector that could be used to recover s(t) from the modulated waveform of part (a).
(f) Sketch the output of the synchronous demodulator if the waveform of part (c) forms the input.

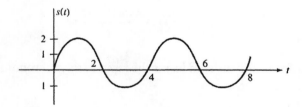

Figure P3.2

3.3 You are given the voltage signals $s(t)$ and $\cos 2\pi f_c t$ and you wish to produce the AM wave. Discuss two practical methods of generating this AM waveform.

3.4 A system is as shown in Fig. P3.4. Note that the system resembles a gated modulator, except that the gating function goes between +1 and -1 instead of between +1 and 0, and the bandpass filter has been replaced with a lowpass filter.

Can this system still produce an AM waveform? If your answer is yes, find the minimum and maximum values of filter cutoff in order for the system to act as a modulator. If your answer is no, show all work that made you arrive at this conclusion.

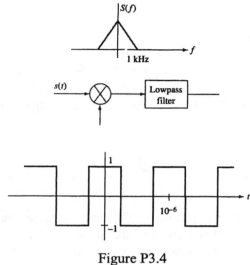

Figure P3.4

3.5 The signal

$$s(t) = \frac{2\sin 2\pi t}{t}$$

modulates a carrier of frequency 100 Hz. A signal of the form

$$n(t) = \frac{\sin 199\pi t}{t}$$

adds to the AM waveform, and the sum forms the input to a synchronous demodulator. Find the output of the demodulator.

3.6 The waveform $v_{in}(t)$ shown in Fig. P3.6 forms the input to an envelope detector. Sketch the output waveform.

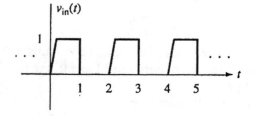
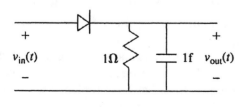

Figure P3.6

3.7 You are given the system shown in Fig. P3.7. An AMTC waveform forms the input. $p(t)$ is the periodic function, and $S(f)$ is as sketched. $|X(f)|$, $|Y(f)|$, and $|Z(f)|$ are the magnitude Fourier transforms of $x(t)$, $Y(t)$, and $z(t)$, respectively. Assume the $f_c \gg f_m$. Assume also that $s(t)$ never goes negative.

(a) Sketch $|X(f)|$.

(b) Sketch $|Y(f)|$.

(c) Sketch $|Z(f)|$.

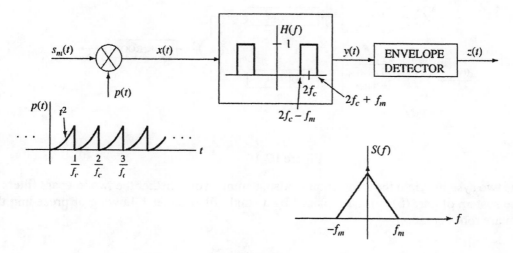

Figure P3.7

3.8 The input to an envelope detector is $r(t)\cos 2\pi f_c t$ where $r(t)$ is always greater than zero.

(a) What is the output of the envelope detector?

(b) What is the average power of the input in terms of the average power of $r(t)$?

(c) What is the average power of the output?

(d) Discuss any apparent discrepancies.

3.9 Replace the local carrier in a synchronous demodulator with a square wave at a fundamental frequency of f_c. Will the system still operate as a demodulator? Will the same be true if periodic signals other than the square wave are substituted for the oscillator?

3.10 An AMTC signal $s_m(t)$ is given by

$$s_m(t) = [A + s(t)]\cos(2\pi f_c t + \theta)$$

This AM waveform is applied to both of the systems shown in Fig. P3.10. The maximum frequency of $s(t)$ is f_m, which is also the cutoff frequency of the lowpass filters. Show that

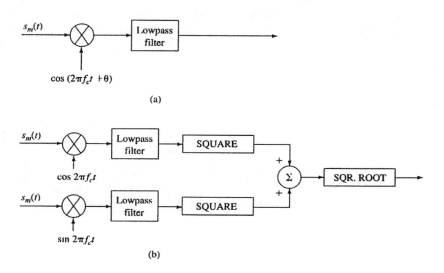

(a)

(b)

Figure P3.10

the two systems yield the same output. Also, comment on whether the two lowpass filters in the system of part (b) can be replaced by a single filter either following or preceding the square root operation

3.11 A synchronous demodulator is used to detect an amplitude-modulated suppressed carrier double-sideband waveform. In designing the detector, the frequency is matched perfectly, but the phase differs from that of the received carrier by $\Delta\theta$, as shown in Fig. P3.11. The phase difference is random and Gaussian distributed with a mean of zero and variance of σ^2. When the phases are matched, the output is $s(t)/2$. What is the maximum variance of the phase error such that the output amplitude is at least 50% of this optimum value 99% of the time?

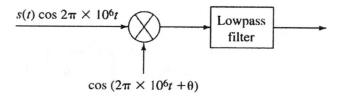

Figure P3.11

3.12 Show that the rectifier detector of Fig. 3.43 demodulates transmitted carrier AM when the rectifier is half wave.

3.13 The system illustrated in Fig. P3.13 performs a simple scrambling operation: reversing frequencies. (That is, *dc* switches to the highest frequency, while the highest frequency switches to *dc*; frequencies near f_m flip to a location near zero.)
(a) Sketch the Fourier transform of the output signal Y(f).
(b) Find the output time signal if

$$r(t) = 5\cos100\pi t + 10\cos200\pi t + 3\cos1000\pi t$$

(c) Design a system that can recover the original signal *r(t)* from the scrambled output of the system.

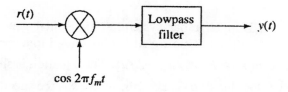

Figure P3.13

3.14 Consider the carrier injection system shown in Fig. 3.10. The information signal is

$$r(t) = \cos4\pi t + \cos20\pi t + \cos40\pi t + \cos60\pi t$$

Suppose that the bandpass filter in the receiver is tuned to the carrier frequency with a bandwidth that passes 3 Hz on either side of this frequency. Thus, the reconstructed carrier is perturbed by the 2-Hz signal. Find the output of the system. Now define the error as the difference between this output and an appropriately scaled version of the input. Find the percentage of error.

3.15 A quadrature receiver is shown in Fig. P3.15. The local oscillators have shifted by 30°, as indicated. Find the output of the receiver if an AMTC signal forms the input as shown.

3.16 You have a friend who is a guitar player. Your friend asks you to design a system that will help in the tuning of a guitar. Specifically, you are asked to design a system that accepts two inputs,

$$\cos2\pi f_o t; \quad \cos[2\pi(f_o + \Delta f)t + \theta]$$

and produces an output which is a *dc* signal that is proportional to the frequency difference Δf. The phase angle θ is unknown. Draw a block diagram of your system.

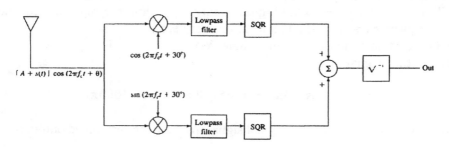

Figure P3.15

3.17 You wish to design a Doppler radar system. A sinusoidal generator continuously generates the signal, $s(t)=A\cos2\pi f_c t$. This signal is transmitted to a speeding car, and the reflected signal is of the form $r(t)=B\cos[2\pi(f_c+\Delta f)t]$. The situation is shown in Fig. P3.17. The frequency difference (Doppler shift) is $\Delta f=10s$, where s is the speed of the car in miles per hour.

You wish to display the speed of the car on a voltmeter that reads from zero to 100 volts, and you wish the voltage reading to be the same as the speed. That is, if the car is traveling at 50 miles per hour, you want the meter to read 50 V. You have a laboratory full of equipment, including filters, multipliers, and any other type of device you need. Design the system.

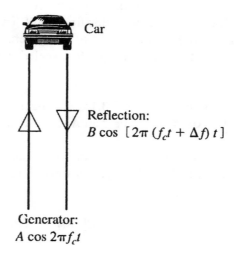

Car

Reflection:
$B\cos[2\pi(f_c t+\Delta f)t]$

Generator:
$A\cos2\pi f_c t$

Figure P3.17

3.18 Design a superheterodyne receiver that operates on AM signals in the band between 1.7 and 2 MHz. As a minimum, your design must include selection of an *if* and design of the heterodyne circuitry.

3.19 Figure P3.19 shows a dual conversion receiver. Assume that the first *if* frequency (f_{LO1}) is 30 MHz and the second (f_{LO2}) is 10 MHz. Assume further that the receiver is designed to demodulate a band of channels between 135 and 136 MHz, each of which is 100 kHz in bandwidth.

(a) Suggest the range of frequencies for the local oscillators.

(b) Determine all possible image station frequencies.

(c) How would you remove the unwanted image stations?

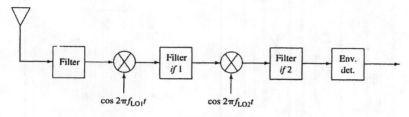

Figure P3.19

3.20 Twenty-five radio stations are broadcasting in the band between 3 MHz and 3.5 MHz. You wish to modify an AM broadcast receiver to receive the broadcasts. Each audio signal has a maximum frequency f_m = 10 kHz. Describe, in detail, the changes you would have to make to the standard broadcast superheterodyne receiver in order to receive the broadcast.

3.21 Sketch the Fourier Transform of the waveform at points (A), (B), and (C) on the diagram of Fig. P3.21. Assume $s(t)$ has the triangular Fourier transform we have used throughout with a maximum frequency of f_m. $p(t)$ is a unit height pulse train with a period of 10 μsec and duty cycle of 50%

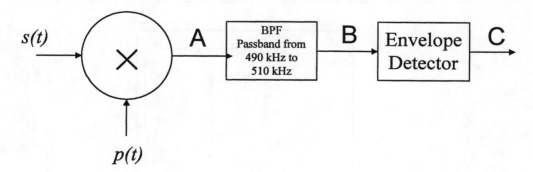

3.22 Figure P3.22 shows five waveforms, labeled (a) through (e). Determine which waveform corresponds to each of the following equations. Write the letter of the waveform next to the equation.

$(2 + \cos 2\pi f_m t)\cos 2\pi f_c t$ _____ $(0.52 + \cos 2\pi f_m t)\cos 2\pi f_c t$ _____

$(12 + \cos 2\pi f_m t)\cos 2\pi f_c t$ _____ $\cos 2\pi f_m t)\cos 2\pi f_c t$ _____

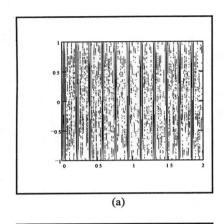

(a)

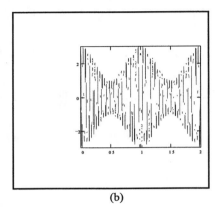

(b)

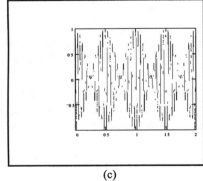

(c)

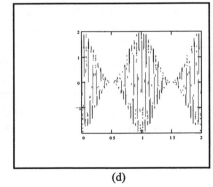

(d)

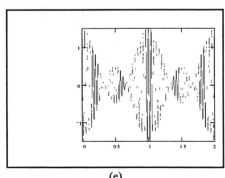

(e)

Figure P3.22

3.23 The system shown in Fig. P3.23 adds two sinusoids together, squares the result and filters that square. Find the time function, $y(t)$, at the output of the filter.

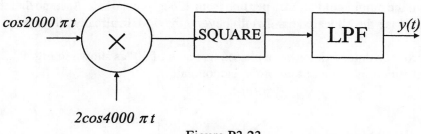

Figure P3.23

3.24 A Broadcast AM receiver has a local oscillator which is set to 1055 kHz. What is the carrier frequency of the station this radio is tuned to? What is the carrier frequency of the image station?

3.25 Two signals, $s_1(t)$ and $s_2(t)$ each have maximum frequencies of 5 kHz and maximum amplitudes of ±1 volt. They are amplitude modulated (transmitted carrier) and added together to form

$$x(t) = [1 + s_1(t)]\cos 2\pi \times 10^6 t + [1 + s_2(t)]\cos 3\pi \times 10^6 t$$

This sum, $x(t)$, forms the input to the system shown in Fig. P3.25. Find the time signal, $y(t)$, at the output of the system.

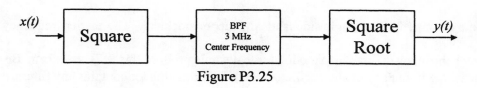

Figure P3.25

3.26 You wish to modify an AM broadcast receiver to receive the following types of broadcasts: 10 stations are broadcasting in the band between 4 MHz and 4.4 MHz. Each audio signal has a maximum frequency of f_m=20 kHz.

   Describe, in detail, the changes you would have to make to the standard broadcast receiver.

3.27 Design a system (i.e., draw a block diagram of the transmitter and receiver) that could be used to send stereo signals using AM radio. The maximum frequency of both the left and right signals is 5 kHz, and each station is allocated 40 kHz of bandwidth (instead of 10 kHz) within the normal AM band (i.e., the first station is allocated 550-590 kHz, the second is allocated 590-530 kHz, and so on).

3.28 Briefly discuss the changes (tradeoffs) you would have to make to increase the resolution of standard television by 20% (i.e., to increase the number of visible lines from 495 to 594 and the number of PIXELS per line from 426 to about 511). Assume that the FCC does **NOT** increase the allocated bandwidth beyond the 6 MHz already allocated.

3.29 A synchronous demodulator has a phase mismatch of 30° as shown in Fig. P3.29. Find the exact time function at the output of the demodulator.

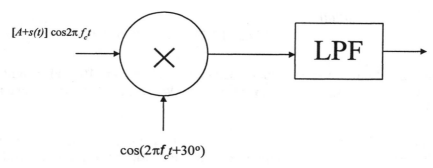

$[A+s(t)]\cos2\pi f_c t$

$\cos(2\pi f_c t+30°)$

Figure P3.29

3.30 The following signal,

$$s(t) = \frac{\sin20\pi t}{\pi t}$$

is to be transmitted using single sideband AM (upper sideband). The carrier frequency is 1 MHz.
(a) Sketch the block diagram of a modulator that would produce this SSB waveform. Be sure to indicate the frequency of any oscillators, and if your modulator contains any filters, make sure to indicate the various cutoff frequencies on the filter characteristic.
(b) Sketch the Fourier transform of the SSB modulated waveform. Make sure to indicate all important frequencies.

3.31 Binary information is transmitted at 100 kbps using OOK. The carrier frequency is 20 MHz, and the received carrier amplitude is 10^{-3} V. The additive noise power is $N_o = 10^{-12}$ watt/Hz.
(a) Design a coherent detector and find the bit error rate.
(b) Design an incoherent detector and find the bit error rate.

3.32 You are given a binary system with

$$s_1(t)=\cos 2\pi f_c t$$
$$s_0(t) = 0$$
$$T_b = \frac{5}{f_c}$$
$$f_c = 100 \ kHz$$

The additive noise has a power spectral density in the range between $N_o = 10^{-2}$ and 10^{-1} watt/Hz.
(a) Plot the probability of error as a function of N_o for coherent detection.
(b) Repeat for an incoherent detector, and compare your answer to that of part (a).

3.33 Find the maximum bit transmission rate for an OOK system in which the peak amplitude of the sinusoidal bursts is 1 V, the additive white Gaussian noise has a power of $N_o = 10^{-7}$ watt/Hz, and the maximum bit error rate is 10^{-5}.

3.34 A binary matched filter detector is used in OOK where the peak amplitude of the sinusoidal burst is 0.5 V and the additive white Gaussian noise has a power of $N_o = 10^{-6}$ watt/Hz. Information is transmitted at 10 kbps.
(a) Find the bit error rate.
(b) A "soft decision" rule is now proposed. To decrease the probability of error, a "dead space" is defined in which no decision is made. That is, if the output is greater than some threshold Δ, we decide that a one was transmitted. If the output is less than -Δ, a decision is made in favor of a zero. However, if the output is between -Δ and Δ, no decision is made. (In some systems, a request for retransmission is sent.) We wish to use this technique to reduce the bit error rate of part (a) by a factor of 2. Design the detector.

3.35 A coherent matched filter detector is used to detect an OOK signal. The carrier frequency is 1 MHz, the additive noise has a power spectral density of 10^{-5} watt/Hz, and the amplitude of the signal is I volt. The bit rate is 10 kbps.
   In designing the matched filter detector, there is a phase mismatch of K^o. Find the maximum value of K such that the coherent detector performs better than an incoherent detector.

3.36 Repeat Problem 3.35 for a bit transmission rate of 1 kbps.

3.37 A random variable x is Ricean distributed with a mean value of zero. Find the probability that $x > x_o$, and compare it to the complementary error function for large x_o.

3.38 Prove Eq. (3.113) for the probability of error of an incoherent detector. [Hint: Use the result of Problem 3.37]

3.39 ASK (OOK) is used to send 100,000 bits per second of binary information. The amplitude of the received signal is 0.2 V. The carrier frequency is 1 MHz. The noise power per Hertz, N_o, ranges from 10^{-8} to 10^{-7} watt/Hz. You are told that the bit error rate must be less than 10^{-3}.

(a) Design the optimum detector.

(b) Check whether your design is capable of meeting the error rate specification for the entire range of noise powers. If it isn't, indicate which part of the range meets the specifications.

3.40 A unit-amplitude bipolar baseband signal is shaped by a lowpass filter that passes frequencies up to $1/T_s$, where the sampling period T_s, is 1 ms. The shaped signal then frequency modules a carrier of frequency 1 MHz. Find the bandwidth of the resulting signal.

3.41 Determine the envelope of the waveform

$$s(t) = \cos 10\pi t + 17\cos 20\pi t \cos 1000\pi t$$

3.42 Starting with the transform of a lower sideband single-sideband wave

$$S_{lsb}(f) = \frac{1}{2}H(f)\big[S(f-f_c) + S(f+f_c)\big]$$

where $H(f)$ is as shown in Fig. P3.42, prove that synchronous demodulation can be used to recover $s(t)$ from $s_{lsb}(t)$.

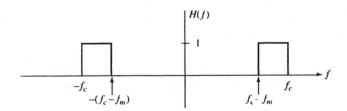

Figure P3.42

3.43 Show that the system of Fig. P3.43 produces a single-sideband (lower sideband) waveform.

What changes would you make to the system to produce the upper sideband waveform?

3.44 A vestigial sideband (VSB) signal is formed by amplitude modulating a carrier with the signal $s(t)$ shown in Fig. P3.44. A pure carrier term of amplitude two is added to the result. The double-sideband signal is then filtered with the system function shown. Assume that f_m is large enough to pass all significant harmonics of the upper sideband. An envelope detector is used for demodulation. Find the minimum value of f_v such that the error is less than 10%.

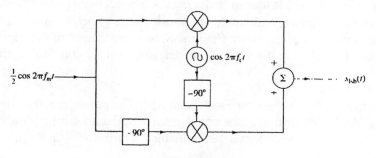

Figure P3.43

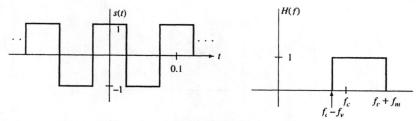

Figure P3.44

Define the error as the difference between the demodulated signal and $s(t)$. The percent error is the ratio of error power to signal power.

3.45 An information signal

$$s(t)=5cos1000\pi t$$

is transmitted using single-sideband suppressed carrier and is demodulated using a synchronous demodulator. Noise with power spectral density $G_n(f) = N_o/2 = 10^{-4}$ adds to the signal during transmission. Find the SNR at the output of the receiver, in dB.

3.46 A signal

$$s(t) = 20cos1000\pi t + 10cos2000\pi t$$

is transmitted using single sideband. Noise of power spectral density $G_n(f) = N_o/2 = 10^{-3}$ adds during transmission.
(a) Sketch a block diagram of the required receiver.
(b) Find the SNR at the output of the receiver.
(c) Suppose a bandpass filter with passband between 400 and 1,100 Hz is added to the output of the receiver. Find the improvement in SNR of this filter.

3.47 A signal $s(t)=4\sin(200\pi t+10^\circ)$ is transmitted using double-sideband transmitted carrier, double-sideband suppressed carrier, and single sideband. Noise of power spectral density $G_n(f) = N_o/2 = 10^{-2}$ adds during transmission. Find the SNR at the output of the appropriate receiver for each case. (Assume that an envelope detector is used for double-sideband transmitted carrier.)

3.48 A signal, $s(t)$, is transmitted using single-sideband AM. The power spectral density of $s(t)$ is as shown in Fig. P3.48. White noise of spectral density $N_o/2$ adds during transmission. Find the SNR at the output of a synchronous demodulator (in terms of N_o and f_m).

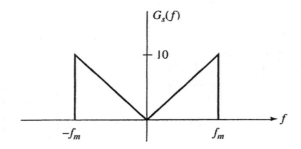

Figure P3.48

3.49 A DSBSC waveform with f_m=5 kHz and f_c=1 MHz is transmitted. Nonwhite noise with power spectral density as shown in Fig. P3.49 adds to the signal prior to detection with a synchronous demodulator. Find the SNR at the output of the detector, assuming that the power of the signal, $s(t)$, is 1 watt.

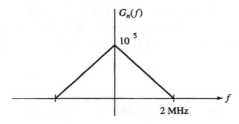

Figure P3.49

CHAPTER 4

FREQUENCY MODULATION

What we will cover and why you should care

In Chapter 3, we introduced the concept of modulation. An information signal is transmitted by varying a particular parameter of a sinusoidal signal (known as the carrier). In Chapter 3, it was the amplitude of the carrier that was varied.

While varying the amplitude seems like a natural thing to do, the performance of such modulated systems is relatively poor. This is true since additive noise affects the amplitude of the transmitted signal. The noise therefore causes errors in reception.

The current chapter explores variation of the carrier's frequency as a means of superimposing the information. This is known as *frequency modulation*. We begin with a study of traditional analog FM as it is used in broadcast radio and in the sound portion of analog television. We then explore the digital equivalent, which is known as *frequency shift keying* (FSK). This represents a very simple form of digital transmission, and it is used in such everyday devices as low-speed MODEMS and touch-tone dialing systems.

After introducing the concept of frequency modulation, we turn our attention to the design of modulators and demodulators. Once we have learned how to configure systems at the transmitter and receiver, we evaluate the performance of these systems. We examine MODEMS as a particularly important application of the theory. We conclude the chapter by exploring a number of contemporary systems.

Necessary Background

As was the case in amplitude modulation, a familiarity with Fourier transform theory is necessary to understand the frequency translation process common to all modulation techniques (See Appendix A). To understand more than the basics of the demodulation and carrier recovery systems presented in Section 4.3, you need to be familiar with feedback control systems, stability, and convergence. Understanding of the performance derivations of Section 4.4 requires an understanding of probability and random processes (see Appendix C).

Although it is helpful to study modulation techniques in the order presented in the text (i.e., amplitude followed by frequency followed by phase), it is not entirely necessary that you understand amplitude modulation prior to studying frequency modulation. However, it is assumed that you understand the basic concept of modulation as discussed in Section 3.1.

4.1 FREQUENCY MODULATION

Before exploring frequency modulation, we must first make sure that we all understand what is meant by frequency. After we agree on the definition, we will examine the modulation process.

4.1.1. Instantaneous Frequency

We approach the definition of frequency by pulling three time functions out of a hat:

$$s_1(t) = A\cos 6\pi t$$

$$s_2(t) = A\cos(6\pi t + 5) \tag{4.1}$$

$$s_3(t) = A\cos(2\pi t e^{-t})$$

The frequencies of $s_1(t)$ and $s_2(t)$ are each clearly 3 Hz. But the frequency of $s_3(t)$ is, at present, undefined. Your traditional definition of frequency does not apply to this type of waveform.

Some thought should indicate that your intuitive definition of frequency (the one you have used since high school) can be stated as follows:

> If $s(t)$ can be put into the form, $A\cos(2\pi f t + \theta)$, where $A, f,$ and θ are constants, then the frequency of $s(t)$ is defined as f Hz.

Unfortunately, $s_3(t)$ cannot be put into this form.

Our concept of frequency must be broadened to apply to the case where the frequency does not remain constant. We must define *frequency* in a manner that can be applied to general waveforms whose frequency is varying with time. To avoid contradictions, we structure this definition so that it agrees with the long-standing concept of frequency for those waveforms which possess constant frequency [e.g., the first two waveforms in Eq. (4.1)].

We begin our definition by expressing the given function in the form of a cosine,

$$s(t) = A\cos\theta(t) \tag{4.2}$$

where A is a constant. *Any signal* can be expressed in this manner by first normalizing, and then defining $\theta(t)$ as the inverse cosine. That is, any signal can be written as

$$s(t) = |s_{max}|\cos\left(\cos^{-1}\frac{s(t)}{|s_{max}|}\right)$$

where s_{max} is the maximum value of $s(t)$. The reason we must divide $s(t)$ by $|s_{max}|$ is that the inverse cosine is only defined for arguments with magnitudes bounded by one (e.g., $\cos^{-1}2$ is undefined).

The *instantaneous frequency* of s(t) is defined as the rate of change of the phase. That is,

$$f_i(t) = \frac{1}{2\pi}\frac{d\theta}{dt}$$

(4.3)

where $f_i(t)$ is the instantaneous frequency in Hz. Note that both sides of Eq. (4.3) have units of Hz (sec^{-1}).

You should take a moment to convince yourself that the instantaneous frequencies of the signals given in Eq. (4.1) are 3 Hz, 3 Hz, and $e^t(1-t)$ Hz respectively.

Example 4.1

Find the instantaneous frequency of the following waveform:

$$s(t) = \begin{cases} \cos 2\pi t & t<1 \\ \cos 4\pi t & 1<t<2 \\ \cos 6\pi t & 2<t \end{cases}$$

Solution: The wave is of the form

$$s(t) = \cos[2\pi t g(t)]$$

where *g(t)* is as shown in Fig. 4.1.

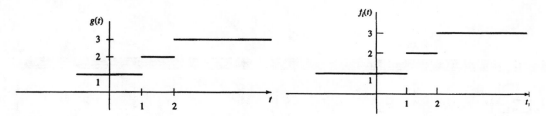

Figure 4.1 - g(t) for Example 4.1   Figure 4.2 - Instantaneous Frequency
for Example 4.1

The instantaneous frequency is given by

$$f_i(t) = \frac{d}{dt}[t g(t)] = g(t) + t\frac{dg}{dt}$$

Note that *dg/dt* is zero everywhere except at the transitions. The instantaneous frequency is then as shown in Fig. 4.2.

Example 4.2

Find the instantaneous frequency of the following waveform.

$$s(t) = 10\cos 2\pi[1000t + \sin 10\pi t]$$

Solution: We apply the definition to this waveform to find,

$$f_i(t) = \frac{1}{2\pi}\frac{d\theta}{dt} = 1000 + 10\pi\cos 10\pi t$$

This frequency is shown in Fig. 4.3.

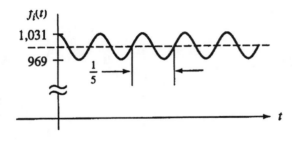

Figure 4.3 - Instantaneous Frequency
for Example 4.2

4.1.2. Narrowband FM

Frequency modulation was invented by Edwin Armstrong (the same engineer who invented the superheterodyne receiver) in 1933. He modulated the instantaneous frequency, $f_i(t)$, with the signal, $s(t)$. This contrasts with amplitude modulation where we modulate the amplitude, A, with the signal. As with AM, we still want efficient transmission and the ability to separate stations. To provide these features, we must shift the frequencies of $s(t)$ up to some carrier frequency. Therefore, rather than vary the instantaneous frequency around zero (which would present a problem since there is no such thing as negative frequency), we vary it around a positive bias level, f_c. We therefore *define* frequency modulation (FM) as being a waveform with the following instantaneous frequency:

$$f_i(t) = f_c + k_f s(t) \tag{4.4}$$

f_c is the (constant) carrier frequency and k_f is a proportionality constant which relates frequency changes to amplitude values of $s(t)$. If $s(t)$ is in volts, k_f has units of Hz/volt, or (volt-sec)$^{-1}$.

You might visualize Eq. (4.4) by thinking of operating a laboratory oscillator, the type with a rotating frequency adjustment. You begin by setting the frequency equal to the carrier, f_c. Then you start wiggling the dial back and forth to follow the waveform, $s(t)$. As $s(t)$ increases, you turn the dial clockwise and the frequency increases. The proportionality constant, k_f, determines the magnitude of your wiggling of the frequency dial. Let's carry this example further. Suppose $s(t)$ is the result of your whistling into a microphone. Thus $s(t)$ is a sinusoid whose frequency depends on the pitch of the sound, and whose amplitude depends on how loudly you whistle. If you whistle louder, your hand will have to swing further in each rotational direction of the frequency adjust dial. If you whistle at a higher pitch, your hand will wiggle back and forth faster. Since FM is conceptually more difficult to understand than is AM, it may help to refer back to this example from time to time. We strongly recommend you pretend you are actually wiggling a dial (only if you are alone when you read this).

Equation (4.4) specifies the instantaneous frequency of a waveform. We now develop the time formula for that waveform. We know that frequency is the time derivative of phase, so the phase is of the form shown in Eq. (4.5).

$$\theta(t) = 2\pi \int_0^t f_i(\tau) d\tau = 2\pi \left(f_c t + k_f \int_0^t s(\tau) d\tau \right) \tag{4.5}$$

We have assumed a zero initial condition for the phase. The FM waveform is then given by Eq. (4.6). We use the Greek letter lambda (λ) for the FM waveform.

$$\lambda_{fm}(t) = A\cos\theta(t) = A\cos 2\pi \left(f_c t + k_f \int_0^t s(\tau) d\tau \right) \tag{4.6}$$

If we turn off the information signal, $s(t) = 0$. The FM waveform of Eq. (4.6) then becomes a pure unmodulated carrier.

The FM waveform of Eq. (4.6) represents a sinusoid of amplitude A. Since the power of a sinusoid does not depend on the frequency, we see that the average power of FM is $A^2/2$ regardless of the form of the information signal, $s(t)$. We shall see this more explicitly later in this section when we have derived an alternate form of Eq. (4.6).

Example 4.3

Sketch the FM and AMSC modulated waveforms for the information signals of Fig. 4.4.

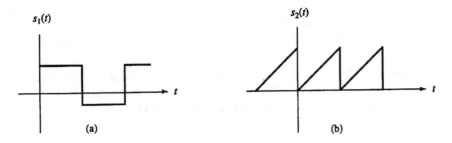

Figure 4.4 - Waveforms for Example 4.3

Solution: The AMSC and FM waveforms corresponding to the information signals of Fig. 4.4 are illustrated in Fig. 4.5. $s_{m1}(t)$ and $s_{m2}(t)$ are the AMSC waveforms and $\lambda_{m1}(t)$ and $\lambda_{m2}(t)$ are the FM waveforms. Since this is meant only as a sketch, we have not worried about exact values of $f_i(t)$.

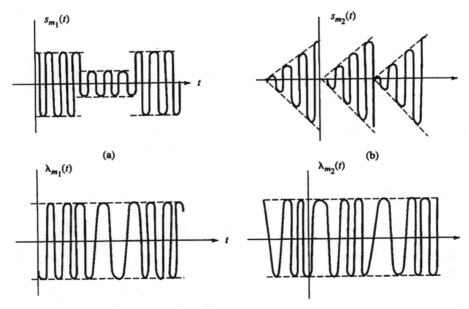

Figure 4.5 - Modulated Waveforms for Example 4.3

The instantaneous frequency of $\lambda_{fm}(t)$ varies from $f_c + k_f[\min s(t)]$ to $f_c + k_f[\max s(t)]$. You might find it useful to refer back to the laboratory oscillator scenario. As you wiggle the frequency dial back and forth, the above expressions represent the maximum excursions from the rest position of f_c. Therefore, by making k_f arbitrarily small, the instantaneous frequency of $\lambda_{fm}(t)$ can be kept arbitrarily close to f_c. You might reason that this could result in a great bandwidth savings. We love to save bandwidth! Recall that in AM we invented new schemes such as single sideband to conserve bandwidth. Here, it appears that we simply have to reduce the constant, k_f. Unfortunately, this reasoning is flawed. The Fourier transform of a waveform is the result of evaluating an integral over all time. Instantaneous frequency does not take the intended time interval into account.

To help illustrate this last point, think of a constant, $s(t) = A$. This has a frequency of zero Hz. On the other hand, a square pulse consists of piecewise constant waveforms. Therefore the instantaneous frequency of the pulse is zero (except at the transitions). However, the bandwidth of the pulse is inversely proportional to its width. Clearly, just because a function's *instantaneous* frequency is confined to a particular band does not mean that the function occupies that same band of frequencies.

We shall soon see that no matter how small k_f is made, the bandwidth can never be less than that of double-sideband AM. Thus, FM cannot be justified on the basis of bandwidth savings. It does, however, possess other advantages over AM, primarily in the area of noise immunity, but we are getting ahead of the game.

We have introduced FM by simply defining the waveform, but have not yet shown that it is of any value. In the introduction to Chapter 3, we established three guidelines for a modulation scheme to be useful: (1) It must be capable of efficient transmission through the air; (2) The modulated signal must be in a form that allows multiplexing; (3) The information signal, $s(t)$, must be uniquely recoverable from the modulated waveform. The demonstration of these properties for FM is much more difficult than it was for AM. This is true since the modulation is *not linear* in $s(t)$. That is, if $s(t)$ is replaced with a sum of signals, the resulting FM wave is *not* the sum of the individual FM waveforms. Indeed, the cosine function is non-linear.

$$\cos(A+B) \neq \cos A + \cos B$$

$$A\cos\left(2\pi\left[f_c t + k_f\int s_1(t) + s_2(t)dt\right]\right) \neq A\cos\left(2\pi f_c t + k_f\int s_1(t)\right) + A\cos\left(2\pi f_c t + k_f\int s_2(t)\right)$$

Because of this non-linear property, we will be forced to make approximations. In the end, we will be satisfied with a far less detailed analysis than was possible for AM.

The first two guidelines, efficient transmission and multiplexing capability, depend on the frequency content of the modulated waveform. Therefore to show that FM is efficient and can be adapted to frequency multiplexing, the Fourier transform of the FM wave must be found. This is not a simple task. We will find it easier to first divide FM into two classes, depending on the size of the constant, k_f. A relatively simple approximation is possible for very small values of k_f. We call this case *narrowband FM* for reasons that will become clear later in this analysis.

We start with the general form of FM wave from Eq. (4.6). To avoid having to rewrite the integral term many times, we define a new time function, $g(t)$, as the integral of the information signal.

$$g(t) = \int_0^t s(\tau)d\tau \tag{4.7}$$

Equation (4.6) then becomes

$$\lambda_{fm}(t) = A\cos\left(2\pi[f_c t + k_f g(t)]\right) \tag{4.8}$$

Now let's expand the cosine using trigonometric identities (cosine of sum).

$$\lambda_{fm}(t) = A\cos\left(2\pi[f_c t + k_f g(t)]\right)$$

$$= A\cos 2\pi f_c t \cos 2\pi k_f g(t) - A\sin 2\pi f_c t \sin 2\pi k_f g(t) \tag{4.9}$$

Since we are assuming that k_f is very small, some simplifying approximations are possible in Eq. (4.9). The cosine of a very small angle is unity while the sine of a very small angle is the angle itself (i.e., we retain the first term in a power series expansion). To justify making these approximations, k_f must be small enough such that $2\pi k_f g(t)$ represents a very small angle. Equation (4.9) then becomes

$$\lambda_{fm}(t) \approx A\cos 2\pi f_c t - 2\pi A g(t) k_f \sin 2\pi f_c t \tag{4.10}$$

This expression is linear in $g(t)$, and is therefore linear in the information signal, $s(t)$. We can find its Fourier transform with little difficulty, as follows.

The Fourier transform of $g(t)$ is related to $S(f)$ by

$$G(f) = \frac{S(f)}{j2\pi f}$$

This is true since an integrator is a linear system with transfer function, $H(f) = 1/j2\pi f$ ($1/s$ in Laplace transform). Then if we assume that $S(f)$ is limited to frequencies below f_m, we find that $G(f)$ is also limited to the same range of frequencies. Integration is a linear process that shifts phase by 90° and divides amplitudes by the frequency. As in any linear operation, integration does not create any new frequencies.

We can now take the Fourier transform of the expression in Eq. (4.10) to find

$$\Lambda_{fm}(f) = \frac{A}{2}[\delta(f-f_c)+\delta(f+f_c)] + \frac{2\pi A k_f}{4\pi}\left[\frac{S(f-f_c)}{f-f_c} - \frac{S(f+f_c)}{f+f_c}\right] \tag{4.11}$$

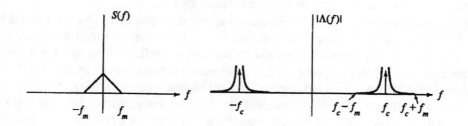

Figure 4.6 - Magnitude Transform of FM Waveform

Figure 4.6 shows a representative $S(f)$ and the magnitude of the transform of the FM waveform. Note that although the shape of $S(f)$ is meant only to be representative of a signal limited to f_m, we have attempted to show the transform of $g(t)$ as being proportional to $S(f)/f$.

Figure 4.6 immediately verifies that narrowband FM meets the first two goals of a modulating system. That is, the range of frequencies can be made as high as necessary for efficient transmission by adjusting f_c to any desired value. If adjacent carrier frequencies are separated by at least $2 f_m$, multiple signals may be transmitted simultaneously on the same channel without interfering with each other. We shall defer consideration of the third requirement, being able to recover $s(t)$ from the modulated waveform, to our study of demodulation later in this chapter. The same demodulators can be used regardless of the size of k_f, so we will cover both the small and large k_f cases together.

The bandwidth of the FM waveform is $2f_m$, just as in the double sideband AM case. This is true regardless of how small k_f is made. This is an important result which does not agree with intuition. Returning to the laboratory oscillator example, suppose a speech waveform frequency modulates a carrier. Further assume the speech waveform has a maximum frequency of 5 kHz. Therefore, you will have to wiggle the frequency control dial back and forth up to 5000 times per second. Suppose k_f is very small such that the maximum deviation away from f_c is only one Hz. You will then be wiggling the dial over a swing of only two Hz, from f_c-1 to f_c+1. Yet the Fourier transform of the resulting waveform will occupy the band between f_c-5000 and f_c+5000. Clearly the instantaneous frequency and the manner in which it changes *both* contribute to the FM bandwidth.

The fact that we call the small k_f case *narrowband* should give you a hint that as k_f is raised, the bandwidth increases from its minimum of $2f_m$. We shall prove this in the next section.

4.1.3. Wideband FM

If k_f is *not* small enough to permit the approximations of the previous section, we have what is known as *wideband FM*. The transmitted signal is still of the form of Eq. (4.8), which we repeat here.

$$\lambda_{fm}(t) = A\cos 2\pi[f_c t + k_f g(t)]$$

(4.12)

Recall that $g(t)$ is the time integral of the information signal, $s(t)$.

If $g(t)$ were a known function, the Fourier transform of this FM waveform could be evaluated. We would simply use the defining integral for the Fourier transform, and although the integral may require numerical techniques to evaluate, it nevertheless yields the Fourier transform. However, we do not wish to restrict ourselves to a particular $s(t)$. Just as in the case of AM, we only want to constrain $s(t)$ to frequencies below f_m. With this as the only given information, it is not possible to find the Fourier transform of the FM waveform. This is true because of the non-linear relationship between $s(t)$ and the modulated waveform. Since we cannot find the Fourier transform of the FM wave in the general case, let's see how much we can possibly say about the modulated signal.

To show that FM can be efficiently transmitted and that multiple channels can be multiplexed, we must only gain some feeling for the *range* of frequencies occupied by the modulated waveform. That is, we don't really care about the exact shape of the Fourier transform of $\lambda_{fm}(t)$–only the range of frequencies it occupies. We have already addressed this for narrowband modulation, so we now concentrate on wideband. The approach we take is significantly different from that used in AM. In the case of AM, we found the Fourier transform of $s_m(t)$, and used this throughout the analysis, including our presentation of the operation of modulators and demodulators. In the case of frequency modulation, we use the Fourier transform for *only one thing* –to determine the *range of frequencies* occupied. All of our other analyses will be performed in the time domain. For this reason, we do not really care about the exact shape of the transform.

Unfortunately, even finding the range of frequencies is impossible in general. To ease our way around this dilemma, we begin by restricting ourselves to a specific type of information signal, a pure sinusoid. This allows us to use trigonometry in the analysis process. Although nobody would go through much effort to transmit a simple sinusoidal information signal, once we analyze this special case, we will generalize the results.

Let the information signal be

$$s(t) = a\cos 2\pi f_m t$$

where a is a constant amplitude. The instantaneous frequency of the FM wave is then given by

$$f_i(t) = f_c + ak_f \cos 2\pi f_m t \tag{4.13}$$

We integrate the frequency to find the FM waveform.

$$\lambda_{fm}(t) = A\cos\left(2\pi f_c t + \frac{ak_f}{f_m}\sin 2\pi f_m t\right) \tag{4.14}$$

The *modulation index*, β, is defined as

$$\beta = \frac{ak_f}{f_m} \tag{4.15}$$

The modulation index is a dimensionless quantity (Hz/Hz). It is the ratio of the amplitude of the time-varying term in Eq. (4.13) to the frequency of this term. Using this definition, Eq. (4.14) can be simplified to

$$\lambda_{fm}(t) = A\cos(2\pi f_c t + \beta\sin 2\pi f_m t) \tag{4.16}$$

We now want to separate the information portion from the carrier so we can isolate the portion due to $s(t)$. To perform this separation, we expand the cosine. We would have to use trigonometric identities (i.e., the *cosine of a sum* identity) to do so, and two terms would have to be tracked throughout the analysis. To simplify the mathematics and allow us to work with only one expression, we convert to exponential notation (This is a trick you should have learned in your study of basic circuit theory).

$$\lambda_{fm}(t) = Re\left\{A\exp(j2\pi f_c t + j\beta\sin 2\pi f_m t)\right\} = Re\left\{Ae^{j2\pi f_c t}\, e^{j\beta\sin 2\pi f_m t}\right\} \tag{4.17}$$

The second term of the product of exponentials is the one that contains the information signal. This term is

$$e^{j\beta\sin 2\pi f_m t}$$

The real part is in the form of a cosine of a sine, and is generally difficult to work with. We can make some progress by noting that the exponential is a periodic function with period $1/f_m$. To prove this, substitute $t+1/f_m$ for t. Doing so does not change the value of the expression. Note that the periodic function is complex. We can expand it in the complex Fourier series with a fundamental frequency of f_m (i.e., the reciprocal of the period).

$$e^{j\beta\sin 2\pi f_m t} = \sum_{n=-\infty}^{\infty} c_n e^{jn2\pi f_m t} \tag{4.18}$$

The Fourier coefficients are given by

$$c_n = f_m \int_{\frac{-1}{2f_m}}^{\frac{1}{2f_m}} e^{j\beta\sin 2\pi f_m t} e^{-jn2\pi f_m t} dt \tag{4.19}$$

Making a change of variables, we let $x = f_m t$. We then find

$$c_n = \int\limits_{-\frac{1}{2}}^{+\frac{1}{2}} e^{j\beta\sin 2\pi x} e^{-jn2\pi x} dx$$

This integral is a real function of n and β (It's real because the imaginary part of the integrand is odd). It is not a function of f_m. The integral cannot be evaluated in closed form. It is tabulated under the name *Bessel function of the first kind* , and its symbol is $J_n(\beta)$. Given n and β, you can simply look this up in a table. To derive more general results, we need to take a moment to explore the properties of Bessel functions.

Bessel Functions

The Bessel function of the first kind is generated as solutions of the following differential equation:

$$x^2\frac{d^2y}{dx^2} + x\frac{dy}{dx} + (x^2-n^2)y(x) = 0$$

Although the Bessel function is defined for all values of n, we are only concerned with the positive and negative real integers. For integer values of n,

$$J_{-n}(x) = (-1)^n J_n(x)$$

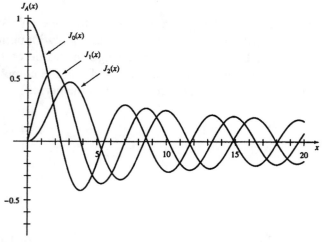

Figure 4.7 - Bessel Functions for n - 0, 1, and 2

Figure 4.7 shows $J_n(x)$ for values of n of 0, 1, and 2. Note that for very small x, $J_0(x)$

approaches unity while $J_1(x)$ and $J_2(x)$ approach zero.

Another useful property of the Bessel function is that the sum of the squares is equal to unity. That is,

$$\sum_{n=-\infty}^{\infty} J_n^2(x) = 1 \qquad for\ any\ x$$

We need to observe the behavior of the Bessel function as n becomes large. To gain insight, we examine a particular point, $\beta=10$. Figure 4.8 is a plot of $J_n(10)$ as a function of n. This function appears to be in underdamped oscillation for negative n, but again note that we are only concerned with integer values of n. For these integer values, symmetry exists and we can focus attention on the positive-n axis.

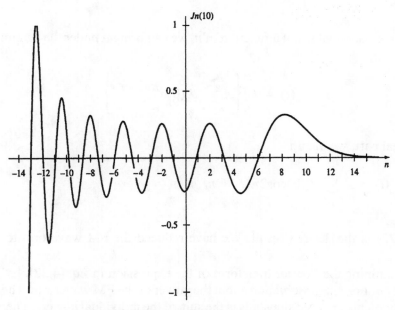

Figure 4.8 - $J_n(10)$ as a Function of n

The important observation to make is that for n greater than about nine, the Bessel function asymptotically approaches zero. In fact, for fixed n and large β, the Bessel function can be approximated by

$$J_n(\beta) \approx \frac{(\beta/2)^n}{\Gamma(n+1)} \qquad\qquad (4.20)$$

where $\Gamma(n+1)$ is the *Gamma function*. Gamma functions approach infinity for arguments greater than 2. For example, the value of the Gamma function for arguments of 2,3,4,5, and

6, is 1,2,6,24, and 120 respectively. Since the Gamma function is in the denominator, the Bessel function gets small rapidly with increasing n beyond the value of the argument, β. This is a critical property for finding the bandwidth of the FM waveform.

Returning now to Eq. (4.19) we stated that the right side of that equation is $J_n(\beta)$. The Fourier coefficients are given by

$$c_n = J_n(\beta)$$

and the FM waveform becomes

$$\lambda_{fm}(t) = Re\left\{Ae^{j2\pi f_c t}\sum_{n=-\infty}^{\infty} J_n(\beta)e^{jn2\pi f_m t}\right\} \tag{4.21}$$

Since the first exponential is not a function of n, we can bring it under the summation sign to get

$$\lambda_{fm}(t) = Re\left\{A\sum_{n=-\infty}^{\infty} J_n(\beta)e^{j2\pi(f_c + nf_m)t}\right\}$$

Taking the real part, we obtain

$$\lambda_{fm}(t) = A\sum_{n=-\infty}^{\infty} J_n(\beta)\cos 2\pi(f_c + nf_m)t \tag{4.22}$$

Equation (4.22) is the desired result. We have reduced the FM waveform to a sum of sinusoids!

Before examining the Fourier transform of the expression in Eq. (4.22), let's use this result to verify our earlier observation about the power of the FM waveform. The power of a sum of harmonically-related sinusoids is the sum of the individual powers. Therefore, the power of the FM wave is given by

$$\overline{\lambda_{fm}^2(t)} = \sum_{n=-\infty}^{\infty} \frac{A^2 J_n^2(\beta)}{2} = \frac{A^2}{2}$$

We used the property that the sum of the squares of Bessel functions is unity. Once again, we see that the power of the FM wave is independent of the modulating signal, $s(t)$.

The Fourier transform of this sum of sinusoids is a train of impulses. This Fourier transform is sketched in Fig. 4.9.

Yuk! We seem to be in big trouble. The transform extends forever in both directions. It would appear that the FM wave has an infinite bandwidth. Indeed, unless $J_n(\beta)$ goes to zero

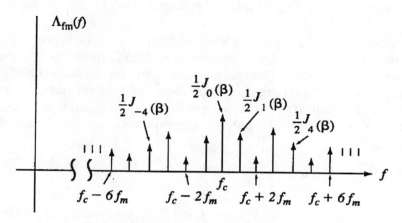

$\Lambda_{fm}(f)$

$\frac{1}{2}J_{-4}(\beta)$

$\frac{1}{2}J_0(\beta)$

$\frac{1}{2}J_1(\beta)$

$\frac{1}{2}J_4(\beta)$

$f_c - 6f_m$ $f_c - 2f_m$ f_c $f_c + 2f_m$ $f_c + 6f_m$

f

Figure 4.9 - Fourier Transform of FM With
Sinusoidal Information Signal

for n above some value, the bandwidth is *not limited* and we can neither transmit efficiently nor multiplex multiple channels.

We are rescued by returning to our previous discussion of Bessel functions. For fixed β, the functions $J_n(\beta)$ eventually do approach zero as n increases. We are not saying that the Bessel functions decrease monotonically for n less than β—only that they *eventually* approach zero. In fact, for certain choices of β, the $J_0(\beta)$ term goes to zero and the carrier term is eliminated from the FM waveform. In the case of AM, elimination of the carrier increases efficiency. In FM, elimination of the carrier does not gain anything since the total power of the modulated waveform remains constant.

To approximate the bandwidth of the FM waveform, we must examine the strengths of the impulses in Fig. 4.9. Let's first choose a small value of β. We see from Fig. 4.7 that if $\beta<0.5$, $J_2(\beta)<0.03$ (actually you can't read Fig. 4.9 that accurately, but this result is nonetheless true). The higher order Bessel functions ($n>2$) are even smaller. At $\beta=0.5$, J_1 is 0.24. For these small values of β, it is therefore reasonable to include only the three impulses centered at the carrier in Fig. 4.9. That is, there is the component at the carrier, and two additional components spaced $\pm f_m$ away from the carrier. This yields a bandwidth of $2f_m$. But we already knew this since very small values of β (i.e., ak/f_m) correspond to the narrowband condition. It's comforting that our result does not contradict the earlier conclusion.

Now suppose that β is *not* small. For example, suppose it is equal to 10. The properties that we have previously discussed would indicate that $J_n(10)$ attenuates rapidly for $n>10$. Referring to Fig. 4.9, we then consider the significant components to be the carrier term and 10 harmonics on each side of the carrier. In general, for large β, the number of terms that must be included on each side of the carrier is β (rounded to an integer). This yields a bandwidth of $2\beta f_m$. Note that these bandwidths are approximations since the Fourier transform of the FM wave is not identically zero beyond any frequency point. However, in practice if the energy outside this band is rejected with a bandpass filter, the resulting distortion is usually not noticeable ("noticeable" is admittedly not a precise engineering term, but we hope you get the idea.).

In many instances, we will be working with modulation indexes between the two extremes. In the early days of FM studies, a person named John Carson proposed a rule of thumb that has been widely adopted. This rule gives the approximate bandwidth of FM as a function of the frequency of the information signal and the modulation index. The reason this rule has been widely accepted is that the amount of energy outside this band has been shown to be negligible for most applications. Carson's rule says the bandwidth is approximately given by

$$BW \approx 2(\beta f_m + f_m) \tag{4.23}$$

This agrees with our two limiting cases. For very small β, the bandwidth is approximately $2f_m$, while for large β, it is approximately $2\beta f_m$.

Carson's approximation can be written in a more meaningful way by replacing substituting $\beta = ak_f/f_m$. Equation (4.23) then be comes

$$BW \approx 2(ak_f + f_m) \tag{4.24}$$

Let's try to give meaning to the terms in Eq. (4.24). Recall that the instantaneous frequency is given by Eq. (4.13) which we repeat here.

$$f_i(t) = f_c + ak_f\cos2\pi f_m t$$

Looking at the two terms in Eq. (4.24), we see that f_m is the rate at which the instantaneous frequency varies while ak_f is the maximum amount that it deviates away from the carrier. Returning to our laboratory oscillator example, f_m is the rate at which your hand must wiggle back and forth while ak_f is the maximum amount it must twist away from the rest position. Doesn't it make intuitive sense that both of these quantities should contribute to the bandwidth of the FM wave?

Example 4.4

Find the approximate band of frequencies occupied by an FM wave with carrier frequency of 5 kHz, $k_f = 10$ Hz/v, and

  (a) $s(t) = 10\cos10\pi t$ volts
  (b) $s(t) = 5\cos20\pi t$ volts
  (c) $s(t) = 100\cos2000\pi t$ volts

Solution: The bandwidth in each case is approximately given by

  (a) BW $2(ak_f+f_m) = 2[10(10) + 5] = 210$ Hz
  (b) BW $2(ak_f+f_m) = 2[5(10) + 10] = 120$ Hz
  (c) BW $2(ak_f+f_m) = 2[100(10) + 1000] = 4$ kHz

The band of frequencies occupied is therefore in the range
  (a) 4895 Hz to 5105 Hz
  (b) 4940 Hz to 5060 Hz
  (c) 3 kHz to 7 kHz

Equation (4.24) was developed for the special case of a sinusoidal information signal. If the modulation were linear in the information signal, we could simply apply this formula to the highest frequency component of $s(t)$ to find the bandwidth. However, FM is non-linear, so this approach is not legal.

It is time for a rather bold generalization. Let's stare at Eq. (4.24) and attempt to develop a similar formula for the general case. To help supply inspiration, we present Fig. 4.10 which shows the instantaneous frequency for the special case of a sinusoidal information signal, and also for the more general case.

Looking first at the special case, the two parameters which feed into Eq. (4.24) are the frequency of $f_i(t)$ and the *maximum frequency deviation*, ak_f, as marked on the figure. This maximum frequency deviation is the largest difference between the instantaneous frequency and the carrier. It's the furthest you turn the frequency dial (if you have been using the laboratory oscillator analogy).

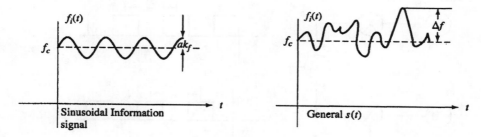

Figure 4.10 - Instantaneous Frequency

In the general case, $f_i(t)$ no longer varies at a single frequency, but contains a continuum of frequencies up to a maximum of f_m. Since we are looking for the *maximum* bandwidth, it would seem intuitive to replace the "frequency of $f_i(t)$" term in Eq. (4.24) with the "maximum frequency of $f_i(t)$". Since we call each of these f_m, there is no need for any change in this portion of the equation. For the sinusoidal information signal case, ak_f is the maximum deviation of frequency away from f_c. Figure 4.10 shows the equivalent *maximum frequency deviation* for the general case. We assign the symbol, Δf, to this frequency deviation (In most real life situations, the maximum positive deviation will be the same as the maximum negative deviation. So it doesn't matter which you use in the equation). The general form of Eq. (4.24) is then given by

$$BW \approx 2(\Delta f + f_m) \tag{4.25}$$

and we have accomplished our goal.

Let's attach an intuitive meaning to the wideband case. If Δf is much larger than f_m, we have wideband FM and the frequency of the carrier varies over a large range but at a slow rate of variation. That is, the instantaneous frequency of the carrier goes from $f_c - \Delta f$ to $f_c + \Delta f$ very slowly. The FM wave therefore approximates a pure sinusoid over long segments of time. We can think of it as a superposition of many sinusoidal bursts with frequencies between the two limits. The Fourier transform is then approximately equal to a superposition of the transforms of each of these many sinusoids, all lying between the frequency limits. The Fourier transform of a long sinusoidal burst approaches an impulse as the length of the burst approaches infinity. It is therefore reasonable to assume that the FM bandwidth is approximately the width of this frequency interval, or $2\Delta f$. Once again, looking at our laboratory oscillator, if we move the dial back and forth very slowly, the Fourier transform of the resulting waveform approximately occupies the same range of frequencies as that over which the dial is varied.

On the other hand, for very small Δf, we have a carrier which is varying over a very small range of frequencies, but doing so relatively rapidly. We can approximate this with two oscillators at the frequency limits, each being gated on for half of the total time. This is shown in Fig. 4.11.

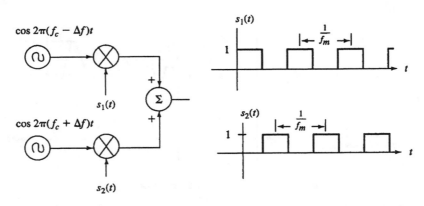

Figure 4.11 - Approximation to Narrowband FM

The approximate band of frequencies occupied by the output signal of Fig. 4.11 is from $f_c - \Delta f - f_m$ to $f_c + \Delta f + f_m$. For small Δf, this yields a bandwidth of $2f_m$. The details of this derivation are left to the problems at the back of this chapter.

We have seen that the bandwidth of an FM waveform increases with increasing values of k_f. At this point, there appears to be no reason to use anything other than narrowband FM, which has the minimum bandwidth of $2f_m$. However, we will see that wideband FM possesses a noise advantage over both narrowband FM and AM.

Example 4.5

A 10 MHz carrier is frequency modulated by a sinusoidal signal of 5 kHz frequency such that the maximum frequency deviation of the FM wave is 500 kHz. Find the approximate band of frequencies occupied by the FM waveform.

Solution: We must first find the approximate bandwidth. This is given by

$$BW \approx 2(\Delta f + f_m)$$

We are given the value of Δf, and the maximum frequency component of the information is 5 kHz. Indeed, this is the *only* frequency component of the information, and therefore, certainly is the maximum. The bandwidth is then approximately

$$BW \approx 2(500 \ kHz + 5 \ kHz) = 1010 \ Hz$$

Thus the band of frequencies occupied is centered around the car rier frequency and ranges from 9495 kHz to 10505 kHz.

The FM signal of this example is wideband. If it were narrowband, the bandwidth would be only 10 kHz.

Example 4.6

A 100 MHz carrier is frequency modulated by a sinusoidal signal of one volt amplitude. k_f is set at 100 Hz/volt. Find the approximate bandwidth of the FM waveform if the modulating signal has a frequency of 10 kHz.

Solution: Again, we use Carson's approximation. That is,

$$BW \approx 2(\Delta f + f_m)$$

Since the information, $s(t)$, has unit amplitude, the maximum frequency deviation, Δf, is given by k_f, or 100 Hz. f_m is simply 10 kHz, the frequency of the modulating signal. Therefore,

$$BW \approx 2(100 \ Hz + 10 \ kHz) = 20{,}200 \ Hz$$

Since f_m is much greater than Δf, this is a narrowband FM signal. The bandwidth necessary to transmit the same information waveform using DSB AM would be 20 kHz, which is approximately the same as the bandwidth of this FM wave.

Example 4.7

An FM waveform is described by

$$\lambda_{fm}(t) = 10\cos[2\times10^7\pi t + 20\cos000\pi t]$$

Find the approximate bandwidth of this waveform.

Solution: f_m is equal to 500 Hz. In order to compute Δf, we first find the instantaneous frequency, $f_i(t)$.

$$f_i(t) = \frac{1}{2\pi}\frac{d}{dt}(2\times10^7\pi t + 20\cos1000\pi t)$$

$$= 10^7 - 10,000\sin1000\pi t$$

The maximum frequency deviation is the maximum value of $10,000\sin1000\,\pi t$, which is simply 10 kHz. The approximate bandwidth is therefore given by

$$BW \approx 2(10,000 + 500) = 21\ kHz$$

This is clearly a wideband FM waveform since Δf is much greater than f_m.

4.1.4 Frequency Shift Keying (FSK)

In analog communication, FM is often used in place of AM because of resulting performance improvement in the presence of additive noise. We will find additional reasons to use FM for digital communication. For example, we will present some extremely simple implementations for the modulators and demodulators.

We start with a system which generates the NRZ baseband signal [See Section 2.3.1]. Let us assume that the baseband signal is composed of piecewise-constant segments. That is, a binary one is transmitted with a square pulse of voltage equal to V_1 and a binary zero is transmitted with a pulse of V_0 volts, as shown in Fig. 4.12.

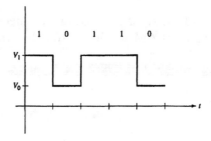

Figure 4.12 - Baseband Digital Signal

When using *frequency modulation*, the frequency of the carrier varies in accordance with the information baseband signal. This contrasts with amplitude modulation where the amplitude of the carrier is controlled by the baseband signal. Since we assume that the baseband signal takes on only one of two values, the frequency of the modulated waveform also takes on one of two values, and the modulation process can be thought of as a *keying* operation. The resulting FM waveform is known as *frequency shift keying* (FSK).

If we denote that baseband signal as $s_b(t)$, the resulting FM waveform is

$$\lambda_{fm}(t) = A\cos\left[2\pi f_c t + 2\pi k_f \int s_b(t)dt\right]$$

This equation makes things look more complicated than they are. $\lambda_{fm}(t)$ reduces to a signal that jumps between two frequencies. When a binary one is being sent, the instantaneous frequency is $f_c + k_f V_1$ and when a binary zero is sent, the instantaneous frequency is $f_c + k_f V_0$. We'll call these f_1 and f_0 in the following.

Figure 4.13 shows a representative binary FSK waveform for the binary sequence 1010.

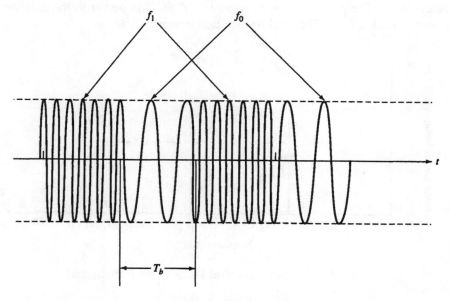

Figure 4.13 - Representative BFSK Waveform

FSK Spectrum

The FSK waveform of Fig. 4.13 can be considered as the superposition of two amplitude-shift-keyed signals. One of these is the ASK signal resulting from modulating a carrier of frequency f_1 using ON-OFF keying from the baseband signal. The other results from amplitude shift keying a carrier of frequency f_0 with the *complement* of the baseband signal. Figure 4.14 shows the system that would accomplish this.

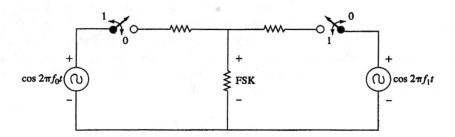

Figure 4.14 - FSK as Superposition of Two ASK Waveforms

Now that we have decomposed the FSK waveform into two ASK waves, we can borrow the ASK results of Section 3.2.5 to find the spectrum of the FSK signal. Each of the two components has a frequency spectrum which is of the general shape of $(\sin^2 f)/f^2$ shifted to the carrier frequency. Thus, the power spectrum of the FSK signal is as shown in Fig. 4.15. We are assuming the zeros and ones are equally probable (i.e., each occurs half of the time on the average). The power of each carrier is $A^2/8$. The power in the sidebands around each carrier is also $A^2/8$. The total transmitted power is $A^2/2$.

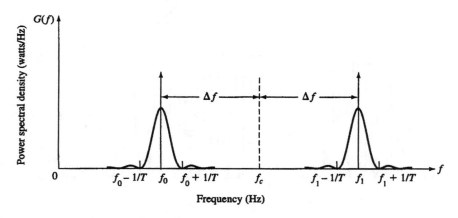

Figure 4.15 - Power Spectral Density for FSK Signal

Figure 4.16 shows a modified version of Fig. 4.15, where we assume that the spacing between the two carrier frequency is equal to the bit transmission rate. This is known as *orthogonal tone spacing*. Its importance will become clear when we view the performance of FSK systems.

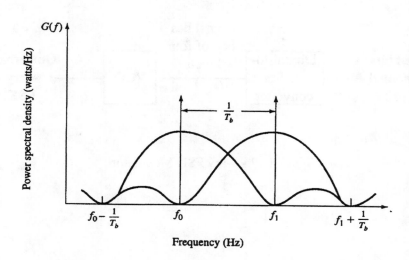

Figure 4.16 - FSK Power Spectral Density for Orthogonal Tone Spacing

The *bandwidth* of the FSK waveform is approximately equal to $f_1 - f_0$, $+ 2R_b$, where we are defining bandwidth out to the first zero of each lobe of the power spectral density. This is often called the *nominal* bandwidth. In the case of orthogonal tone spacing, the nominal bandwidth is $3R_b$. This contrasts with $2R_b$ for the ASK waveform.

4.1.5. MFSK

Until now, we have been discussing *binary* FSK. Let us suppose that in a particular binary communication system (e.g., baseband, ASK, or FSK) the bandwidth of the signal is too large for the channel. One way to reduce the signal bandwidth is to lower the bit rate. However, this may be unacceptable if information must be transmitted at a specified rate.

A second option is to leave the realm of binary communications by combining bits into groupings. For example, if we combine bits in pairs, each pair can take on one of four possible values. These four possible values can be transmitted using FSK with four different frequencies. This is known as *4-ary* (or *quaternary*) FSK. The generator can be visualized as a D/A converter operating on pairs of bits. The D/A converter is followed by a VCO (voltage-controlled oscillator or voltage-to-frequency converter), as shown in Fig. 4.17. The output of the D/A takes on one of four values, zero, one, two or three. This drives the VCO to one of four frequencies.

The rate at which the frequencies change is one-half of the original bit rate. The power spectral density of an FSK system with four frequencies is the sum of the four waveforms shown in Fig. 4.18. We are illustrating this for orthogonal tone spacing, where in this case, the frequencies are separated by $R_b/2$. Note that the bandwidth is $5R_b/2$, as compared to the $3R_b$ required for binary FSK.

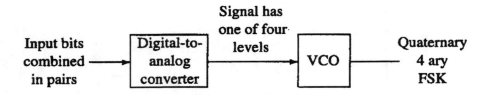

Figure 4.17 - 4-ary FSK Modulator

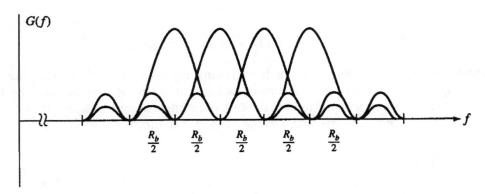

Figure 4.18 - Power Spectral Density of 4-ary Orthogonal FSK

4.2. MODULATORS

Now that we know what frequency modulation is, we must learn how to build a modulator. This is the subject of the current section. Then the next section explores demodulators, followed by a detailed analysis of the performance of these devices.

4.2.1. Analog Modulators

We have shown that FM waveforms are approximately bandlimited to a range of frequencies around f_c, the carrier frequency. The first two criteria of a useful modulation system are therefore satisfied. We can transmit efficiently by choosing f_c in the proper range, and we can frequency multiplex separate signals. We simply need to make sure that the adjacent carrier frequencies are separated by a sufficient amount such that the transforms of the FM waveforms do not overlap in frequency.

We must still show that the original signal, $s(t)$, can be recovered from the angle-modulated waveform, and also to show that modulators and demodulators can be constructed simply.

We start by re-examining narrowband FM. The waveform is approximately expressed by Eq. (4.26)

$$\lambda_{fm}(t) = A\cos2\pi f_c t - 2\pi Ag(t)k_s\sin2\pi f_c t \qquad (4.26)$$

Recall that $g(t)$ is defined as the integral of the information signal, $s(t)$. The block diagram of Fig. 4.19 directly implements the expression of Eq. (4.26).

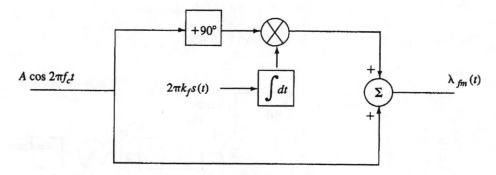

Figure 4.19 - Narrowband FM Modulator

The instantaneous frequency of the output of the system of Fig. 4.19 is

$$f_i(t) = f_c + k_s s(t)$$

The reason this represents a narrowband signal is that the maximum value of $k_s s(t)$ [i.e., the frequency deviation] is small compared to the frequencies present in $s(t)$. Suppose we put the output narrowband FM waveform through a non-linear device that multiplies all frequencies by a constant, C. The resulting waveform has an instantaneous frequency given by

$$f_i(t) = Cf_c + Ck_s s(t)$$

The frequency deviation of this new waveform is C times that of the old, while the rate at which the instantaneous frequency varies has not changed. This is shown in Fig. 4.20. Therefore, for high enough values of C, frequency multiplication changes narrowband FM into wideband FM. It also moves the carrier frequency, but the carrier has no effect upon whether an FM wave is narrowband or wideband. If the bandwidth of the FM wave is significantly larger than $2f_m$, the signal is wideband. If the new higher carrier is not desired, we can shift (heterodyne) the result to any part of the axis without affecting the bandwidth.

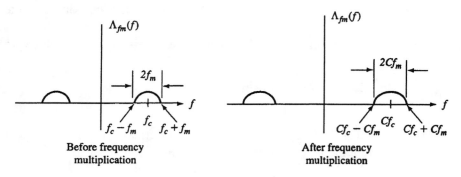

Figure 4.20 - Frequency Multiplication

The resulting FM modulator is shown in Fig. 4.21.

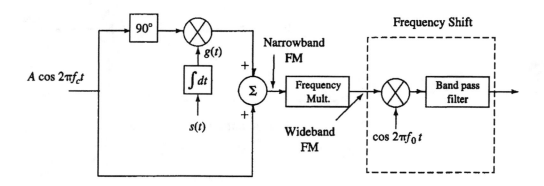

Figure 4.21 - Wideband FM Modulator

You may be wondering how to build a frequency multiplier. Any nonlinear device acts as a frequency multiplier. One example is a squaring device. If we assume that the input to the squarer is an FM wave, the output of the squarer is given by

$$\left[A\cos2\pi\left(f_c t + k_f g(t)\right)\right]^2$$

$$= \frac{A^2}{2}\left[1 + \cos4\pi\left(f_c t + k_f g(t)\right)\right]$$

The bandpass filter eliminates the dc term. The instantaneous frequency of the resulting waveform is

$$f_i(t) = 2f_c + 2k_f s(t)$$

so the squaring device acts as a frequency doubler.

There are more direct ways to generate wideband FM. An electronic oscillator has an output frequency which depends on the size of energy storage devices. There are a wide variety of oscillators whose frequencies depend on a particular capacitor value. By varying the capacitor value, the frequency of oscillation varies. If the capacitor variations are controlled by $s(t)$, the result is an FM waveform. One traditional way to accomplish this is with *varactor diodes*. When this type of diode is back biased, it acts as a capacitance with a value that depends on the magnitude of the back biasing voltage. A similar operation can be performed digitally. Timer chips can be configured so an external voltage controls the pulse rate. The output pulse train is bandpass filtered to extract the fundamental.

The FM modulator is essentially a *voltage controlled oscillator* (VCO, also known as voltage-to-frequency converter). VCOs are available as integrated circuits. The 566 is a representative device which is used for frequencies below 1 MHz. The block diagram of the 566 VCO is shown in Fig. 4.22.

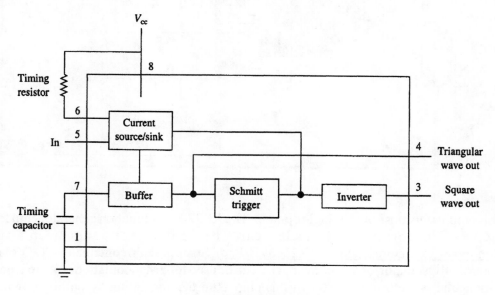

Figure 4.22 - 566 VCO

The timing capacitor receives a constant charging or discharging current from the *current source/sink*. The voltage on the *modulation input* controls the amount of current produced by the current source/sink. The timing capacitor output forms the input to a Schmitt trigger. When the Schmitt trigger output changes level, the current source/sink reverses its operating mode. Thus the slope of the voltage ramp on the timing capacitor is proportional to the output frequency of the IC.

Simulation Example

Open the file called *FM Modulator* in the Tina software on the CD included with this text. See Appendix F if this is the first time you are using Tina. You screen should look like this:

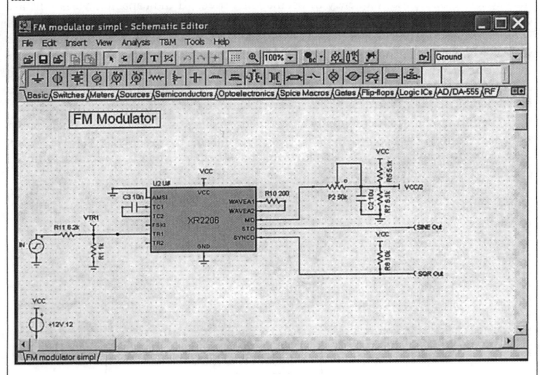

This modulator consists of a single op-amp. The XR2206 is a function generator chip from EXAR Corporation. Details can be found at the web site: http://www.exar.com/products/XR2206v103.pdf. Since the chip contains a VCO (Voltage Controlled Oscillator, of V/F converter), it can be used for FM modulation. We are feeding a triangular waveform into TR1 of this chip. Run the simulation by pulling down the *Analysis* menu and selecting *Transient* and you will get plots of the input and the output. (If the curves display on top of each other, simply pull down the *View* menu and select *Separate Curves*.)

4.2.2. Digital (FSK) Modulators

Digital FM can be considered to be a special case of analog FM where the information signal takes on one of two discrete values (for binary). For this reason, we can consider a digital modulator to be no different from an analog modulator, where the input is a digital baseband signal. However, a baseband signal which jumps instantaneously between two values is not frequency limited.

Figure 4.23(a) illustrates a simple modulator consisting of two oscillators and a switch (key). If M-ary FSK were required, you just increase the number of oscillators.

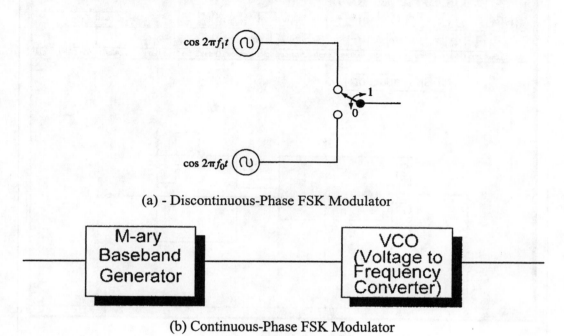

(a) - Discontinuous-Phase FSK Modulator

(b) Continuous-Phase FSK Modulator

Figure 4.23 - FSK Modulators

So far we have been presenting FSK as if the frequency abruptly changes between the two possible values. That is, taking the baseband modulator approach, we have been assuming that the baseband is composed of perfect square pulses. This produces an FSK signal with discontinuous phase. That is, at the moment the switch moves from one position to the other, we have very little control over matching the phase of the sinusoid just before switching with that of the (different frequency) sinusoid just after switching.

In practice, the pulses are often shaped (conditioned) prior to modulating the carrier. The frequency transitions are therefore smooth rather than instantaneous. If the modulator of Fig. 4.23(b) is used instead of that of Fig. 4.23(a), the phase will be continuous. The baseband generator can be binary or M-ary.

Simulation Example

Open the file called *FSK Modulator* in the Tina software on the CD included with this text. You screen should look like the following:

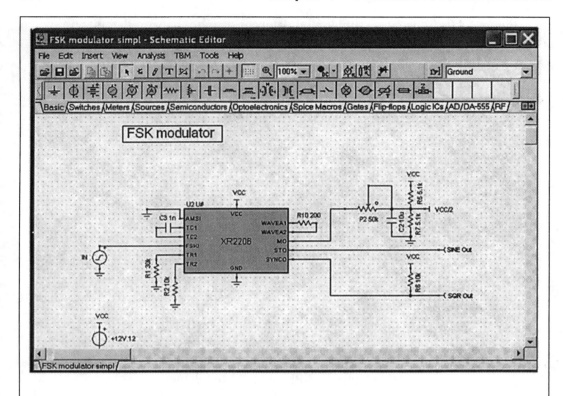

The FSK modulator uses the same IC (the XR2206) as the FM modulator. Note that we use an input of a square wave so that only two frequencies are produced at the output. Run the simulation by pulling down the *Analysis* menu and selecting *Transient*. You will see the FSK waveform produced.

4.3 DEMODULATORS

We separate our study of demodulators into the analog and digital. Within these two groupings, we will have both coherent and incoherent detectors. Just as in the case of amplitude modulation, the incoherent detectors will be easier to build, but in the next section we will find that they do not perform as well as the coherent receivers.

4.3.1. Analog Demodulators

The challenge of analog FM demodulation can be stated as follows. Given $\lambda_{fm}(t)$ as in Eq. (4.27), recover the information signal, $s(t)$.

$$\lambda_{fm}(t) = A\cos\left(2\pi f_c t + k_f \int_{-\infty}^{t} s(t)\ dt \right) \qquad (4.27)$$

Demodulators fall into one of two broad classifications. The first of these employ discriminators–devices which discriminate one frequency from another by converting frequency variations into amplitude variations. The resulting amplitude changes are detected just as was done in AM. The second category of demodulator uses a phase lock loop to match a local oscillator to the modulated carrier frequency.

Discriminator Detector

If you think back to your calculus classes, you will find that you already know one way of translating frequency variations into amplitude variations. Time differentiation of a sinusoid multiplies the sinusoid by its instantaneous frequency. Therefore we start by investigating the derivative of the FM waveform.

$$\frac{d\lambda}{dt} = -2\pi A[f_c + k_f s(t)]\sin 2\pi \left(f_c t + k_f \int_0^t s(\tau)d\tau \right) \tag{4.28}$$

This derivative is sketched in Fig. 4.25 for a representative $s(t)$.

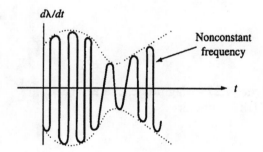

Figure 4.25 - Derivative of FM Waveform

If we now assume that the frequency of the sinusoid in Eq. (4.28) is always much greater than f_m (a reasonable assumption in real life), this "carrier" term fills in the area between the amplitude and its mirror image. We have exaggerated the carrier in sketching Fig. 4.25. Actually, the area between the upper and lower outline should be shaded due to the extremely high carrier frequency. Even though the carrier frequency is not constant [i.e., it's equal to $f_c+k_f s(t)$], the envelope of the waveform is still clearly defined by

$$|2\pi A[f_c + k_f s(t)]| \tag{4.29}$$

The slight variation in the frequency of the carrier would not even be noticed by an envelope detector.

In a practical system, f_c is much larger that $k_f s(t)$. Therefore the quantity in brackets in Eq. (4.28) is normally non-negative, and we can eliminate the absolute value sign in Eq. (4.29). The result is a constant added to a scaled version of the original signal. A differentiator followed by an envelope detector can therefore be used to recover $s(t)$ from the FM waveform. This is shown in Fig. 4.26.

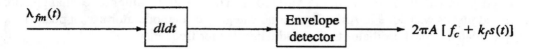

Figure 4.26 - FM Demodulator

We note that the above analysis did not make any assumptions about the size of k_f. The demodulator of Fig. 4.26 therefore works for either wideband or narrowband FM signals.

The appearance of the envelope detector should give one clue that AM is somehow occurring in this system. Indeed, this is the case as the following analysis reveals.

The transfer function of a differentiator is given by

$$H(f) = 2\pi jf \qquad (4.30)$$

The magnitude characteristic is shown in Fig. 4.27.

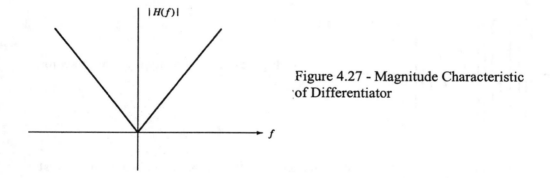

Figure 4.27 - Magnitude Characteristic of Differentiator

The magnitude of the output of the differentiator is linearly proportional to the frequency of its input. The differentiator therefore changes FM into AM. When a differentiator is used in this manner, it is called a *discriminator*.

A differentiator is not the only system that can be used as a discriminator. Any system which has a transfer function magnitude that is approximately linear with frequency will change FM into AM. The system characteristic need only be linear over the range of frequencies of the system's input. Even a practical (non-ideal) bandpass filter can work as a discriminator if operated over a limited frequency range relative to the filter bandwidth. This is shown in Fig. 4.28.

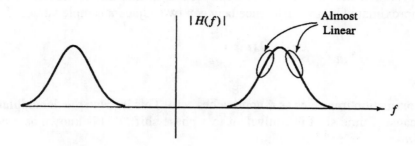

Figure 4.28 - Bandpass Filter as a Discriminator

We can improve the linearity of the bandpass filter discriminator in a manner similar to that applied to the balanced AM modulator. We subtract the characteristic from a shifted version of itself, as shown in Fig. 4.29. That is, we take the difference between the output of two bandpass filters with center frequencies separated as shown.

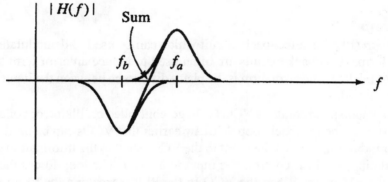

Figure 4.29 - Improved Linearity Discriminator

If properly designed, this differencing process effectively eliminates even powers in the power series expansion of the filter rolloff function. The circuit of Fig. 4.30 accomplishes this. The tuned circuit consisting of the upper half of the output winding of the transformer and C_1 is tuned to f_a, and the tuned circuit consisting of the other half of the output winding and C_2 is tuned to f_b. This circuit is known as a *slope demodulator* since it is using the sloped portion of the filter characteristic as part of the demodulation process.

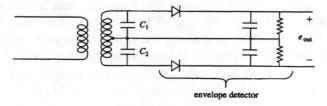

Figure 4.30 - Slope Demodulator

If we return to the mathematics, we can see one other related approach. The derivative can be approximated by the difference between two adjacent sample values. That is,

$$t_o \frac{d\lambda}{dt} \approx \lambda(t) - \lambda(t-t_o) \tag{4.31}$$

The approximation improves as t_o approaches zero. This leads to the demodulator of Fig. 4.31. Because a time shift is equivalent to a phase shift, this is known as a *phase shift demodulator*.

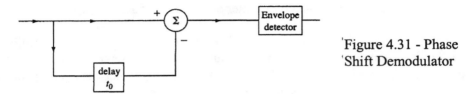

Figure 4.31 - Phase Shift Demodulator

Phase Lock Loop

The *phase lock loop* (PLL) is a feedback circuit which can be used to demodulate angle modulated waveforms. Feedback circuits are often used to reduce an error term toward zero. In the case of the PLL, the error term is the phase difference between the input signal and a reference sinusoid.

The phase lock loop incorporates a VCO (voltage-controlled oscillator, or voltage-to-frequency converter) in the feedback loop. We saw earlier that VCOs can be used as FM modulators. For modulating, we set the input to the VCO equal to the information signal, $s(t)$. For demodulating, the PLL controls the input to a VCO. The loop forces the VCO output to match the FM wave. When the VCO in the PLL is tracking the received FM wave, the input to that VCO tracks the information, $s(t)$.

Before formalizing this discussion, we must learn a little about PLLs. This important feedback system finds application both in analog and digital communication. A typical loop configuration is shown in Fig. 4.32. The loop compares the phase of the input signal to the phase of the VCO signal. If the phase difference between these two signals is anything other than zero, the output frequency of the VCO is adjusted in a manner which forces the difference toward zero.

The output of the phase comparator, $v_o(t)$, forms the input to the VCO. The output of the VCO is an FM waveform with an instantaneous frequency that is proportional to the phase difference between the input and the VCO output. If, for example, the input frequency is higher than the VCO frequency, the phase difference between the two increases linearly with time causing the VCO frequency to increase until it matches that of the input. Thus, the VCO attempts to follow the input frequency. Since the frequency of the VCO output is proportional to the voltage at its input, it is this input voltage which

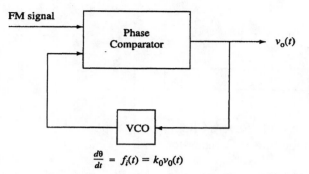

FM signal

Phase
Comparator

$v_o(t)$

VCO

$$\frac{d\theta}{dt} = f_i(t) = k_0 v_0(t)$$

Figure 4.32 - Phase Lock Loop

also tries to follow the frequency of the loop input signal. Thus, monitoring the input to the VCO yields a demodulated version of the FM waveform.

Now that we have an intuitive feel for the PLL operation, we need to get more sophisticated and take a mathematical approach. We begin by examining the phase comparator which is a part of the PLL. The simplest method of phase comparison consists of a product operation followed by a lowpass filter, as shown in Fig. 4.33.

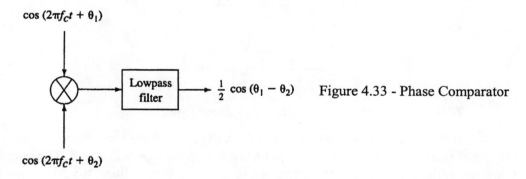

$\cos(2\pi f_c t + \theta_1)$

Lowpass
filter

$\frac{1}{2}\cos(\theta_1 - \theta_2)$ Figure 4.33 - Phase Comparator

$\cos(2\pi f_c t + \theta_2)$

Suppose that the two inputs to the comparator are $\cos(2\pi f_c t + \theta_1)$ and $\cos(2\pi f_c t + \theta_2)$. The output of the lowpass filter is then $\frac{1}{2}\cos(\theta_1 - \theta_2)$. This output is zero when the phase difference is 90°. Since the phase comparator generates the loop error function (the quantity that changes the frequency of the VCO), the loop is in lock when the input and the VCO output are at the same frequency and 90° out of phase.

It is sometimes desirable to have a comparator output which varies *linearly* with the phase difference rather than sinusoidally. This provides for more uniform adjustment as the phase difference approaches 90°. If the input sinusoidal signals are severely clipped (or amplified to the point where they severely saturate the electronics), they can be thought of as square waveforms. A representative situation is illustrated in Fig. 4.34.

The product of the two square waves is now averaged by the low pass filter. This average is proportional to the fraction of time that the square waves are equal. Therefore, the output amplitude is linearly related to the phase difference.

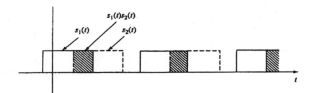

Figure 4.34 - Severe Clipping
in the Linear Comparator

Using the multiplier as a phase comparator, we redraw the PLL with FM input and a general *loop filter* in place of the low pass filter. The purpose of this filter is to allow the loop to track the input phase but not be too sensitive to noise variations. This is shown in Fig. 4.35.

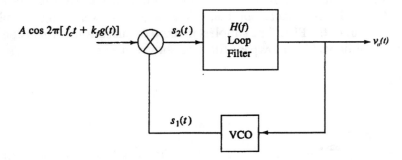

Figure 4.35 - PLL With Multiplier as Phase Comparator

When the loop is in lock, the output of the VCO follows the input FM wave, but is 90° out of phase with it. The VCO output has an amplitude we will call B. The output frequency varies from the nominal setting of f_c by an amount proportional to $v_o(t)$, the input to the VCO. We shall call this VCO proportionality constant, k_o. Then, $s_1(t)$ is given by

$$s_1(t) = B\sin 2\pi\left(f_c t + k_o\int_0^t v_o(\tau)d\tau\right) \qquad (4.32)$$

It is not enough to know the lock condition of the loop. It is also important to know the time behavior so that we can determine if the loop is capable of tracking a time-varying input. One way to analyze the loop is to find its step response. That is, suppose the loop is in lock and the input phase changes instantaneously.

We shall perform this transient analysis for the *first-order loop*, where the loop filter is a lowpass filter. The signal at the output of the multiplier is

$$s_2(t) = ABcosw\pi[f_ct+k_fg(t)]sin2\pi\left[f_ct + k_o\int_0^t v_o(\tau)d\tau\right]$$

$$= \frac{AB}{2}sin2\pi\left[k_fg(t) - k_o\int_0^t v_o(\tau)d\tau\right] \tag{4.33}$$

$$+ \text{ higher-frequency terms}$$

We don't bother writing the higher order terms in Eq. (4.33) since they will not get through the lowpass filter.

Let us define the following two phase factors:

$$\theta_{fm}(t) = 2\pi k_f g(t)$$

$$\theta_o(t) = 2\pi k_o\int_0^t v_o(\tau)d\tau \tag{4.34}$$

In terms of these factors, the output of the lowpass filter is

$$v_o(t) = \frac{ABsin[\theta_{fm}(t) - \theta_o(t)]}{2} \tag{4.35}$$

If a linear phase detector is used (or if the phase factors are small enough to approximate the sine by its angle), Eq. (4.35) becomes

$$v_o(t) = \frac{AB[\theta_{fm}(t) - \theta_o(t)]}{2} \tag{4.36}$$

To find the transient loop response, we take the derivative of both sides of Eq. (4.36) to get

$$\frac{dv_o}{dt} = \frac{AB}{2}\left[\frac{d\theta_{fm}}{dt} - \frac{d\theta_o}{dt}\right]$$

$$= AB[k_f s(t) - k_o v_o(t)] \tag{4.37}$$

Finally, the differential equation is given by

$$\frac{dv_o}{dt} + \pi k_o ABv_o(t) = \pi k_f ABs(t) \tag{4.38}$$

The steady-state solution of Eq. (4.38) is found by setting the derivative equal to zero. Then

$$v_o(t) = \frac{k_f}{k_o}s(t) \tag{4.39}$$

The transient response of this first-order differential equation is an exponential. The time constant of the transient response is $1/\pi k_o AB$. This determines the response time of the loop.

Integrated Circuit PLL

Phase lock loops are available as integrated circuits. A typical device is the LM565. The block diagram is shown in Fig. 4.36 configured for frequency demodulation. The first page of the data sheet is shown as Fig. 4.37.

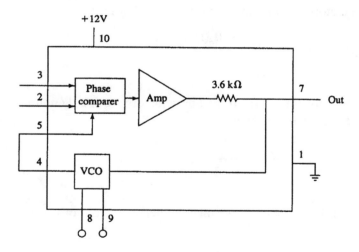

Figure 4.36 - LM565 IC PLL

The amplifier coupled with an external resistor and capacitor form a lowpass filter. The lowpass filter output is the demodulated output of the circuit. The 565 PLL operates in a range of frequencies up to 500 kHz. Higher frequency operation requires a more complex IC. For example, the NE564 is capable of operation up to 50 MHz.

In selecting the proper PLL for a particular application, the engineer encounters a variety of descriptive terms. We now define these terms, for we will be using them again when we use PLLs in digital circuitry.

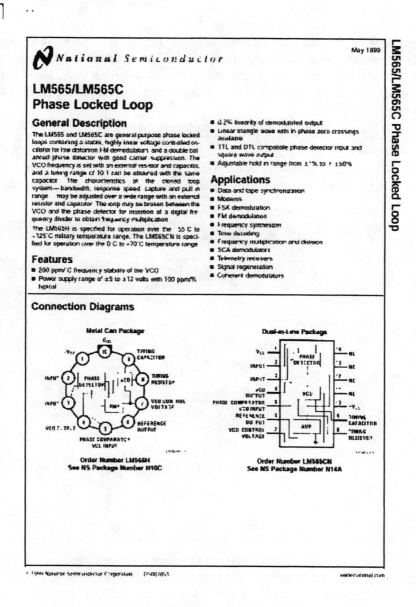

Figure 4.37 - LM565 PLL Data Sheet

The *free-running frequency*, or center frequency is the frequency of the VCO without any input.

The *lock range* is the range of frequencies over which the loop is capable of staying in lock. It is normally centered around the free-running frequency. This is similar to the *tracking range* except for a factor of two. The lock range is the total frequency change between the highest and lowest operating frequency. The tracking range is the maximum deviation permitted from the free-running frequency.

The *capture range* is the range of frequencies over which the loop is capable of locking in. It is normally smaller than the lock range. That is, once the loop is in lock, the frequency can deviate a certain maximum amount without losing lock. However, if the system is not yet in the lock condition and is searching for the correct frequency, the allowable deviation is less.

The *lock-up time* is the length of the transient response.

Simulation Example

Open the file called *FM Deodulator* in the Tina software on the CD included with this text. See Appendix F if this is the first time you are using Tina. You screen should look like this:

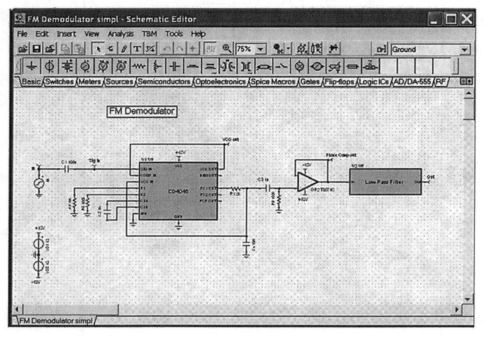

The input to this demodulator is a user defined function. You can double click on *IN* and then click on the user defined function portion of the window, then on "...". You will see the equation of this source. The carrier is 100 kHz and the modulating signal is a sinusoid at 5 kHz. This demodulator consists of one op-amp, the CD4046 and a sub-circuit which acts as a lowpass filter. The CD4046 chip is a phase locked loop. Details can be found at http://www.fairchildsemi.com/ds/CD/CD4046BC.pdf. The op-amp at the output of the phase locked loop is a buffer where the output waveform is the same as the input waveform. Run the transient analysis by pulling down the *Analysis* menu and selecting *Transient*. You will see the FM input waveform, several intermediate waveforms, and the output. The output contains a transient response since the frequencies are riding around a large carrier. You can see the steady state response by increasing the simulation time, but this will make the simulation run slower.

4.3.2. Digital Demodulators

FSK detectors can be either coherent (keep track of phase) or incoherent. We begin with the coherent matched filter detector of Fig. 4.38.

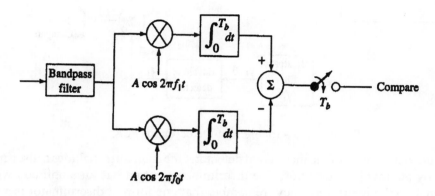

Figure 4.38 - Matched Filter Detector for FSK

As we have done throughout this text, we have implemented the detector using correlators instead of matched filters. We should note, however, that the matched filter for a sinusoidal burst is a filter whose impulse response is a sinusoidal burst. This is approximated by a bandpass filter tuned to the frequency of the sinusoid.

The coherent detector requires reconstruction of the two separate carriers at the receiver. Please recall that FSK can be viewed as a superposition of two ASK waveforms. Two bandpass filters or two phase lock loops can be used to reconstruct sinusoids for use in the detector of Fig. 4.38. However, as in the case of ASK, the carrier is not always present. For example, during periods when a one is being transmitted, the sinusoid at f_0 is not being transmitted. Likewise, when zeros are transmitted, the carrier at f_1 is "shut off". During such periods, the phase lock loop can drift or, if bandpass filters are used, a transient will be present. These degrading effects get worse as the number of consecutive ones or zeros increases. This situation is known as *static data*.

Since the coherent detector suffers from the problems we have just discussed, we are motivated to look for another approach. We now consider the incoherent detector. One simple implementation of the incoherent detector is shown in Fig. 4.39. The detector contains two bandpass filters, one tuned to each of the two frequencies used to send zeros and ones. The output of each filter forms the input to an envelope detector and the envelope detector output is baseband detected using an integrate and dump operation or a single-sample detector (the single-sample detector is illustrated in the figure). In intuitive terms, this detector is simply evaluating which of two possible sinusoids is stronger at the receiver. If we take the difference of the outputs of the two envelope detectors, the result is bipolar baseband (please take a moment to verify this).

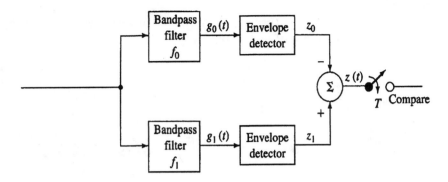

Figure 4.39 - Incoherent Detector for FSK

There are other forms of the incoherent detector. One such form includes a discriminator followed by an envelope detector. The discriminator output has an amplitude which is proportional to the input frequency (remember that one form of discrminator is a simple differentiator, and when a sinusoid is differentiated, we multiply by the frequency). The output of the envelope detector is therefore a baseband signal consisting of two different positive levels. This signal can be detected using any of the baseband techniques.

Simulation Example

Open the file called ***FSK Demodulator*** in the Tina software on the CD included with this text. Your screen should look like this:

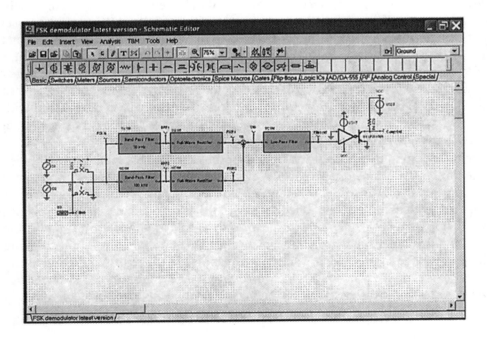

Note that we have used the software to simulate the incoherent detector of Fig. 4.39, but instead of envelope detectors we are using rectifier detectors. Since the output of the lowpass filter is a smoothed version of the baseband waveform, we feed it into a comparator to turn it into a square-type wave. We could simulate the entire detector by sampling the filter output and comparing that sample to a threshold. However, to illustrate the detector operation, we wish to reproduce the baseband waveform. That is why we are using the comparator.

Run the transient analysis to produce six waveforms (the program is performing many calculations, so depending on the speed of your computer, this could take a while to run). We plot the FSK time function, the outputs of the two bandpass filters, the output of the lowpass filter, and the overall input and output of the system.

Comp Out is the comparator output, and *Mod* is the baseband waveform used at the input to create the FSK signal. Ideally, these should be identical. Note that the output contains a time shift due to the fact that we are simulating practical filters.

Please note that we have set U3 to be a 2 kHz square waveform. Thus, our FSK waveform has a bit rate of 4 kbps. We have chosen our two frequencies to be 30 kHz and 100 kHz. This is a much wider separation that would be used in most practical situations. Indeed, for orthogonal tone spacing, we would use 30 kHz and 34 kHz. We used this wide separation for illustrative purposes only so we could relax the constraints on the bandpass filter rolloff.

MFSK

The MFSK detector is an extension of the binary detectors of Figs. 4.38 and 4.39. In the coherent matched filter detector case, we increase the number of correlators to match the number of frequencies, and the outputs of the various correlators are compared to each other in order to pick the largest. For the incoherent detector, we increase the number of filters and choose the largest output. This is shown for 4-ary FSK in Fig. 4.40.

4.4 PERFORMANCE

Just as we did in chapters two and three, we now examine the performance of the various detectors introduced in the previous section. The major subdivisions are between analog and digital, and between coherent and incoherent detectors.

4.4.1. Analog

Since FM is a non-linear form of modulation, analysis of noisy systems proves much more difficult than in the AM case. We have to be content with some approximations and generalizations.

Let us first assume that a noise-free FM waveform is transmitted.

$$\lambda_{fm}(t) = A\cos 2\pi [f_c t + k_f g(t)] \tag{4.40}$$

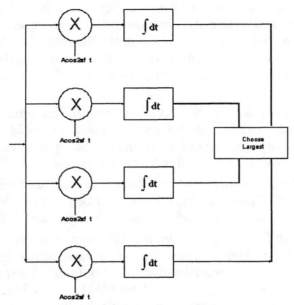

(a) Coherent Detector

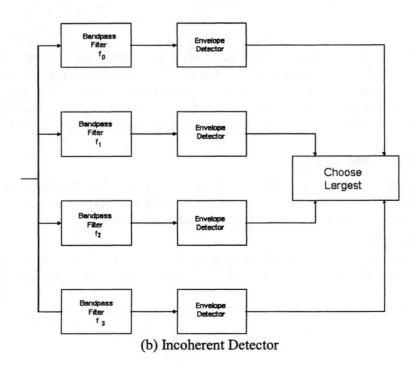

(b) Incoherent Detector

Figure 4.40 - Detectors for MFSK

Recall that $g(t)$ is the time integral of the information signal. Noise is added to this FM signal along the path between the transmitter and receiver. The additive noise is assumed to be white with a power of N_o watts/Hz. The received signal, $r(t)$, which is a combination of signal and noise, enters the receiver antenna as shown in Fig. 4.41. Although the figure indicates that the noise is added prior to the receiver input, this model can also be used for the case of front end receiver noise.

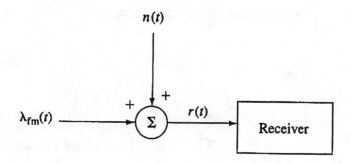

Figure 4.41 - FM Plus Noise into Receiver

FM Receivers

An FM waveform carries its information in the form of frequency. The amplitude of the FM wave is constant. Another way to view this is to assume the signal information is contained in the zero crossings of the wave. The FM waveform can be clipped at a low level (i.e., effectively changed from a continuous waveform to a square-type waveform) without loss of signal information. Additive noise can significantly affect the amplitude of the FM wave, but it perturbs the zero crossing to a lesser degree. Receivers therefore often clip, or limit, the amplitude of the received waveform prior to frequency detection. This provides a constant amplitude waveform as input to the discriminator. One effect of clipping is to introduce higher harmonic terms. A *post-detection* lowpass filter is used to reject these higher harmonics.

A block diagram of the simplified FM receiver is shown in Fig. 4.42. This receiver does not include the heterodyning and *if* strip since these do not change the signal to noise ratio.

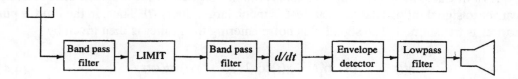

Figure 4.42 - Simplified FM Receiver

Signal to Noise Ratio

The analysis of noise in FM systems is facilitated by a phasor presentation. Figure 4.43 illustrates a phasor diagram of the received FM signal plus additive noise.

The signal phasor has length A and angle $2\pi k_f g(t)$. The phasor wiggles around according to $2\pi k_f g(t)$, while its amplitude remains constant. The noise phasor can be added to the signal if we expand the noise in quadrature components (see Appendix C).

$$n(t) = x(t)\cos 2\pi f_c t - y(t)\sin 2\pi f_c t \tag{4.41}$$

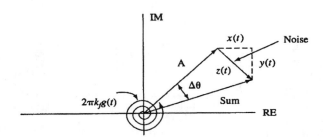

Figure 4.43 - Phasor Diagram of FM Signal Plus Noise

$x(t)$ and $y(t)$ are low-frequency random processes. It is the bandpass filter in the receiver which allows us to use this bandlimited quadrature representation for the noise.

The angle of the resultant phasor contains all of the useful information. The length of the vector can be disregarded, since the receiver's limiter eliminates any amplitude variations.

As long as the magnitude of the noise vector does not exceed A, the maximum angular difference between the resultant and the signal vector, $\Delta\theta$, is 90°. The limiting value of 90° would only occur when the noise amplitude reaches A, and the vectors are colinear and 180° out of phase. As the noise gets smaller, the angular variations around the signal vector are limited to decreasing amounts less than 90°.

The angle of the signal vector is $2\pi k_f g(t)$. For wideband FM, this angle can be considerably larger than 360°. Thus, the proportional degradation caused by the noise can be made arbitrarily small by increasing k_f. Try to picture the vector of Fig. 4.43 wildly winding and unwinding around the original as $g(t)$ varies positively and negatively.

As a first step in quantifying these observations, let's analyze the special case of a large carrier to signal ratio. That is, assume A is much larger than $g(t)$. Thus, in the limit, a pure carrier is transmitted. The signal plus noise entering the limiter is then given by

$$r(t) = A\cos 2\pi f_c t + x(t)\cos 2\pi f_c t - y(t)\sin 2\pi f_c t \tag{4.42}$$

To trace this through the limiter, we combine terms and then eliminate the amplitude variations. We rewrite $r(t)$ as

$$r(t) = \sqrt{[A + x(t)]^2 + y^2(t)} \; \cos\left[2\pi f_c t - \tan^{-1}\left(\frac{y(t)}{A + x(t)}\right)\right]$$

(4.43)

Since the output of the limiter has constant amplitude, we eliminate the amplitude term in Eq. (4.43). The limiter output can then be written as

$$\cos\left[2\pi f_c t - \tan^{-1}\left(\frac{y(t)}{A + x(t)}\right)\right] + higher \; harmonics$$

(4.44)

The higher harmonics are rejected by the bandpass filter. The input to the discriminator is therefore the first term in Eq. (4.44). The output of the discriminator-envelope detector combination is given by Eq. (4.45). We have taken the derivative of the inverse tangent and we have also eliminated the constant term arising from the carrier frequency.

$$\frac{d\theta}{dt} = \frac{[x(t) + A]dy/dt - y(t)dx/dt}{y^2(t) + [x(t) + A]^2}$$

(4.45)

The signal of Eq. (4.45) enters the final lowpass filter of Fig. 4.42. To find the power at the output of this filter, we must examine the frequency content of Eq. (4.45). We know the power spectral densities of $x(t)$ and $y(t)$ and differentiation operations can be modeled as linear systems with $H(f)=j2\pi f$. However, before doing the mathematics we can considerably simplify Eq. (4.45) under the assumption of high carrier to noise ratios. That is, we assume that A is much larger than $x(t)$ and $y(t)$. Equation (4.45) therefore becomes

$$\frac{d\theta}{dt} \approx \frac{dy/dt}{A}$$

(4.46)

Our goal is to find the output noise power, so we investigate the power spectral density of the output noise as given in Eq. (4.46). Note that the signal is not present since we assumed that the received waveform was a pure carrier plus noise. The power spectral density of $y(t)$ is N_o for frequencies between $-BW/2$ and $+BW/2$, where BW is the bandwidth of the receiver input filter. Even though the bandwidth (BW) can be many times f_m for wideband FM, the final lowpass filter will cut off any components above f_m. The power spectral density of dy/dt is $(2\pi f)^2$ times the density of $y(t)$. Therefore, the power spectral density of the output noise of Eq. (4.46) is as shown in Fig. 4.44.

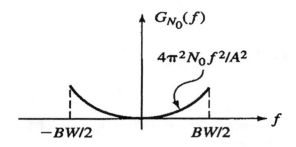

Figure 4.44 - Spectral Density
of Output Noise

The output noise power is found by integrating the power spectral density over the frequency range of the lowpass filter.

$$P_N = 2\int_0^{f_m} G_{N_o}(f)df = \frac{8\pi^2}{3}\frac{N_o}{A^2}f_m^3 \tag{4.47}$$

Since the signal waveform at the output of the FM receiver is $2\pi k_f s(t)$, the output signal power is $4\pi^2 k_f^2 P_s$, where P_s is the power of the information signal, $s(t)$. The output signal to noise ratio for high CNR is then given by

$$SNR_o = \frac{4\pi^2 k_f^2 P_s}{P_N} = \frac{3A^2 k_f^2 P_s}{2N_o f_m^3} \tag{4.48}$$

We can give some intuitive interpretation to Eq. (4.48) by considering the special case of a sinusoidal information signal. That is, assume $s(t) = a\cos 2\pi f_m t$. Then $P_s = a^2/2$, and Eq. (4.48) becomes

$$SNR_o = \frac{3\beta A^2/2}{2\beta f_m N_o}\left(\frac{ak_f^2}{f_m}\right) = 3\beta^3 SNR_i \tag{4.49}$$

We are defining SNR_i as the power of the received carrier divided by the received noise power in the FM bandwidth of $2\beta f_m$. Equation (4.49) shows the effect of wideband FM. As β increases, the bandwidth increases but so does the output signal to noise ratio. The equation makes it look as if the improvement follows the third power of β. This is misleading since the input noise power to the receiver is linearly proportional to β. Therefore, if we referenced output signal to noise ratio to a signal to noise ratio that uses noise in a fixed bandwidth, the relationship would follow the square of β. This leads to a trade-off decision in the design of systems.

Threshold Effect

Let us now return to Fig. 4.43 to more fully explore the effect of noise on wideband FM. We made the assumption that the signal vector magnitude, A, was larger than the noise vector magnitude. If the reverse is true, the diagram could be redrawn as in Fig. 4.45. Alas, the receiver doesn't know the difference between signal and noise. The wideband FM receiver improves noise performance only when the signal is larger than the noise. If the noise gets larger than the signal, the receiver locks onto the noise and suppresses the signal. The poor receiver acts as if the noise is the signal and the signal is the noise! This phenomenon is known as *noise capture*. It is clearly seen in FM broadcast radio when one listens to a distant station (e.g., in an automobile as you approach the distance limits of reception). The station occasionally drops out and is replaced by static.

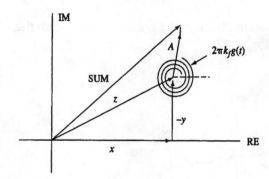

Figure 4.45 - Noise Larger Than Signal

The signal amplitude remains constant at A. As the average length of the noise vector approaches A, we can expect this random quantity to sometimes be larger than and sometimes be smaller than A. Figure 4.46 shows a possible trajectory of the noise vector, and the resultant sum as the noise vector hovers around a magnitude of A.

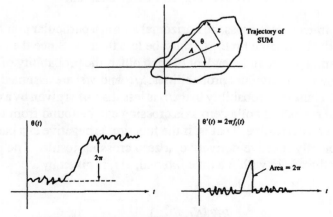

Figure 4.46 - Trajectory of Sum Vector

A counter-clockwise rotation around the origin is illustrated. Also shown in the figure are the resulting angle and its derivative (the frequency). Note that the pulse in frequency is positive and has an area of 2π. If the encirclement of the origin had been in the clockwise direction, the pulse would have been negative. In cases where the origin is not encircled, the average value (total area) of the angular variations is zero. The spike in the frequency waveform would result in an audible *click* in the receiver audio output. Since the noise vector is sweeping around as the quadrature components [$x(t)$ and $y(t)$] change, we can expect occasional clicks as z approaches A. The larger the average value of z, the more frequent will be the clicks until noise capture takes effect. At this point, the audio output would simply be approximately that of the random noise–a rushing sound in the speaker.

Since the noise is random, the clicks cannot be characterized deterministically. We can only derive probability averages. One quantity of interest is the average number of clicks per second.

For simplicity, the phasor diagram will be rotated to align the horizontal axis with the signal vector. The new diagram is redrawn as Fig. 4.47.

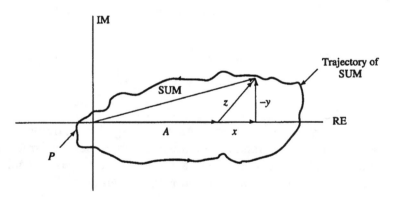

Figure 4.47 - Realigned Phasor Diagram

In order for the trajectory to cross the horizontal axis at a particular point, $y(t)$ must have a zero crossing and, at the same time, $x(t)$ must be less than $-A$. Since the noise quadrature components, $x(t)$ and $y(t)$, are independent of each other, the probability of this joint event is the product of the two individual probabilities. $x(t)$ and $y(t)$ are assumed to be Gaussian random processes. Thus, the probability that $x(t)$ is less than $-A$ is given by a complementary error function. The probability of a zero axis crossing can be found from the joint density of the function and its derivative. That is, if the function is negative at a particular point in time with a sufficiently positive derivative, a zero crossing results. The probability of a trajectory circling the origin within a time interval, Δt, is given by

$$Pr(click) = \frac{BW\ \Delta t}{2\sqrt{3}}\ erfc\left(\sqrt{\frac{A^2}{2N_o BW}}\right) \qquad (4.50)$$

In Eq. (4.50), BW is the bandwidth of the FM waveform. The average number of clicks per second is then given by

$$CPS_{avg} = \frac{BW}{2\sqrt{3}} \; erfc \left(\sqrt{\frac{A^2}{2N_o BW}} \right) \qquad (4.51)$$

Note that the argument of the complementary error function is the square root of the signal to noise ratio. Thus, if we define this signal to noise ratio as

$$SNR = \frac{A^2}{2N_o BW} \qquad (4.52)$$

the average number of clicks per second becomes

$$CPS_{avg} = \frac{BW}{2\sqrt{3}} \; erfc \left(\sqrt{SNR} \right) \qquad (4.53)$$

As the signal to noise ratio increases, the complimentary error function decreases and the number of clicks per second also decreases. As the bandwidth of the FM wave increases, the number of clicks per second increases since additional noise is being admitted to the system.

The area under each pulse in Fig. 4.46 is 2π. We can therefore get an approximation to the output signal to noise ratio near threshold by adding the power of the clicks to the noise power found in the high CNR case. The power of the clicks is estimated in a manner similar to that used for impulse noise analysis. The final result is

$$SNR = \frac{3 BW(SNR) k_f^2 P_s / f_m^3}{1 + (BW)^3 \; 4\sqrt{3} \, erfc(\sqrt{SNR}) f_m^3} \qquad (4.54)$$

As the average number of clicks per second approaches zero, the denominator of Eq. (4.54) approaches unity and the SNR becomes that of Eq. (4.49) for the high CNR case.

Pre-emphasis and De-emphasis

Equation (4.49) indicates that the signal to noise ratio of an FM waveform increases with increasing k_f. Thus, a noise advantage is realized by increasing the bandwidth of the FM signal.

An even greater noise advantage can be realized by observing that the noise at the output of the FM receiver is *not white*, even though the input noise is white. Figure 4.44 illustrates that the output noise power spectral density increases parabolically with increasing frequency. Therefore, the high-frequency parts of the original signal are more severely affected by noise than are the lower frequencies. A more efficient use of the band of frequencies would occur if these higher signal frequencies were *emphasized* and the lower frequencies were *de-emphasized* in a manner that keeps the total power constant. The signal could therefore be shaped by a filter with $H(f)$, as shown in Fig. 4.48.

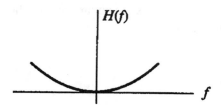

Figure 4.48 - Pre-emphasis Filter

This shaping is done prior to transmission. All signal frequencies would then be equally affected by the noise. The filter represents a signal distortion. The filtering operation must therefore be undone at the receiver using an inverse filter known as the *de-emphasis filter*. This filter would have transfer function $1/H(f)$. The composite filter transfer function of the two filters is unity.

   The de-emphasis filter changes the noise power spectral density from that shown in Fig. 4.44 to a density which is approximately white. This decreases the output noise power. It also assures equal noise disturbance over all signal frequencies. Total noise power is not the only consideration. If the noise power is concentrated at high frequencies (as is the case without de-emphasis), the overall SNR may seem acceptable while high-frequency reception is very poor. Many information signals in real life drop off as frequency increases. It may therefore be desirable to overemphasize the signal with a filter characteristic which increases more rapidly with frequency than that shown in Fig. 4.48.

   Commercial broadcast FM uses pre-emphasis. In the interest of economy, the exact function shown in Fig. 4.48 is not used. A simple *RC* lowpass filter in the receiver provides the de-emphasis. The inverse filter is used in the transmitting station for pre-emphasis.

4.4.2. Digital

We now leave the analog world and turn our attention to digital FSK performance. As with AM (of the previous chapter) we could consider digital as a special case of analog where the baseband signals consist of a sequence of waveforms used to send the various symbols. However it will prove easier to analyze the performance of digital frequency modulation by starting from scratch using the specialized detectors presented in the previous section.

Coherent Detection

We begin by analyzing the performance of the coherent binary matched filter, or correlation detector. The bit error rate is given by

$$P_e = \frac{1}{2} erfc\left(\sqrt{\frac{E(1-\rho)}{2N_o}}\right) \tag{4.55}$$

E is the average energy of the two signals and ρ is the correlation coefficient.

We need simply find E and ρ. The transmitted signal is a sinusoid of amplitude A regardless of whether a one or a zero is being transmitted. The average energy per bit is then $A^2 T_b/2$. The correlation coefficient is given by

$$\rho = \frac{A^2 \int\limits_0^{T_b} \cos(2\pi f_0 t)\cos(2\pi f_1 t)dt}{A^2 T_b/2} \tag{4.56}$$

$$= \frac{1}{T_b}\left(\int\limits_0^{T_b}\cos 4\pi(f_0 + f_1)t\,dt + \int\limits_0^{T_b}\cos 2\pi(f_1 - f_0)t\,dt\right)$$

Since $f_1 + f_0 \gg f_1 - f_0$, we normally assume that the first integral of Eq. (4.56) is negligible (i.e., when a sinusoid is integrated, we divide by the frequency). We then have

$$\rho \approx \frac{1}{T_b}\int\limits_0^{T_b}\cos 2\pi(f_1 - f_0)t\,dt = \frac{\sin 2\pi(f_1 - f_0)T_b}{2\pi(f_1 - f_0)T_b} \tag{4.57}$$

Recall that we want to make ρ as small as possible. To minimize ρ, we evaluate the integral and then take its derivative with respect to $f_1 - f_0$ to find the best choice of frequency spacing. The result is that

$$2\pi(f_1 - f_0)T_b = \tan[2\pi(f_1 - f_0)T_b]$$

$$(f_1 - f_0) = \frac{0.715}{T_b} = 0.715 R_b \tag{4.58}$$

We substitute this frequency separation back into Eq. (4.56) to get $\rho = -0.22$. Other values of frequency separation lead to larger values of ρ. The separation of Eq. (4.58) would therefore seem to be a good design choice.

This was too easy. There is some associated bad news. We now show that the resulting error probability is sensitive to phase with this selection of frequency deviation. We assumed in the derivation that the two sinusoids were matched in phase. This is extremely difficult to achieve in real life. The signals may take different paths with different path lengths, and therefore different phase shifts. Controlling the phase at signal transitions (e.g., between zeros and ones) is also very difficult since the phase is continuous at these transitions. If the phases are *not* perfectly matched, ρ increases, and can even increase beyond zero to become positive.

To illustrate this, let's assume that the two signals are:

$$
\begin{aligned}
s_0(t) &= A\cos 2\pi f_0 t \\
s_1(t) &= A\cos(2\pi f_1 t + \theta)
\end{aligned}
\tag{4.59}
$$

We solve for the correlation coefficient, ρ, as follows.

$$
\begin{aligned}
\rho &= \frac{A^2 \int_0^{T_b} \cos(2\pi f_0 t)\cos(2\pi f_1 t + \theta)dt}{A^2 T_b/2} \\[2ex]
&= \frac{1}{T_b}\left(\int_0^{T_b} \cos 2\pi(f_0+f_1)t+\theta)dt + \int_0^{T_b} \cos(2\pi(f_1-f_0)t+\theta)dt \right) \\[2ex]
&\approx \frac{1}{T_b}\int_0^{T_b} \cos(2\pi\times 0.715/T_b t+\theta)dt = \frac{1}{T_b}\left.\frac{\sin 2\pi\times 0.715/TT_b t+\theta)}{2\pi\times 0.715/T_b}\right|_0^{T_b} \\[2ex]
& \frac{1}{2\pi\times 0.715}[\sin(2\pi\times 0.715 + \theta) - \sin\theta]
\end{aligned}
\tag{4.60}
$$

Plotting ρ as a function of θ yields the curve of Fig. 4.49. Note that although the correlation coefficient is -0.22 when the phase difference is zero and the frequency spacing is $0.714R_b$, the coefficient can become positive as the phase deviates from zero.

Now suppose that we choose a different frequency separation. Let's set this separation equal to the bit rate. That is, $f_1 - f_0 = R_b$. For this choice of frequency spacing, Eq. (4.60) becomes,

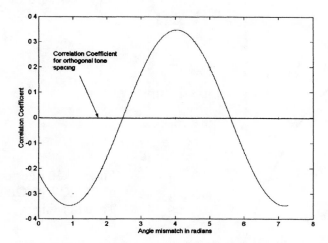

Figure 4.49 - Correlation Coefficient
as Function of Phase Mismatch

$$\rho \approx \frac{1}{T_b}\int_0^{T_b}\cos(2\pi R_b t \;+\; \theta)dt \;=\; \frac{1}{T_b}\left[\frac{\sin(2\pi T_b t + \theta)}{2\pi R_b}\right]_0^{T_b}$$

$$= \frac{1}{2\pi}\Big[\sin(2\pi R_b t T_b \;+\; \theta) \;-\; \sin\theta\Big] = \frac{1}{2\pi}\Big[\sin(2\pi + \theta) \;-\; \sin\theta\Big] = 0$$

(4.61)

This is true regardless of the phase relationship between the two signals. *The performance of the detector is insensitive to phase differences between $s_0(t)$ and $s_1(t)$.* Thus in Fig. 4.49, the x axis presents the $f_1\text{-}f_0=R_b$ case. This therefore represents the more conservative approach to design, and a frequency spacing equal to the bit rate is almost always used.

You can draw an analogy to automobiles. You can have a high-performance automobile that achieves maximum power for a given engine displacement. However, it that engine gets slightly out of tune, its performance degrades rapidly. On the other hand, you can have a very conservative car that doesn't get very good performance, but that performance will hardly change at all as the engine gets out of tune.

Since ρ is equal to zero, the signals are uncorrelated–they have nothing in common with each other. It is like two vectors that are at right angles to each other. For this reason, this choice of frequency spacing is known as *orthogonal tone spacing.* For orthogonal tone spacing, the performance is then given by

$$P_e = \frac{1}{2}\; erfc\left(\sqrt{\frac{E(1-\rho)}{2N_o}}\right) = \frac{1}{2}erfc\left(\sqrt{\frac{A^2 T_b}{4N_o}}\right)$$

(4.62)

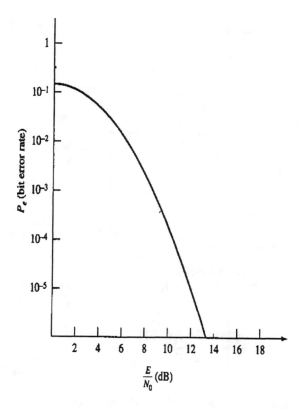

Figure 4.50 - Performance of BFSK With Orthogonal Tone Spacing

We plot this result in Fig. 4.50. The abscissa of this plot is $A^2T_b/2N_o$, the signal to noise ratio, expressed in dB.

MFSK

We examine the coherent detector for MFSK which was shown in Fig. 4.40(a). We assume that the signals are orthogonal. That is, the frequencies are spaced so that the integral of the product of two different symbol waveforms is zero.

The probability of a correct decision is the probability that the output of the correlator corresponding to the transmitted signal is the largest of all the correlator outputs. The probability of symbol error is the complement of the probability of correct decision–the sum of the two is equal to one.

Because of the orthogonality assumption, the outputs off all M - 1 "unmatched" correlators are identically distributed zero-mean Gaussian random variables. They are also independent of each other. The probability that all of these unmatched correlator outputs are less than some number, say y_o, is therefore given by

$$\left[\frac{1}{\sqrt{2\pi}\sigma} \int_{-\infty}^{y_o} \exp\left(-\frac{x^2}{2\sigma^2} \right) dx \right]^{M-1}$$

The output of the one correlator matched to the transmitted signal is also a Gaussian random variable, but it has a mean of $A^2 T_s/2$ instead of zero. The probability of symbol error is then given by the complicated-looking equation,

$$P_M = \frac{1}{\sqrt{2\pi}\sigma} \int_{-\infty}^{\infty} \left\{ 1 - \left[\frac{1}{\sqrt{2\pi}\sigma} \int_{-\infty}^{y} \exp\left(-\frac{x^2}{2\sigma^2} \right) dx \right]^{M-1} \right\} \exp\left(\frac{(y-A^2 T_s/2)^2}{2\sigma^2} \right) dy$$

Suppose we wish to plot the probability of bit error as a function of the bit signal to noise ratio. We observe that the signal to noise ratio per symbol is

$$SNR_{symbol} = \frac{A^2 T_s}{2N_o}$$

The signal to noise ratio per bit is simply the symbol ratio divided by the number of bits per symbol. Therefore,

$$SNR_{bit} = \frac{A^2 T_s}{2N_o \log_2 M}$$

The relationship between symbol error rate and bit error rate is more complicated than it was for MASK. In our analysis of MASK in Section 3.5.1, we assumed that a symbol error usually results in only one bit error (we assumed a Gray code). But because MFSK is orthogonal, there is no reason to assume symbol errors result in adjacent symbols. For example, in 4-FSK, it is just as likely for $s_0(t)$ to change to $s_1(t)$ as it is to change to $s_3(t)$.

Let's start by examining the case of M=4. We are combining bits in pairs. There are four symbols corresponding to 00, 01, 10, and 11. If we transmit any one of these four symbols and experience an error, there are only three ways this can occur. Two of these result in a single bit error, and the third results in two bit errors. The average number of bit errors per symbol error is therefore 4/3. Since there are two bits per symbol, the bit error rate is 2/3 of the symbol error rate.

Suppose we increase M to 8, representing combinations of three bits. Of the seven possible symbol errors, three represent a single bit error, three represent two bit errors, and one represents three bit errors. The average number of bit errors per symbol error is then 12/7, and since there are three bits per symbol, the bit error rate is 4/7 times the symbol error rate.

In general, the relationship between bit error rate and symbol error rate is

$$P_b = \frac{2^{(\log_2 M - 1)}}{2^{\log_2 M} - 1} P_M$$

Putting all of this together yields the curve of Fig. 4.51. The bit error rate decreases as M increases. This may seem counter to your intuition. As additional signal waveshapes are needed, it might seem that they would have to be "closer together" and therefore lead to increased errors. However, remember we are assuming orthogonal tone spacing. Therefore as the number of frequencies increase, the bandwidth of our MFSK signal increases. So we are trading off bit error rate for bandwidth. This was not the case for M-ary baseband or MASK, nor will it apply to MPSK (which we cover in the next chapter).

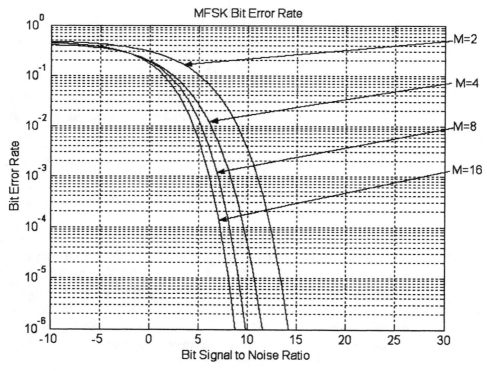

Figure 4.51 - Bit Error Rate for MFSK

Incoherent Detection

We now turn our attention to the performance of the incoherent FSK detector. The nonlinearities in the incoherent detector make the analysis more complex than that of the coherent detector. This is true since the various random quantities are no longer Gaussian distributed.

The probability analysis of the coherent detector exhibits symmetry, but the incoherent

ASK detector probability curves do not contain such symmetries. Due to symmetry of coherent detectors, the probability of erroneously detecting a transmitted one as a received zero (miss) is the same as the probability of detecting a transmitted zero as a received one (false alarm). Thus, the probability of bit error is equal to either one of these probabilities. We need only calculate one of the error probabilities. When we sample the envelope detector outputs, the resulting random variable distributions are Rayleigh and Ricean, depending on whether or not a signal is present. We will hint at the derivation of the error rate. Please refer back to Fig. 4.39 for the incoherent detector.

If the signal is absent from one leg of the detector, the envelope detector output follows a Rayleigh density. That is,

$$p(z) = \frac{2z}{N_o} e^{z^2/N_o} U(z) \tag{4.63}$$

In contrast, if a signal of amplitude A is present, the envelope detector output follows a Ricean density given by

$$p(z) = \frac{2z}{N_o} \exp\left(\frac{-(z^2+A^2)}{N_o} \right) I_o\left(\frac{2Az}{N_o} \right) U(z) \tag{4.64}$$

Now let's suppose a one is sent. An error will be made if z_0 is larger than z_1. The probability of this happening is equal to the double integral,

$$P_e = \int_{z_1=0}^{\infty} \frac{2z_1}{N_o} e^{-(z_1^2+A^2)/N_o} I_o\left(\frac{2Az_1}{N_o} \right) \int_{z_0=z_1}^{\infty} \frac{2z_{09}}{N_o} e^{-z_0^2/N_o} dz_0 dz_1 \tag{4.65}$$

With some difficulty, this integral can be evaluated to yield,

$$P_e = \frac{1}{2}\exp\left(-\frac{A^2 T_b}{4N_o} \right) \tag{4.66}$$

This result is much easier to derive if we make the following assumptions: (a) The signal to noise ratio is high enough so that we are operating far out on the tails of the Ricean density and (b) the bandpass filters of the incoherent detector block the incorrect signal totally. That is, the filter tuned to f_1 totally blocks a signal at a frequency of f_0.

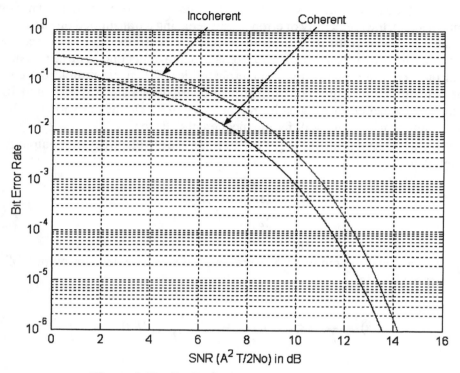

Figure 4.52 - Performance Comparisons for FSK

We now wish to compare the performance of the incoherent detector to that of the coherent detector. That is, we compare Eq. (4.62) with Eq. (4.66). Figure 4.52 shows the two error probabilities as a function of the signal to noise ratio. The curve for the coherent detector is a repeat of Fig. 4.50.

MFSK

We will not give a detailed derivation of the error rate for incoherent MFSK detection. Instead, we simply present an upper bound on this error. The bit error rate for the incoherent MFSK detector is bounded by

$$P_e \le \frac{M-1}{2} \exp\left(-\frac{E}{2N_o}\right)$$

Example 4.6

Find the probability of error of the incoherent detector when used for FSK where the bit period is 2 sec and the signals have amplitude of 0.4 volt. The two frequencies are 1 kHz and 2 kHz. The additive noise has power of 10^{-2} watts/Hz. Compare the incoherent detector performance with that of the coherent detector.

Solution: We must first determine if the results of Fig. 4.49 apply to this problem. Within the derivation of those results, the only assumptions (in the incoherent case) were that the signal frequencies are sufficiently separated that the bandpass filters reject the alternate frequency, and that the signal to noise ratio is high enough to yield a small bit error rate. The bandwidth of the filters would be set to 1 Hz (i.e., $2R_b$), and the frequencies are separated by 1 kHz, so the first assumption is clearly valid. The second assumption will be tested after we derive a bit error rate.

The signal to noise ratio, E/N_o, is given by $A^2 T_b/2N_o$, or 16 (12 dB). We then read from Fig. 4.51 [or Eq. (4.27)] that the probability of error for the incoherent detector is 1.7×10^{-4}.

We now view the curve for the coherent case. The assumption to derive that curve was orthogonal tone spacing. The power spectral density of each signal passes through zero at spacings of 1/2 Hz away from the carrier. Since the spacing between the two frequencies is 1 kHz, and this is a multiple of 1/2, the orthogonal assumption is valid. In actuality, the spacing is so large that, even if it were not a multiple of R_b, the correlation between signals would still be essentially zero. Thus, the probability of error is read from Fig. 4.50 [or Eq. (4.62)] as 3.8×10^{-5}. This is 4.5 times better than that of the incoherent detector.

4.5 Applications

We now examine three important applications of frequency modulation. We begin with classic FM broadcast and FM stereo. Then we examine some basic MODEMs that use frequency modulation (FSK). Finally, we look at Dual Tone Multifrequency Dialing as one simple application that uses a variation of FSK.

4.5.1 Broadcast FM and FM Stereo

If you pick up any FM program guide, you will notice that adjacent stations are separated by 200 kHz. The FCC has assigned carrier frequencies of the type 101.1 MHz, 101.3 MHz, 101.5 MHz, ... to the various transmitting stations within the United States. Thus, while the bandwidth allocated to each station in AM is only 10 kHz, it is an impressive 200 kHz for FM stations. This gives each broadcaster lots of room to work with.

The maximum frequency of the information signal, f_m, is set at 15 kHz. This is three times the figure specified for AM, accounting for the fact that the FM generally sounds much better than AM. The higher frequencies provide much fuller music sounds, and even voice signals sound much clearer.

To transmit this signal using narrowband FM would require a bandwidth of $2f_m$, or 30 kHz. Since 200 kHz is available, we see that broadcast FM is wideband.

Using the approximate formula developed for the bandwidth of an FM signal, we see that a maximum frequency deviation of about 85 kHz is possible. The actual figure used is a conservative 75 kHz.

The FM broadcast receiver looks very much like the superheterodyne AM receiver. The only differences are the range of frequencies of the image rejection filter and local oscillator (the FM band extends from 88 MHz to 108 MHz), the *if* frequency (which is set at 10.7 MHz), and the addition of the discriminator prior to the envelope detector. This is illustrated in the block diagram of Fig. 4.53.

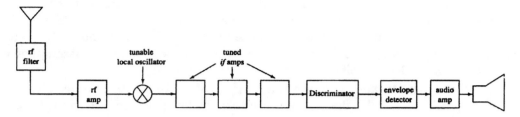

Figure 4.53 - Block Diagram of FM Receiver

FM Stereo

FM stereo is the process of sending two independent audio signals simultaneously within the same FM channel. But the FCC does not allocate any additional bandwidth. So what do we do? Unlike AM, we do not have to look very hard to find room for additional channels. We have already noted that only 30 kHz of bandwidth is needed to send the (monaural) signal by narrowband FM. Since 200 kHz is allocated, there is plenty of room, though we will have to use narrowband instead of wideband.

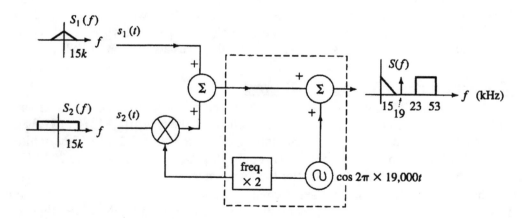

Figure 4.54 - FM Stereo Transmitter

We begin with a system that multiplexes the two individual audio channels. This system is illustrated in the left portion of Fig. 4.54. We are using two different shapes to represent the two Fourier transforms, $S_1(f)$ and $S_2(f)$. Each of these is meant to represent a general low-frequency limited signal. The reason for the two different shapes is to make it easier to track these through the system. We amplitude modulate a carrier at 38 kHz with $s_2(t)$. This shifts the signal into the range between 23 kHz and 53 kHz. This does not overlap the frequencies of $s_1(t)$, so we can now add the two signals together confident in the thought we can pull them apart later. The Fourier transform of the sum is shown on the

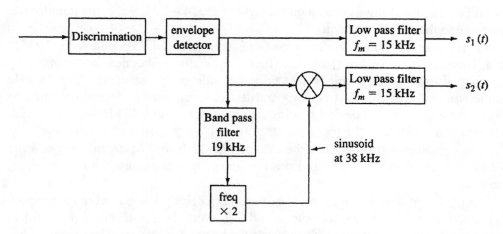

Figure 4.55 - FM Stereo Receiver

right side of the figure. Do not worry about the impulse in the transform nor the circuit in the dashed box. We will return to these later.

The *composite* signal,

$$s_1(t) + s_2(t)\cos Z(2\pi \times 3.8 \times 10^4 t) \tag{4.67}$$

represents a time function with an upper frequency of 53 kHz. We can frequency modulate our carrier with this time function. If we use narrowband FM, the resulting waveform uses only 106 kHz of the available 200 kHz.

At the receiver, we demodulate the FM waveform to recover the composite signal. This is shown on the left side of Fig. 4.55.

A lowpass filter recovers $s_1(t)$ while rejecting the second term. A bandpass filter would separate the second term from the first, but we then must recover $s_2(t)$ from the modulated waveform. Even if we chose to add a carrier to make this AMTC, we could still not use an envelope detector to receive $s_2(t)$. This is true since the carrier frequency of 38 kHz is about 2.5 times the maximum frequency of $s_2(t)$. Proper operation of the envelope detector requires that the envelope be at a *much lower* frequency than the carrier. We *must* use synchronous demodulation. This is shown in Fig. 4.55 where the composite signal is multiplied by a carrier at 38 kHz and then lowpass filtered to recover $s_2(t)$.

How do we assure that the 38 kHz sinusoid in the receiver is perfectly synchronized to the received carrier? We could transmit some of the carrier and use a phase lock loop to recover it at the receiver. However, a much simpler option exists. Returning to the transmitter of Fig. 4.54 note that the 38 kHz carrier is developed by doubling the frequency of a 19 kHz oscillator. The 19 kHz signal is added to the composite signal. Thus, the actual composite signal that goes to the narrowband FM modulator is

$$s_1(t) + s_2(t)\cos(2\pi \times 3.8 \times 10^4 t) + A\cos(2\pi \times 1.9 \times 10^4 t) \tag{4.68}$$

This has a Fourier transform as shown on the right of Fig. 4.54. Note that the impulse has appeared due to the 19 kHz sinusoid.

At the receiver, the output of the envelope detector of Fig. 4.55 contains this 19 kHz sinusoid. This is separated using a bandpass filter. The bandpass filter need not be too sharp since no signal energy lies within 4 kHz on either side of this sinusoid[1]. The 19 kHz reconstructed signal is doubled in frequency and used to synchronously demodulate the AM waveform. We are assured of perfect synchronization since the two 38 kHz sinusoids (that in the transmitter and that in the receiver) are derived from the same 19 kHz source.

Only one problem remains. When the FCC set standards for FM stereo, *compatibility* was a requirement. That is, a monaural receiver must not receive just the left (or right) channel.

The upper leg of Fig. 4.55 represents a monaural receiver. The output of the receiver would be $s_1(t)$. If $s_1(t)$ were the left signal and $s_2(t)$ were the right, this would lead to violating the compatibility standard. Thus, we do not want $s_1(t)$ and $s_2(t)$ to represent the left and right channels. Instead, we let $s_1(t)$ be the sum of the left and right signals, and $s_2(t)$ be the difference. The monaural receiver then recovers the sum of the left and right. The stereo receiver must do an additional linear operation. That operation is to add $s_1(t) + s_2(t)$ to get one channel, and take the difference to get the other. This is known as a *matrix* operation. Thus

$$s_1(t) = s_L(t) + s_R(t)$$
$$s_2(t) = s_L(t) - s_R(t)$$
$$s_1(t) + s_2(t) = 2s_L(t)$$
$$s_1(t) - s_2(t) = 2s_R(t)$$

The unused portion of the allotted band between 53 kHz and 100 kHz deviation from f_c is used for so-called SCA (*Subsidiary Communications Authorization*) signals. These include the commercial-free music heard in some restaurants, transmitted on a subcarrier of 67 kHz. It is also licensed for special applications such us utility load management and paging systems.

This discussion illustrates the flexibility of FM broadcast systems. This flexibility is due to the fact that the bandwidth of an FM signal depends on more than just the bandwidth of the information signal.

---

[1] This filter provides us with a bonus. If monaural FM forms the input to the receiver, the output of this 19 kHz BPF is zero (i.e., all of the signal information is limited to 15 kHz). The output of the filter can therefore be used to light a "stereo indicator" light on the front panel of the receiver.

4.5.2 MODEMS

FSK is the most common form of digital communication in use with low-speed data on telephone transmission systems. When a voice channel is used for transmitting digital information, the form of modulation must be compatible with the characteristics of the voice channel.

Modulator-Demodulators, or *MODEMS* are used to transform the digital signal into a modulated signal (waveform) which can be transmitted through the channel. The channel is typically a telephone channel. The data input to the modem is converted into audio frequency signals which are coupled into the phone lines. Modems are classified as either *asynchronous* or *synchronous*. Asynchronous modems do not have a clock, and data rates need not be constant. This is the least complex type of modem. In fact, most modems operating below 1800 bps are asynchronous. Alternatively, synchronous modems send data at a fixed periodic rate. They can operate at higher data rates than asynchronous modems.

As may be expected, standards have been developed in an attempt to make systems compatible. "Type 103" modems (originated as part of the Bell System 103 and 113 Series) include asynchronous modems which operate up to (a very slow) 300 bps. This series of modem have been reduced to integrated circuit chips (the International Semiconductor Technologies SC11002 is one example). These are *full-duplex modems*[2]. The two directions of transmission are known as *originate* and *answer*. The frequencies are assigned as follows:

	SPACE (0)	MARK (1)
ORIGINATE	1070 Hz	1270 Hz
ANSWER	2025 Hz	2225 Hz

A different, but related standard is the *ITU recommendation V.21*. This standard provides for the following frequency assignments:

	SPACE	MARK
ORIGINATE	1180 Hz	980 Hz
ANSWER	1850 Hz	1650 Hz

Besides consisting of different frequencies, note that the lower frequency in each pair is used for the MARK, in contrast to the 103 series where the reverse is true.

Figure 4.56 shows a block diagram of a low-speed asynchronous modem. The FSK modulator can be a VCO with signal conditioning at the input. The bandpass filter restricts out of band signals prior to coupling the signal to the phone line.

---

[2]*Full-duplex* permits simultaneous two-way communication; *half-duplex* permits two-way communication, but only one way at a time.

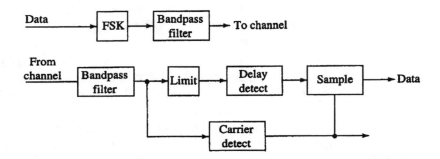

Figure 4.56 - Low-Speed Asynchronous Modem

The receive portion starts with a *bandpass filter* to decrease noise outside of the band of signal frequencies. An *amplitude limiter* is used to decrease noise effects since the signal information is contained in the frequency of the incoming waveform. The *delay detector* compares the signal to a delayed version of itself. In this manner, it approximates the derivative of the signal, which is proportional to the frequency. It therefore acts as a discriminator yielding an output voltage which is proportional to frequency. The *carrier detect* portion of the system is required since the modem is asynchronous. We would not want the system deciding between ones and zeros when nothing is being sent.

To achieve higher speeds, modems can be made synchronous instead of asynchronous. Additional speed improvements are possible by going from full-duplex to half-duplex, and by changing from binary to M-ary transmission. Some improvements are also possible by using forms of modulation other than FSK. These are discussed in the next two chapters.

The *Bell 202 series modem* is a half-duplex modem capable of higher speeds than the 103 series. This modem can send data at 1200 bps on dial-up lines, and at 1800 bps on leased lines with C2 conditioning. A space is sent using 2200 Hz and a mark is transmitted using 1200 Hz.

Because we want to be able to communicate with others throughout the world, standards become important. We saw the V.21 standard in the above discussion. These standards are proposed and maintained through the *International Telecommunications Union* (ITU). This was formerly known as the CCITT (*Comité Consultatif Internationale de Telegraphique et Telephonique*). The modem standards are contained in Section V, with the major ones listed in the following table.

Standard	Date ratified	Speed (bps)	HDX/FDX	PSTN/ private	Modula- tion
V.21	1964	200	FDX(FDM)	PSTN	FSK
V.22	1980	1200	FDX(FDM)	PSTN	PSK
V.22 bis	1984	2400	FDX(FDM)	PSTN	QAM
V.23	1964	1200	HDX	PSTN	FSK
V.26	1968	2400	HDX	private	PSK
V.26 bis	1972	2400	HDX	PSTN	PSK
V.26 ter	1984	2400	FDX(EC)	PSTN	PSK
V.27	1972	4800	HDX	private	PSK
V.27 bis	1976	4800	HDX	PSTN	PSK
V.27 ter	1976	9600	HDX	private	QAM
V.32	1984	9600	FDX(EC)	PSTN	QAM
V.32 bis	1991	14400			TCM
V.32 ter		19200			TCM
V.34 (V.fast)	1994	28800			TCM
V.90		56000			
V.92					

HDX - Half Duplex

FDX - Full Duplex

FDM - Frequency Division Multiplexing

EC - Echo Canceller

PSTN - Public Switched Telephone Network

FSK - Frequency Shift Keying

PSK - Phase Shift Keying

QAM - Quadrature Amplitude Modulation

TCM - Trellis Coded Modulation

4.5.3 Dual-Tone Multifrequency (DTMF) Dialing

Touch-tone dialing permits a standard telephone keypad to be used for dialing a telephone number by generating a pair of frequencies or tones for each key pressed. The TP5087DTMF tone generator, which is shown in Fig. 4.57 supplies four row frequencies, R_i, and four column frequencies, C_i, as shown below.

R_1	697 Hz	C_1	1209 Hz
R_2	770 Hz	C_2	1336 Hz
R_3	852 Hz	C_3	1477 Hz
R_4	941 Hz	C_4	1633 Hz

This leads to 16 possible dual-tone combinations. Contemporary phones use only 12 of these. The block at the left of the figure indicates the tone combinations. For example, when the number seven is pressed, the two sinusoidal frequencies of 852 Hz (R_3) and 1209 Hz (C_1) are simultaneously generated by the IC.

The basic frequency is generated from a low-cost 3.579545 MHz crystal (a standard NTSC TV color burst crystal), which feeds the IC. The row and column inputs, which are energized when a key is pressed, are decoded to develop the high-frequency tone and the low-frequency tone. The two tones are summed in a mixing amplifier and power amplified in an *npn* emitter follower. Note that the tones are derived by dividing the reference frequency by fixed numbers. For example, 697 Hz results from dividing the reference by 5136, and 1633 Hz results if we divide by 2192.

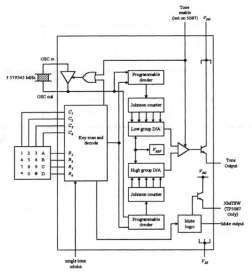

Figure 4.57 - TP50-87DTMF Tone Generator

Problems

4.1 (a) Find the instantaneous frequency of the following waveform:

$$17\cos(5t + 3t^2 + t^3)$$

(b) What is the instantaneous frequency at time t=2 seconds?

4.2 Find the approximate band of frequencies occupied by an FM waveform of carrier frequency 5 MHz, k_f=0.02 Hz/V, and

$$s(t) = 200\cos400\pi t + 300\cos800\pi t \quad volts$$

4.3 An angle modulated waveform is described by

$$\lambda(t) = 100\cos(2\pi\times10^7t + .05\sin2\pi\times500t + 0.25\sin2\pi\times1500t + 0.25\sin2\pi\times2500t)$$

(a) What is $f_i(t)$, the instantaneous frequency of $\lambda(t)$?
(b) What is the approximate bandwidth of $\lambda(t)$?
(c) If $\lambda(t)$ represents an FM signal, what is $s(t)$, the information signal?

4.4 The amplitude of an FM waveform is 5, the carrier frequency is 10 MHz, and the instantaneous frequency is given by

$$f_i(t) = 10^5 + 100\pi\cos(100\pi t) + 50\pi\sin(200\pi t)$$

Write the expression for the FM waveform.

4.5 Find the instantaneous frequency of

$$s(t) = 10[\cos(10t)\cos(30t^2) - \sin(10t)\sin(30t^2)]$$

4.6 Find the instantaneous frequency of

$$s(t) = 2e^{-t}U(t)$$

4.7 An information signal is as shown in Fig. P4.7. The carrier frequency is 1 MHz and k_f=10^4. Sketch the FM waveform and label significant frequencies and values.

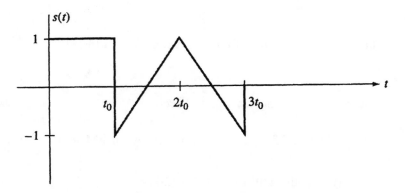

Figure P4.7

4.8 In Eq. (4.19), we found that

$$c_n = \frac{1}{T} \int_{-\frac{T}{2}}^{\frac{T}{2}} e^{j\beta \sin 2\pi f_m t} e^{-jn2\pi f_m t} dt$$

where $T = 1/f_m$. Show that the imaginary part of c_n is zero.

4.9 Show that c_n, as given in Eq. (4.19) [see Problem 4.8], is not a function of f_m or T. That is, show that c_n depends only on β and n. [Hint: Let $x = 2\pi f_m t$ and make a change of variables.]

4.10 Find the approximate Fourier transform of

$$s(t) = \cos\left(\frac{ak_f}{f_m} \sin 2\pi f_m t \right)$$

Do this by expanding the cosine in a Taylor series and retaining the first three terms.

4.11 Repeat Problem 4.10 for

$$s(t) = \sin\left(\frac{ak_f}{f_m} \sin 2\pi f_m t \right)$$

4.12 Use the results of Problems 4.10 and 4.11 to find the approximate Fourier transform of the FM waveform

$$\lambda_{fm}(t) = \cos\left(2\pi f_c t + \frac{ak_f}{f_m}\sin2\pi f_m t\right)$$

4.13 Find the approximate band of frequencies occupied by an FM waveform of carrier frequency 2 MHz, where $k_f = 100$ Hz/V and:
(a) $s(t) = 100\cos2\pi\times150t$ *volts.*
(b) $s(t) = 200\cos2\pi\times300t$ *volts.*

4.14 Find the approximate band of frequencies occupied by an FM waveform of carrier frequency 2 MHz, where $k_f = 100$ Hz/V and $s(t)=100\cos2\pi\times150t+200\cos2\pi\times300t$ *volts.* Compare this with your answers to Problem 4.13. Contrast the result with AM.

4.15 Consider the system shown in Fig. P4.15, where the output is alternately switched between a source at frequency $fc-\Delta f$ and a source at frequency $f_c+\Delta f$. The switching is done at a frequency f_m. Find and sketch the Fourier transform of the output $y(t)$. You may assume that the switch takes zero time to go from one position to another.

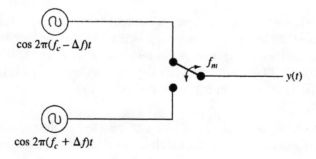

Figure P4.15

4.16 A 10-MHz carrier is frequency modulated by a sinusoid of unit amplitude and $k_f = 10$ Hz/V. Find the approximate bandwidth of the FM waveform if the modulating signal has a frequency of 10 kHz.

4.17 A 100-MHz carrier is frequency modulated by a sinusoid of frequency 75 kHz such that $\Delta f = 500$ kHz. Find the approximate band of frequencies occupied by the FM waveform.

4.18 Find the approximate band of frequencies occupied by the waveform

$$\lambda(t) = 100\cos(2\pi \times 10^5 t + 35\cos 100\pi t)$$

4.19 A frequency modulated waveform is described by

$$\lambda(t) = 50\cos(2\pi \times 10^6 t + 0.001\cos 2\pi \times 500 t)$$

(a) What is $f_i(t)$, the instantaneous frequency of $\lambda(t)$?
(b) What is the approximate bandwidth of $\lambda(t)$?
(c) Is $\lambda(t)$ a narrowband or wideband frequency modulated signal?
(d) What is $s(t)$, the information signal?

4.20 Find the approximate pre-envelope and envelope of

$$s(t)\cos\left(2\pi f_c t + k_f \int_0^t s(\tau)d\tau \right)$$

where the maximum frequency of $s(t)$ is much less than f_c.

4.21 A phase lock loop is used to demodulate an FM waveform. The information signal is a voice waveform with maximum frequency 5 kHz. The comparator has an output that varies linearly with input phase difference. An input phase difference of 90° causes a 25-V change in the output. The VCO output varies linearly and changes by 10 kHz for each 1-V change in input. Will the loop act as an effective demodulator?

4.22 Show that the system of Fig. P4.22 can be used to demodulate an FM waveform. Analyze the performance of this demodulator as a function of the time delay T.

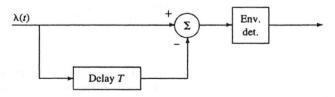

Figure P4.22

4.23 You are given the sum (L + R) and the difference (L - R) signals in a stereo receiver system. Design a simple resistive circuit that could be used to produce the individual left and right signals from the sum and difference waveforms.

4.24 An FM stereo system is being designed for planet X, where the people have hearing that goes up to 25 kHz. The XCC (equivalent to FCC in the US on the earth) has specified the band of frequencies between 100 MHz and 200 MHz for their broadcast FM. Each channel is allocated 200 kHz within this band. You are asked to design the FM stereo system. Describe, in detail, the changes you would make to the system we use on earth in the United States.

4.25 An information signal $s(t)=5\cos 2000\pi t$ frequency modulates a carrier of frequency 1 MHz. $A = 100$ and $k_f = 100$ Hz/V. Noise of power spectral density $N_o = 10^{-3}$ is added during transmission.
(a) Find the SNR in dB at the output of an FM receiver.
(b) If the output is put through a bandpass filter with passband from 900 to 1,100 Hz, find the improvement in SNR of the filter.
(c) Repeat parts (a) and (b) for AM double-sided transmitted carrier. Compare the results.

4.26 Repeat Problem 4.25 for $s(t)=5\cos 2000\pi t+3\cos 4000\pi t$.

4.27 Repeat Problem 4.25 for

$$s(t) = \frac{100\sin 2000\pi t}{\pi t}$$

4.28 You are given the FM signal

$$\lambda(t) = 100\cos(2\pi\times 10^5 t + 35\cos 100\pi t)$$

This signal is received along with additive white noise of power spectral density $No/2 = 0.1$. Find the average number of clicks per second in the output of the detector.

4.29 An FM signal is given by

$$\lambda(t) = 50\cos(2\pi\times 10^6 t + 0.001\cos 2\pi\times 500 t)$$

Noise of power spectral density $N_o/2 = 0.05$ is added to the signal. The sum of the signal and noise enters the detector. Is the detector operating near threshold? If your answer is yes, find the average number of clicks per second at the output of the detector.

4.30 An FM signal is given by

$$\lambda(t) = A\cos(2\pi\times 10^5 t + 35\cos 100\pi t)$$

White noise with power of $N_o = 0.01$ watt/Hz is added to the signal. The sum forms the input to a detector.

(a) Find the value of A so that the average number of clicks per second is 5.

(b) For the value of A in part (a), find the signal to noise ratio at the detector output.

4.31 A first-order phase-locked loop has the following parameter values:

$A = B = 1$; $k_o = 5$ Hz/V: $f_o = 10$ kHz.

Plot the output frequency of the loop as a function of time if a sinusoidal input of frequency 10,010 Hz is applied at time zero.

4.32 Repeat Problem 4.31 with the input changed from a sinusoid to a square wave.

4.33 The clock in a certain system has jitter. Its frequency is given by $f(t) = 10^4 + 100\cos2\pi \times 300t$. Design a phased-lock loop that will track the frequency within 10 Hz.

4.34 The input to a phase-locked loop is a pulse train of very narrow pulses at a frequency of 10 kHz. The pulse frequency varies slightly by a maximum amount of 5 Hz. This variation occurs at a maximum frequency of 100 Hz. That is, the time it takes for the frequency to deviate by the maximum amount of 5 Hz away from 10 kHz is no shorter than 100 msec. Design a first-order phase-locked loop to track the input frequency accurately.

4.35 An FSK signal consists of bursts of frequency of 800 and 900 kHz, with the higher frequency used to transmit a binary one. The bit rate is 2 kbps. Find the FSK bandwidth.

4.36 Find the bandwidth of an FSK signal resulting from binary modulation where the pulses are shaped according to a raised cosine. Assume that an alternating train of zeros and ones is transmitted and that bipolar transmission is used.

4.37 What is the nominal bandwidth of 8-ary FSK if orthogonal tone spacing is used and the bit rate is $R_b = 10^4$ bps?

4.38 Find the bit error rate for an FSK system with

$$s_1(t) = \cos(1100\pi t + 30^\circ)$$
$$s_0(t) = \cos(1000\pi t + 30^\circ)$$
$$\frac{N_o}{2} = 0.1; \quad T_b = 10 \text{ sec}$$

(a) Using a coherent detector.

(b) Using an incoherent detector.

4.39 FSK is used to transmit binary information. The frequency deviation is set at the optimum value, that is, a frequency spacing of 0.715 R_b, so the correlation coefficient is about -0.22. Now suppose that the two signals are mismatched in phase by $\Delta\theta$. Plot the correlation coefficient as a function of the phase mismatch, and find the maximum value of the correlation.

4.40 FSK is used to transmit binary information. The frequency deviation is set for orthogonal tone spacing, that is the tones are separated in frequency by R_b, so the correlation coefficient is zero. Now suppose that the two signals are mismatched in phase by $\Delta\theta$. Show that the correlation coefficient is zero regardless of the phase mismatch.

4.41 Derive an expression for the maximum allowable phase mismatch between two frequency bursts such that a tone spacing of $0.715/T_b$, is preferable to orthogonal tone spacing. Assume that a coherent detector is used.

4.42 Find the probability of error for the incoherent FSK detector of Fig. 4.39 where $A = 5$, $f_0= 1{,}200$ Hz, $f_1=2{,}200$ Hz, and the sampling period is 0.1 msec. Noise of $N_o/2 = 10^{-6}$ watt/Hz is added during transmission.

4.43 Find the probability of error of the incoherent detector when used for FSK where the bit period is 1 msec. The two signals have an amplitude of 2 V, and the frequencies are 1.2 kHz and 2.2 kHz. The additive noise has a power of $N_o = 10^{-5}$ watt/Hz. Compare the performance of the incoherent detector with that of the coherent detector.

4.44 You wish to modify the standard FM broadcast stereo system to increase the maximum signal frequency from 15 kHz to 20 kHz. Each station is still allocated a 200 kHz band and the stations all occupy the range between 88 MHz and 108 MHz. You are asked to design the FM stereo system.. Describe, in detail, the changes you would make to the system presented in the text. A block diagram of the transmitter and receiver would help.

4.45 A coherent FSK detector is used to detect FSK where the two signals have amplitudes of 100 microvolts and the two frequencies are 3000 kHz and 3002 kHz. The bit transmission rate is 2 kbps.
(a) Design the detector (draw a block diagram).
(b) What is the bandwidth of the FSK signal?
(c) What is the maximum value of noise power per Hertz if the bit error rate must be less than 3×10^{-3} ?
(d) Using the value of noise power found is part (c), assume that the modulation scheme changes from FSK to ASK. The maximum amplitude and bit rate stays the same. What is the new value of probability of bit error?

4.46 A paging system transmits an audio tone to a paging device. The audio signal is limited to 10 kHz upper frequency. The system transmits on an FM stereo station with carrier of 101.7 MHz a shown in the block diagram in Fig. P4.46. Design (draw a block diagram) a receiver that would recover $s_{paging}(t)$. Make sure to indicate the cutoff frequency of any filters in your system.

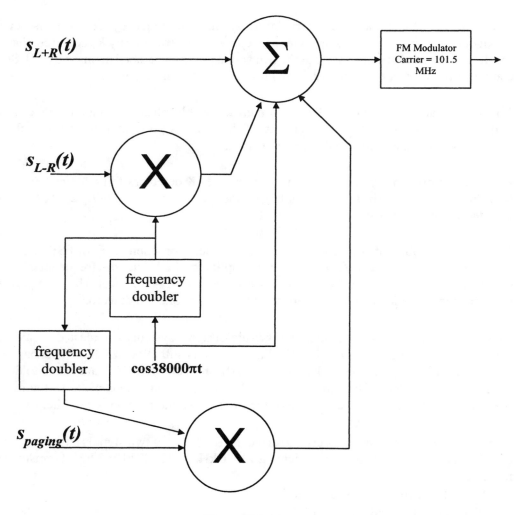

Figure P4.46

4.47 A coherent FSK detector (i.e., matched filter detector) is used to detect FSK where the two signals have amplitude of A volts and the frequencies are 2000 Hz and 2010 Hz. The bit transmission rate is 10 bits per second and the noise power per Hertz is $N_o=10^{-4}$.

(a) Design the detector (draw a block diagram).

(b) What is the correlation coefficient, ρ.

(c) What is the minimum signal amplitude, A, if the bit error rate must be less than 10^{-5}?

4.48 Shown in Fig. P4.48 is a curve for probability of bit error as a function of $A^2T/2N_o$ in dB for FSK. [Note that this does not necessarily agree with curves in the text since I am not telling you the frequency spacing nor the value of ρ. Therefore you must answer this question using the curve and not the formulas.]

(a) You are told that the noise power per Hertz is 10^{-5} watts/Hz and that the signal amplitude is 1 millivolt. Find the maximum bit transmission rate, R_b, such that the probability of error is less than 10^{-4}.

(b) Using the bit transmission rate of part (a), the noise power per Hertz now increases by a factor of 10 to 10^{-4} watts/Hz. Find the new probability of error.

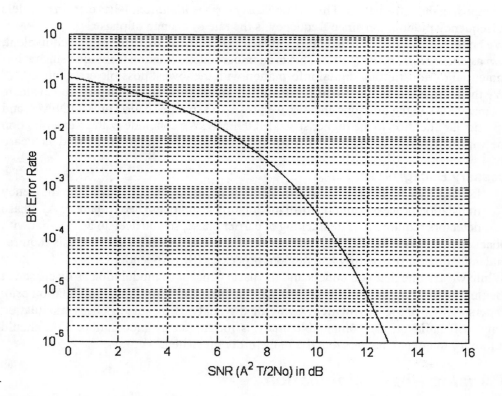

CHAPTER 5

PHASE MODULATION

What we will cover and why you should care

This is the last of three parallel chapters covering the basic modulation techniques. In Chapter 3, we used amplitude variation to carry the information. In Chapter 4, it was the frequency that was varied.

The current chapter explores variations of the carrier's phase as a means of superimposing the information. This is known as *phase modulation*. It has certain parallels with frequency modulation since frequency is the rate of change of phase.

We begin with a study of analog PM and then extend the study to the digital equivalent, known as *phase shift keying* (PSK). This is an important modulation technique since it has the potential of transmitting digital information with the lowest possible error rates.

We then turn our attention to the design of modulators and demodulators, and evaluate performance. As in the previous chapter, we present a separate section on MODEMS, and pay particular attention to the higher speed MODEMS. We conclude with an examination of the various tradeoff decisions in the design process. AM Stereo is selected as a case study.

Necessary Background

A familiarity with Fourier transform theory is necessary to understand the frequency translation process common to all modulation techniques (See Appendix A). Phase detectors require accurate tracking of carrier phase, so you need to be familiar with feedback control systems, stability, and convergence. The performance analysis requires an understanding of probability and random processes (See Appendix C).

While we assume that you understand the basic concept of modulation as discussed at the beginning of Chapter 3, it is not necessary that you know amplitude modulation prior to understanding phase modulation. Some of the analysis of analog phase modulation systems will build on the comparable analysis for frequency modulation. You should therefore try to cover Chapters 4 and 5 in the order presented.

5.1 Analog Phase Modulation

There is no basic difference between analog phase modulation and analog frequency modulation. In fact, the terms are often used interchangeably or the two types are collectively referred to as *angle modulation*. Modulating a phase with a particular time signal also modulates the derivative of that phase, frequency, with a related time function. We therefore need not spend a great deal of time developing phase modulation. Indeed, we will be able to extend our FM results to PM with little difficulty. If you have forgotten the material on FM (Section 4.1 of the previous chapter), we suggest you review that before going forward.

We have a modulated waveform of the type

$$\lambda_{pm}(t) = Acos\Theta(t) \tag{5.1}$$

$\theta(t)$ is modulated with $s(t)$. In FM, the derivative of $\theta(t)$ follows the signal. For phase modulation, it is $\theta(t)$ itself that follows the signal. Therefore,

$$\theta(t) = 2\pi[f_c t + k_p s(t)] \tag{5.2}$$

The proportionality constant, k_p, has units of volts^{-1}. The PM wave is then given by

$$\lambda_{pm}(t) = Acos\Theta(t)$$
$$= Acos2\pi[f_c t + k_p s(t)] \tag{5.3}$$

We have included the $2\pi f_c t$ factor in the phase definition of Eq. (5.2) so that when $s(t)=0$, the PM wave is a pure carrier. Had we not included this term, the PM wave would become *dc* when the signal goes to zero.

We can relate the PM wave to an FM wave by finding the instantaneous frequency of the signal of Eq. (5.3).

$$f_i(t) = \frac{1}{2\pi}\frac{d\theta}{dt} = f_c + k_p\frac{ds}{dt} \tag{5.4}$$

This looks very similar to the FM case [Eq. (4.4)] which we repeat here.
$$f_i(t) = f_c + k_f s(t)$$
In fact there is no difference between frequency modulating a carrier with $s(t)$ and phase modulating that same carrier with the integral of $s(t)$. Alternatively, there is no difference between phase modulating a carrier with $s(t)$ and frequency modulating the same carrier with the derivative of $s(t)$. We will use this observation in designing modulators and demodulators.

5.1.1. Narrowband Phase Modulation
As we did for FM, we separate PM into two categories depending on the relative size of the constant, k_p. Equation (5.3) can be expanded as follows:

$$\lambda_{pm}(t) = Acos2\pi[f_c t k_p s(t)]$$
$$= Acos(2\pi f_c t)cos\big(2\pi k_p s(t)\big) - Asin(2\pi f_c t)sin\big(2\pi k_p s(t)\big) \tag{5.5}$$

If the constant, k_p, is sufficiently small, Eq. (5.5) can be approximated as

$$\lambda_{pm}(t) \approx Acos2\pi f_c t - A2\pi k_p s(t)sin2\pi f_c t \tag{5.6}$$

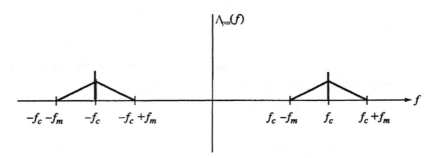

Figure 5.1 - Fourier Transform of Narrowband PM Waveform

Equation (5.6) represents an AM waveform where the modulating signal is $A2\pi k_p s(t)$. The magnitude of the Fourier transform of this wave is sketched as Fig. 5.1.

This signal therefore occupies a range of frequencies between f_c-f_m and f_c+f_m. The bandwidth is $2f_m$.

This result is not surprising since phase modulating with $s(t)$ is exactly the same as frequency modulating with the derivative of $s(t)$. Since the derivative of $s(t)$ contains the same range of frequencies as does $s(t)$, the derivation leading to Fig. 4.6 of the previous chapter would reach the same bandwidth conclusion for PM. That is, a narrowband PM waveform occupies the range of frequencies between f_c-f_m and f_c+f_m, with a bandwidth of $2f_m$. This is the same as the bandwidth for double sideband AM.

For PM to be narrowband, $2\pi k_p s(t)$ must be a very small angle. This permits approximating cosine and sine by the first term in a series expansion.

To gain insight into the phase modulation process, we present several vector plots. We want to compare vector plots from AM with those of PM. The AMTC waveform for a sinusoidal information signal $[s(t) = \cos 2\pi f_m t]$ is

$$\begin{aligned} s_m(t) &= A\cos2\pi f_c t + \cos2\pi f_m t\cos2\pi f_c t \\ &= [A + \cos2\pi f_m t]\cos2\pi f_c t \end{aligned} \qquad (5.7)$$

We can write this as the real part of a complex exponential.

$$s_m(t) = RE\left\{[A + \frac{1}{2}e^{j2\pi f_m t} + \frac{1}{2}e^{-j2\pi f_m t}]e^{j2\pi f_c t}\right\} \qquad (5.8)$$

The phase of the term in braces in Eq. (5.8) has a large angular term due to the $2\pi f_c t$ phase factor. Recall that f_c is usually much larger than f_m. We use the common technique of assuming that a "stroboscopic" picture of this phasor is taken with a flash every $1/f_c$ seconds. As such, we only plot the phasor of the inner bracket in Eq. (5.8). This is shown as Fig. 5.2.

Note that the resultant is in phase with the carrier term. This is as expected since there is no phase variation in AM.

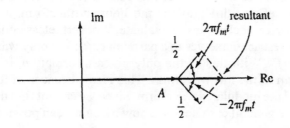

Figure 5.2 - Phasor Plot for AMTC

We now repeat this approach for narrowband PM. Equation (5.6) can be rewritten in the following form, where we have let $s(t)$ be $\cos 2\pi f_m t$.

$$\lambda_{pm}(t) = Re\left\{ \left[A - 2\pi A k_p e^{j(2\pi f_m t - 90°)} + 2\pi A k_p e^{-j(2\pi f_m t - 90°)} \right] e^{j2\pi f_c t} \right\}$$

(5.9)

The quantity in the inner brackets of Eq. (5.9) is plotted in Fig. 5.3.

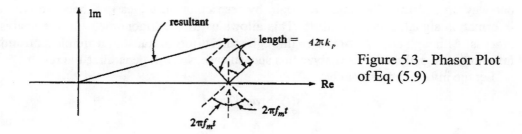

Figure 5.3 - Phasor Plot of Eq. (5.9)

Note that the resultant has almost the same amplitude as A. It is not exactly equal to A because Eq. (5.6) is an approximation. Had we not used an approximation, the amplitude would be constant.

5.1.2. Wideband PM

If k_p is not small enough to permit the approximations of the previous section, we have what is known as *wideband PM*. The transmitted signal is of the form

$$\lambda_{pm}(t) = A\cos 2\pi [f_c t + k_p s(t)]$$

(5.10)

where $s(t)$ is the information signal. The following analysis exactly parallels that of FM in the previous chapter.

If $s(t)$ were a known function, the Fourier transform of this FM waveform could be evaluated. We would simply use the defining integral for the Fourier transform, and although the integral may require numerical techniques to evaluate, it nevertheless could be solved. However, we do not wish to restrict ourselves to a particular $s(t)$. We only want to constrain $s(t)$ to frequencies below f_m. With this as the only given information, it is impossible to find the Fourier transform of the PM waveform. This is true because of the non-linear relationship between $s(t)$ and the modulated waveform. Since we cannot find the Fourier transform of the PM wave in the general case, let's see how much we can possibly say about the modulated signal.

To show that PM can be efficiently transmitted and that multiple channels can be multiplexed, we must only gain some feeling for the *range* of frequencies occupied by the modulated waveform. We have already done this for narrowband modulation, so we now concentrate on wideband. The approach we take is significantly different from that used in AM. In the case of AM, we found the Fourier transform of $s_m(t)$, and used this throughout the analysis, including our presentation of the operation of modulators and demodulators. In the case of phase modulation, we use the Fourier transform for *only one thing*–to determine the range of frequencies occupied. All of our other analyses will be performed in the time domain. For this reason, we do not really care about the exact shape of the transform.

Unfortunately, even finding the range of frequencies is impossible in general. To ease our way around this dilemma, we begin by restricting ourselves to a specific type of information signal, a pure sinusoid. This allows us to use trigonometry in the analysis process. Although nobody would go through much effort to transmit a simple sinusoidal information signal, once we analyze this special case, we will generalize the results.

Let the information signal be

$$s(t) = a\sin 2\pi f_m t$$

where a is a constant amplitude. We are using sine instead of cosine so the results will be similar to those of FM. Phase modulating with sine is the same as frequency modulating with cosine. The PM waveform is then given by

$$\lambda_{pm}(t) = A\cos\left(2\pi f_c t + 2\pi a k_p \sin 2\pi f_m t\right) \tag{5.11}$$

Let us define the *modulation index* as β:

$$\beta \triangleq 2\pi a k_p \tag{5.12}$$

Note that the modulation index represents the maximum amount that the phase varies from what it would be for a pure carrier (i.e., a linear function). The modulation index is a dimensionless quantity since a has units of volts, and k_p has units of volts^{-1}. Using this definition, Eq. (5.11) can be simplified to

$$\lambda_{pm}(t) = A\cos(2\pi f_c t + \beta\sin 2\pi f_m t) \tag{5.13}$$

We now want to separate the information portion from the carrier so we can isolate the portion due to *s(t)*. We would have to use the "cosine of sum" trigonometric identity to do so, and two terms would have to be tracked throughout the analysis. To simplify the mathematics and allow us to work with only one expression, we convert to exponential notation.

$$\lambda_{pm}(t) = Re\{A\exp(j2\pi f_c t + j\beta\sin 2\pi f_m t)\} \tag{5.14}$$

The exponential of a sum as it appears in Eq. (5.14) can be split into a product of exponentials. The second term of this product is the one that contains the information signal. This term is

$$e^{j\beta\sin 2\pi f_m t}$$

The real part is in the form of a cosine of a sine, and is generally difficult to work with. However, note that the exponential is a periodic function with period $1/f_m$. To prove this, substitute $t+1/f_m$ for t. Doing so does not change the value of the expression. Note that the periodic function is complex. We can expand it in the complex Fourier series with a fundamental frequency of f_m.

$$e^{j\beta\sin 2\pi f_m t} = \sum_{n=-\infty}^{\infty} c_n e^{jn2\pi f_m t} \tag{5.15}$$

The Fourier coefficients are given by

$$c_n = f_m \int_{\frac{-1}{2f_m}}^{\frac{1}{2f_m}} e^{j\beta\sin 2\pi f_m t} e^{-jn2\pi f_m t} dt \tag{5.16}$$

The integral of Eq. (5.16) cannot be evaluated in closed form. We discussed this in detail in Chapter 4. The integral converges to a real value. That real value is a function of both n and β. It is not a function of f_m. The integral is tabulated under the name *Bessel function of the first kind*, and its symbol is $J_n(\beta)$. Given n and β, you can simply look this up in a table. We examined Bessel functions in Chapter 4.

Comparing Eq. (5.16) to the definition of the Bessel function, we see that the Fourier coefficients are given by

$$c_n = J_n(\beta)$$

and the PM waveform becomes

$$\lambda_{pm}(t) = Re\left\{Ae^{j2\pi f_c t}\sum_{n=-\infty}^{\infty} J_n(\beta)e^{jn2\pi f_m t}\right\} \tag{5.17}$$

Since the first exponential is not a function of n, we can bring it under the summation sign to get

$$\lambda_{pm}(t) = Re\left\{A\sum_{n=-\infty}^{\infty} J_n(\beta)e^{j2\pi(f_c+nf_m)t}\right\} \tag{5.18}$$

Taking the real part, we obtain

$$\lambda_{pm}(t) = A\sum_{n=-\infty}^{\infty} J_n(\beta)\cos 2\pi(f_c + nf_m)t \tag{5.19}$$

We have reduced the PM waveform to a sum of sinusoids.

The Fourier transform of the sum of sinusoids in Eq. (5.19) is a train of impulses. This Fourier transform is sketched in Fig. 5.4.

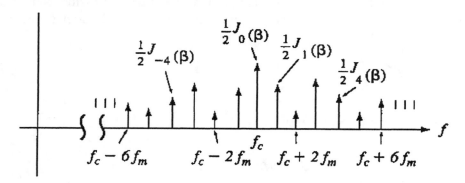

Figure 5.4 - Fourier Transform of PM with
Sinusoidal Information Signal

The transform extends forever in both directions. It would appear that the PM wave has an infinite bandwidth. Indeed, unless $J_n(\beta)$ goes to zero for n above some value, the bandwidth is *not limited* and we can neither transmit efficiently nor multiplex multiple channels.

We are rescued from this dilemma by returning to our previous discussion of Bessel functions. For fixed β, the functions $J_n(\beta)$ eventually *do* approach zero as n increases. Note that the Bessel functions do not decrease monotonically for n less than β. In fact, for certain choices of β, the $J_0(\beta)$ term goes to zero and the carrier term is eliminated from the PM waveform. In the case of AM, elimination of the carrier increases efficiency. In PM, elimination of the carrier does not gain anything since the total power remains constant.

To approximate the bandwidth of the PM waveform, we must examine the strengths of the impulses in Fig. 5.4. Let's first choose a small value of β. Looking back at Fig. 4.7 we see that if $\beta<0.5$, $J_2(\beta)<0.03$. The higher order Bessel functions ($n>2$) are even smaller. At $\beta=0.5$, J_1 is 0.24. For these small values of β, it is therefore reasonable to include only the three impulses centered at the carrier in Fig. 5.4. That is, there is the component at the carrier, and two additional components spaced $\pm f_m$ away from the carrier. This yields a bandwidth of $2f_m$. But we already knew this since very small values of $\beta(2\pi a k_p)$ correspond to the narrowband condition.

Now suppose that β is not small. For example, suppose it is equal to 10. The properties that we have previously discussed would indicate that $J_n(10)$ attenuates rapidly for $n>10$. Referring to Fig. 5.4, we then consider the significant components to be the carrier term and 10 harmonics on each side of the carrier. In general, for large β, the number of terms that must be included on each side of the carrier is β (rounded to an integer). This yields a bandwidth of $2\beta f_m$. These bandwidths are approximations since the Fourier transform of the FM wave is not identically zero beyond any frequency point. However, in practice if the energy outside this band is rejected with a bandpass filter, the distortion is not noticeable. (Here's that non-precise wording again!)

In many instances, we will be working with modulation indexes between the two extremes. In the early days of FM studies, John Carson proposed a rule of thumb that has been widely adopted. This rule gives the approximate bandwidth of FM as a function of the frequency of the information signal and the frequency deviation. The reason this rule has been widely accepted is that the amount of energy outside this band has been shown to be negligible for most applications. Carson's rule says the bandwidth is approximately given by

$$BW \approx 2(\beta f_m + f_m)$$

This agrees with our two limiting cases. For very small β, the bandwidth is approximately $2f_m$, while for large β, it is approximately $2\beta f_m$.

Carson's approximation can be written in a more meaningful way by replacing substituting $\beta=2\pi a k_p$. Equation (5.19) then be comes

$$BW \approx 2(2\pi a k_p + f_m) \tag{5.20}$$

Let's try to give meaning to the terms in Eq. (5.20). Recall that the instantaneous frequency is given by

$$f_i(t) = \frac{1}{2\pi}\frac{d\theta}{dt}$$

$$= f_c + 2\pi a k_p \cos 2\pi f_m t \tag{5.21}$$

Looking at the two terms in Eq. (5.21), we see that f_m is the rate at which the instantaneous frequency varies while $2\pi a k_p$ is the maximum amount that it deviates away from the carrier. It makes intuitive sense that both of these quantities should contribute to the bandwidth of the PM wave.

Example 5.1

A phase modulated signal has a carrier of 100 MHz and the modulating signal is a 1 kHz sinusoid of amplitude 0.01 volts. The proportionality constant, k_p, is 10^4 volts^{-1}. Find the band of frequencies occupied by the PM waveform.

Solution: We could simply plug the numbers into Eq. (5.20) to find the bandwidth. We are given all three parameters, a, k_p and f_m. However, to emphasize the fundamentals, we will not simply plug into this formula. The definition of phase modulation is given by Eq. (5.2).

$$\theta(t) = 2\pi[f_c t + k_p s(t)]$$
$$= 2\pi[10^8 t + 10^4 \times 0.01\cos 2\pi \times 10^3 t]$$

Once we know the instantaneous frequency, we can immediately identify the two parameters that are needed to find the bandwidth. The instantaneous frequency is given by

$$f_i(t) = \frac{1}{2\pi}\frac{d\theta}{dt} = 10^8 - 2\pi \times 10^5 \sin 2\pi \times 10^3 t$$

The frequency deviation is $2\pi \times 10^5$ and the maximum rate of change of the instantaneous frequency is 1 kHz. The bandwidth is therefore approximately

$$BW \approx 2(2\pi \times 10^5 + 10^3) \approx 630 \ kHz$$

The band of frequencies occupied is therefore between 100MHz±315 kHz.

5.2 Digital Phase Modulation

We have just finished showing that for *analog communication*, frequency modulation and phase modulation are very similar. The frequency of a waveform is the time derivative of the instantaneous phase, so varying frequency also varies phase.

For *digital communication*, the distinction between frequency modulation and phase modulation is more significant. This is true since digital information signals are drawn from a discrete set of waveforms.

5.2.1. PSK

In phase modulation, just as in amplitude and frequency modulation, we start with a sinusoidal carrier of the form

$$s_c(t) = A\cos\theta(t) \qquad (5.22)$$

In frequency modulation, the derivative of $\theta(t)$, which is proportional to the instantaneous frequency, follows the baseband or information signal. In phase modulation, it is the phase itself which follows this baseband signal. Thus,

$$\theta(t) = 2\pi[f_c t + k_p s(t)] \tag{5.23}$$

where f_c is the carrier frequency, k_p is the proportionality factor relating phase shift to signal voltage, and $s(t)$ is the baseband information signal.

In *binary frequency shift keying (BFSK)*, (see Section 4.1.4) we switch back and forth between sinusoids of two different frequencies, depending on whether a one or a zero is being transmitted. In *binary phase shift keying (BPSK)*, the frequency of the carrier stays constant while the phase shift takes on one of two constant values.

The two signals used to transmit a zero and one are expressed as:

$$s_0(t) = A\cos(2\pi f_c t + \theta_0)$$
$$s_1(t) = A\cos(2\pi f_c t + \theta_1) \tag{5.24}$$

where θ_0 and θ_1 are constant phase shifts.

We'll return to Eq. (5.24) in a few minutes. But first, let's look at this in an alternate way to gain a different perspective. Given any values of θ_0 and θ_1, we define an average phase, θ, and a deviation, Δ, which is one-half of the difference. That is,

$$\theta = \frac{\theta_0 + \theta_1}{2}$$

$$\Delta = \frac{\theta_1 - \theta_0}{2} \tag{5.25}$$

$$s_0 = A\cos(2\pi f_c t + \theta - \Delta)$$

$$s_1 = A\cos(2\pi f_c t + \theta + \Delta)$$

Therefore, we can express this signal as follows:

$$s_i(t) = A\cos[2\pi f_c t + \theta + d_i(t)\Delta] \tag{5.26}$$

where $d_i(t)$ is the data sequence consisting of +1 or -1 and Δ is the phase deviation, also known as the *modulation index* [compare Eq. (5.26) to Eq. (5.12)]. We use trigonometric expansions to express Eq. (5.26) as follows:

$$s_i(t) = A\cos(2\pi f_c t + \theta)\cos[\Delta d_i(t)] - A\sin(2\pi f_c t + \theta)\sin[\Delta d_i(t)] \tag{5.27}$$

We use the even and odd properties of the cosine and sine to reduce this further to the expression shown in Eq. (5.28).

$$s_i(t) = A\cos(\Delta)\cos(2\pi f_c t+\theta)-Ad_i(t)\sin(\Delta)\sin(2\pi f_c t+\theta)$$

$$(5.28)$$

The first term in Eq. (5.28) is the *residual carrier* (it doesn't depend on the data being sent). It has a power of

$$P_c = \frac{A^2\cos^2(\Delta)}{2} \tag{5.29}$$

The second term represents the *modulated information signal*, or *sidebands*, and the power of this term is

$$P_d = \frac{A^2\sin^2(\Delta)}{2} \tag{5.30}$$

The total transmitted power is the sum of these two terms, or $A^2/2$. This result is as expected since we transmit a pure sinusoid of amplitude A regardless of the specific data being sent. The energy contained in the signal used to transmit each bit is

$$E_b = \frac{A^2 T_b}{2} \tag{5.31}$$

where T_b is the bit period. For this reason, the amplitude is sometimes written in terms of the bit energy as

$$A = \sqrt{\frac{2E_b}{T_b}} \tag{5.32}$$

If the modulation index, Δ, is set equal to $\pi/2$ (as we will see is the most common value), then Eq. (5.28) reduces to

$$s_i(t) = Ad_i(t)\cos(2\pi f_c t) \tag{5.33}$$

We have arbitrarily set θ to -90° to simplify the expression. Had we not done so, we would just have to carry the phase throughout the analysis. The PSK waveform of Eq. (5.33) represents the *suppressed carrier* case and the two signals are the negative of each other.

$$s_1(t) = -s_0(t) \tag{5.34}$$

We call such signals *antipodal*. This is the best choice of phase angle since it achieves a minimum bit error rate (We prove this later in Section 5.4.2 of this chapter). In this case, the two signals are 180° out of phase with each other. They can be represented as shown in Fig. 5.5.

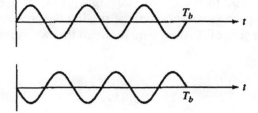

Figure 5.5 - Binary PSK (BPSK) Waveforms

As we study complex phase shift keying systems, and in particular as we generalize from binary to M-ary, an alternate signal representation will often prove useful. This representation is known as *signal space*. A signal space diagram is a vector representation which illustrates the complex projection of the transmitted signal in the direction of two orthogonal normal signals (i.e., generalized unit vectors). Returning to the general case, the two signals are

$$s_0(t) = A\cos(2\pi f_c t + \theta_0)$$

$$s_1(t) = A\cos(2\pi f_c t + \theta_1)$$

(5.35)

These can be expressed in complex notation as

$$s_0(t) = RE\ \{Ae^{j2\pi f_c t}e^{j\theta_0}\}$$

$$s_1(t) = RE\ \{Ae^{j2\pi f_c t}e^{j\theta_1}\}$$

(5.36)

We now shift the frequency down to baseband. This is sometimes known as "despinning" since you can think of the complex vector [the portion in braces in Eq. (5.36)] as composed of variations around a periodic vector which is spinning around the origin with frequency f_c. The signal space diagram is a plot of the complex part of Eq. (5.36) with the periodic component suppressed.[1]

We illustrate the signal space representation in Fig. 5.6(a) where we are using the notation of Eq. (5.25) and aligning the real axis with θ. Figure 5.6(b) shows the signal space representation for the suppressed carrier case. The distance of each point from the origin

---

[1] The concept of signal space relates to orthonormal expansions of the various signals. The signal space diagram is a vector plot of the coefficients of the expansion. We take a simpler approach in this text.

is the square root of the signal energy per bit. The energy per bit is $A^2 T_b/2$ where T_b is the bit period–the length of each sinusoidal burst used to send a single bit. The distance between the two points in the signal space diagram is $2\sqrt{E}$, or $A\sqrt{2T_b}$. This proves to be an important parameter in measuring system performance. As you might guess, the greater the distance between points in the signal space representation, the smaller the probability of bit error. The distance represents the degree of *dissimilarity* between the two signals.

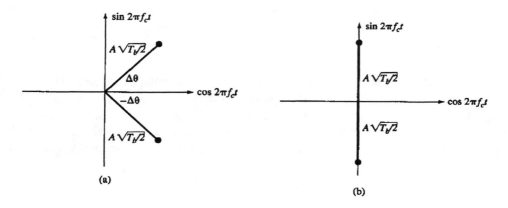

Figure 5.6 - Signal Space Representation

BPSK Spectrum
The BPSK signal can be considered as a superposition of two ASK waveforms, and the bandwidth of the resulting waveform can be found by examining the component parts. In

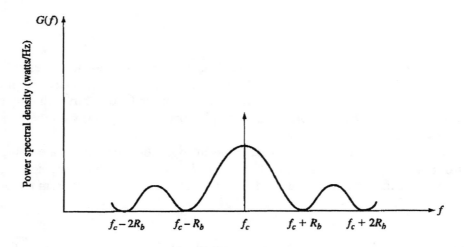

Figure 5.7 - BPSK Power Spectral Density

the suppressed carrier case, PSK is equivalent to taking the difference between two OOK signals. The first is the OOK signal resulting from amplitude shift keying with the data signal. The second results from OOK using the complement of the data signal. The frequency spectrum is therefore found in the same manner as illustrated in Chapter 3. The power spectral density of the BPSK waveform, assuming random data, is then as shown in Fig. 5.7.

Note that the first zero of the spectrum occurs at a distance of R_b, the bit rate, away from the carrier frequency. Therefore, the *nominal bandwidth* of PSK is $2R_b$.

5.2.2. MPSK

The previous section presented binary PSK. The nominal bandwidth of a BPSK signal is twice the bit rate. Suppose this bandwidth is too large to transmit through the channel. One approach is to reduce the bit rate, R_b. But this is often not acceptable. We can effect a bandwidth reduction without reducing the bit rate if we leave the realm of binary communication. For example, if bits are combined in pairs, each grouping can take on one of four possible values. We can transmit these four possible values using *quadrature phase shift keying* (QPSK), also known as 4-ary PSK. Since the phase of the transmitted carrier changes once every $2T_b$ seconds, the nominal bandwidth is *half that of BPSK*.

QPSK can be viewed as the superposition of two BPSK signals, one modulating a cosine carrier and one modulating a sine carrier. The four transmitted signals are separated by 90° from each other. Figure 5.8(a) shows one technique for generating QPSK.

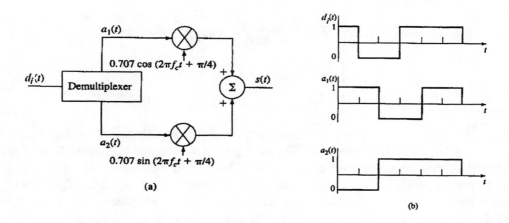

Figure 5.8 - QPSK Generator

The demultiplexer is a type of decommutator which splits the incoming data waveform into an odd and an even part. That is, the first bit takes the top path, the second takes the bottom, the third takes the top, and so on. We assume a bipolar baseband signal, so the a's are +1 or -1. The demultiplexer contains a delay so that the pairs of bits are simultaneously fed to the two modulators. Figure 5.8(b) shows the operation of the demultiplexer for a data signal, $d_i(t)$, representing the binary sequence 100111 (Since the block diagram requires a bipolar signal, think of this as +1 -1 -1 +1 +1 +1). $a_1(t)$ represents the odd bits (i.e., first, third, fifth, etc.) and $a_2(t)$ the even. The output signal, $s_i(t)$, has one of four possible phases. Trigonometric identities should convince you that the output takes on the following form:

a_1	a_2	$s_i(t)$
+1	+1	$+\cos 2\pi f_c t$
+1	-1	$-\sin 2\pi f_c t$
-1	-1	$-\cos 2\pi f_c t$
-1	+1	$+\sin 2\pi f_c t$

Although the output amplitude is one, we could scale this to any desired amplitude, A. The signal space diagram of QPSK is shown in Fig. 5.9.

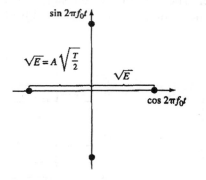

Figure 5.9 - Signal Space Diagram for QPSK

The phase changes after every two input bits. At each transition, the phase can change by either $\pm 90°$ or $180°$. Transitions of $180°$ occur if both bits change from one $2T_b$ interval to the next. Abrupt phase changes of as much as 180 ° are difficult for a demodulator to track. A variation of QPSK is therefore often used. The even bits are *not* delayed, therefore allowing the phase to change after *every bit interval* instead of after every second bit interval. This is known as *offset QPSK* (OQPSK). It is also known as staggered QPSK. The generator for OQPSK is shown in Fig. 5.10. Typical waveforms are illustrated in Fig. 5.11 for the data stream $d_i(t) = 1, 1, -1, -1, -1, 1, 1, 1$.

At each transition, only one of the component signals can change. Thus, the phase can only change by $\pm 90°$ at a transition and not by $180°$. This suppression of large phase jumps leads to better performance under certain practical considerations. In particular, the modification can significantly improve performance if the channel is bandlimited or nonlinear (e.g., when a hard limiter is present). This is true since the envelope of the waveform does not go to zero in the OQPSK case, as it does in the QPSK case.

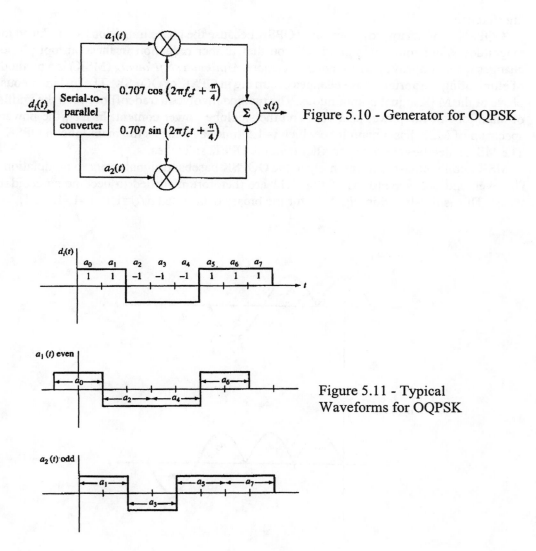

Figure 5.10 - Generator for OQPSK

Figure 5.11 - Typical Waveforms for OQPSK

Although the phase in OQPSK changes at a rate of once every T_b, the bandwidth of the resulting signal is still one half of that of binary PSK. OQPSK is the superposition of two "half-rate" PSK signals. The fact that one of the two signals is offset in time does not affect the power spectral density.

Minimum Shift Keying

In either QPSK or OQPSK, abrupt phase changes occur at bit interval transitions. In the case of QPSK, these changes are of 90° or 180° magnitude, and they occur at one half of the bit rate. In OQPSK, the magnitude of the changes is 90°, and they occur at a rate equal to the bit rate. These abrupt changes result in sidelobes of the power spectral density. Such signals are distorted when passing through a bandlimited channel (thus causing intersymbol interference).

OQPSK exhibits improvement over QPSK because the phase transitions are reduced in magnitude. Your intuition should tell you that further reduction in these abrupt phase changes would improve performance even more. *Minimum shift keying* (MSK) is a method of eliminating the abrupt phase changes occurring in QPSK or OQPSK. The instantaneous phase of the MSK signal is continuous. The method involves a tradeoff between the width of main lobe of the power spectrum and the sidelobe power content. In fact, the power spectrum of MSK has a main lobe which is 1.5 times as wide as the main lobe in QPSK. The MSK sidelobes are much smaller than the QPSK sidelobes.

MSK applies sinusoidal weighting to the OQPSK baseband signals prior to modulation. The even and odd waveforms of Fig. 5.11 are therefore modified to become sinusoidal pulses. This is illustrated in Fig. 5.12 for the binary data signal $d_i(t)=1,1,-1,-1,-1,1,1,1$.

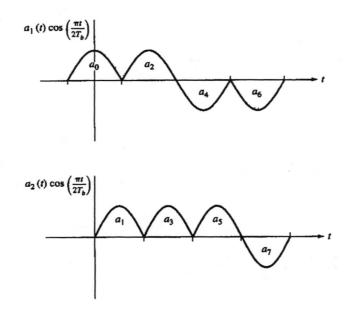

Figure 5.12 - MSK Generation

We then use the system of Fig. 5.10 to obtain the MSK waveform where d_e and d_o are the odd and even data functions.

$$s_i(t) = d_e(T)\cos\left(\frac{\pi t}{2T_b}\right)\cos 2\pi f_c t$$

$$+ d_o(t)\sin\left(\frac{\pi t}{2T_b}\right)\sin 2\pi f_c t$$

(5.37)

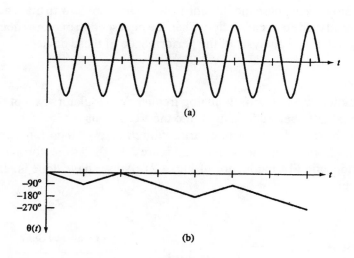

(a)

(b)

Figure 5.13 - MSK Waveform

Figure 5.13 shows the waveform represented by Eq. (5.37). Also shown in the figure is a sketch of the instantaneous phase of this waveform. The instantaneous phase is found by combining the two terms of Eq. (5.37) into a single sinusoid as in Eq. (5.38).

$$s_i(t) = d_e(t)\cos\left(2\pi f_c t - d_e(t)d_o(t)\frac{\pi t}{2T_b}\right) \tag{5.38}$$

The instantaneous phase is therefore linear (a ramp), as shown in Fig. 5.13(b).

5.2.3. DPSK

Demodulation of suppressed carrier PSK is difficult as we will see in Section 5.4.2. An alternative is to modify the signal so we can demodulate by comparing the received waveform with a delayed version of itself. This is known as *differential PSK* (DPSK). In this manner, we are using the carrier in the modulator to observe changes from one bit interval to the next. We therefore are checking for *changes* in the phase of the received signal. This technique does not allow us to observe the absolute phase of the arriving signal. To use this technique, the information must be superimposed upon the carrier in a *differential* manner.

If the original signal is coded using one of the differential forms (e.g., NRZ-M or NRZ-S), then the digital information is contained in changes between adjacent bit intervals. If we can detect changes in phase between adjacent bit intervals, we can decode such differential information.

We'll see how this simplifies the design of detectors when we get to Section 5.4.

5.3 Modulators

Now that we know what phase modulation is, we must learn how to build a modulator. That is explored in the current section. Then the next section explores demodulators, followed by a detailed analysis of the performance of these devices.

5.3.1. Analog

Analog phase modulators are the same as analog frequency modulators except that we first differentiate the signal, $s(t)$, before feeding it into the VCO. This is true since the phase of a signal will vary with $s(t)$ if the frequency varies with ds/dt,. We therefore can adapt any of the FM modulators from the previous chapter. Figure 5.14 shows one form of modulator derived from the wideband FM modulator of Fig. 4.21. All we have done is eliminate the integrator at the input.

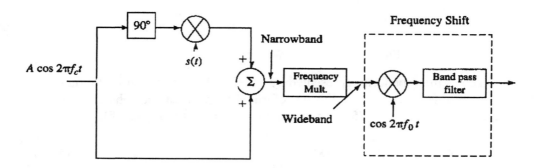

Figure 5.14 - Wideband PM Modulator

Figure 5.15 shows a simplified version using a VCO.

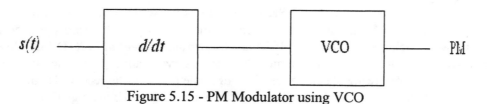

Figure 5.15 - PM Modulator using VCO

Example 5.2

Design a PM modulator for a signal with maximum amplitude of ±5 volts. Use a 74LS624 VCO, and choose a carrier frequency of 5 MHz and an index of modulation of five.

Solution: We must first do some research on the specified VCO. Some web browsing will show that Texas Instruments offers this chip at a very low price (the chip contains two identical VCOs). You would then access the data sheets on the web, and find typical circuits

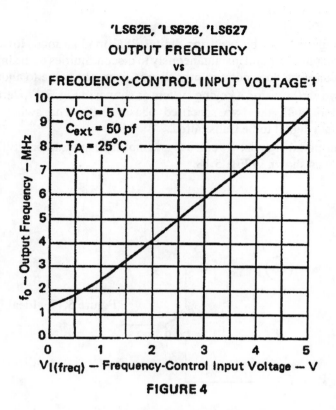

'LS625, 'LS626, 'LS627

OUTPUT FREQUENCY

vs

FREQUENCY-CONTROL INPUT VOLTAGE †

FIGURE 4

Figure 5.16 - 74LS5624 VCO Curve

and characteristics. We have selected one representative curve for a particular value of external capacitor, and this is shown as Figure 5.16. Note that varying the frequency-control input voltage between 0 and 5 volts, the output frequency of the VCO varies from about 1.2 MHz to about 9.5 MHz. Since we want a carrier frequency of 5 MHz, we would bias this input around a constant of about 2.5 volts. The frequency-control input voltage would therefore be $2.5+ks(t)$ where $s(t)$ is our input information which was specified as having an amplitude of 5 volts and frequency of 5 kHz. Our only remaining task is to choose the constant, k. The slope of the characteristic of Figure 5.16 is about 1.66 MHz per volt. If the modulation index is 5 and the input signal has a maximum frequency of 5 kHz, the maximum frequency deviation is 25 kHz. We therefore want the VCO output frequency to vary between 5 MHz - 25 kHz and 5 MHz + 25 kHz. To get a variation of ±25 kHz with a slope of 1.66 MHz per volt requires that the controlling voltage vary up and down by 15 mV. The constant, k, would then have to be $15 \times 10^{-3}/5 = 3 \times 10^{-3}$.

The design has been reduced to conditioning the input by shifting it up 2.5 and scaling it by 0.003. Then we simply connect the VCO according to the data sheet with an external capacitor of 50 pf.

5.3.2. Digital

Analog phase modulators can be used for digital PSK modulation. However, practical analog modulators cannot respond instantaneously to discontinuities in the input waveform. The output phase therefore does not abruptly change from one value to another.

Conceptually, we can envision a keying system as shown in Fig. 5.17.Here we show two sinusoidal generators with the two specified phase shifts. The switch is positioned depending on the data signal to be transmitted.

In the case of antipodal signals, we can use an amplitude modulator with an input of $+V$ or $-V$ (bipolar NRZ) as shown in Fig. 5.18.

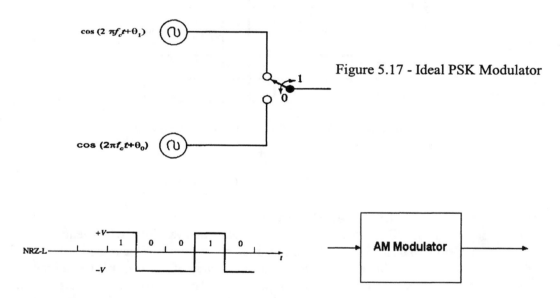

Figure 5.17 - Ideal PSK Modulator

Figure 5.18 - AM Modulator Used for Antipodal PSK

The operation shown in Fig. 5.18 can be accomplished using a gating circuit and a phase splitter. The phase splitter provides two sinusoidal signals of opposite polarity. One simple way to construct a phase splitter is to feed a sinusoid into a transistor amplifier. One output is taken at the collector and the other at the emitter. The emitter output is in phase with the input (as is the case with an emitter follower). The collector output is 180 degrees out of phase with the input.

Software Simulation Example

Start running the Tina simulation software included on the CD. If this is the first time you are using this, please see Appendix F for information.

Open the program called *PSK Modulator*. You screen should look like the picture shown below.

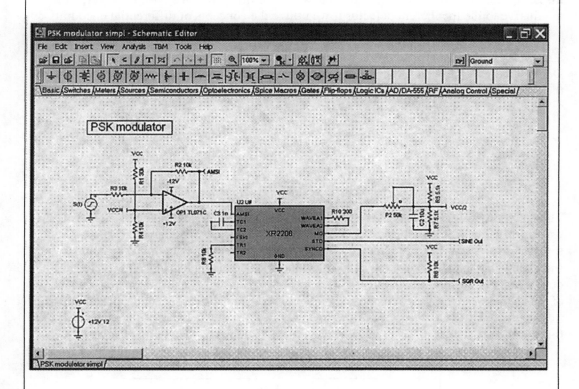

This modulator consists of a single op-amp. The XR2206 is a function generator chip from EXAR Corporation. Details can be found at http://www.exar.com/products/XR2206v103.pdf. Since the chip contains a VCO (Voltage Controlled Oscillator, of V/F converter), it can be used for phase modulation. We are feeding a square waveform into FSK1 of this chip. Run the simulation and you will get plots of the input and the output. (If the curves display on top of each other, simply pull down the *View* menu and select *Separate Curves*.) Note that the frequency of the output waveform remains constant. You can only observe the phase variations by looking at the transition points.

5.4 Demodulators

We separate our study of demodulators into analog and digital. When we examined AM in Chapter 3 and FM in Chapter 4, we saw that within each group, there were coherent and incoherent detectors. But for phase modulation, there is no incoherent detector. This is true since, if we throw away phase information, we have discarded the only parameter that contain signal information. We will, however, find an alternative to the incoherent detector if we utilize the concepts of differential encoding.

5.4.1. Analog

Demodulators for analog PM are identical to those for analog FM except that the output of the FM demodulator is integrated to yield the phase of the carrier.

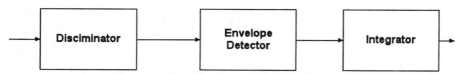

Figure 5.19 - The Discriminator Detector

The discriminator detector is shown in Fig. 5.19. We begin by processing the received PM waveform using a discriminator. One form of discriminator is a differentiator. Alternatively, a discriminator can be any linear circuit with an approximately linear amplitude vs. frequency characteristic over the band of frequencies occupied by the PM waveform. The output of the discriminator has an amplitude which varies with the instantaneous frequency of the PM waveform. The envelope detector yields an output that follows this instantaneous frequency. If we block the dc value and integrate the variations, the result is a signal which follows the phase variations of the PM waveform. We thus have a demodulator.

The phase-lock loop can also be used as a demodulator as shown in Fig. 5.20.

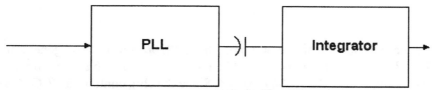

Figure 5.20 - PLL PM Demodulator

The output of the PLL is a signal that follows the frequency of the received waveform. We need simply subtract out the constant and then integrate to form the PM demodulator.

5.4.2. Digital

We examine the coherent detector for PSK. We again emphasize that there *is no incoherent form of detector for PSK*. This is true since incoherent detectors maintain only amplitude information–they throw away phase information. If, for example, a PSK waveform forms the input to an envelope detector, the output is a constant *independent of the data being transmitted.*

The general form of the coherent detector is the matched filter detector shown in Fig. 5.21.

The challenge in building the detector of Fig. 5.21 is to reconstruct the correctly-phased carriers at the receiver. If the transmitted PSK signal is of the form of transmitted carrier [i.e., $\Delta \neq \pi/2$ in Eq. (5.26)] the carrier can be recovered using a very narrowband bandpass filter or a phase-lock loop. This is shown in Fig. 5.22.

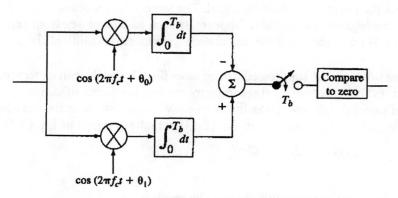

Figure 5.21 - Matched Filter Detector for BPSK

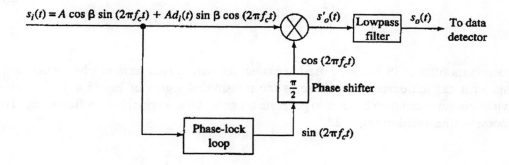

Figure 5.22 - Carrier Recovery for Transmitted Carrier Case

The input to the detector is

$$s_i(t) = A\cos\beta\sin(2\pi f_c t) + Ad_i(t)\sin\beta\cos(2\pi f_c t) \tag{5.39}$$

Equation (5.39) is a repeat of Eq. (5.28). Once again, please note that we have set the constant phase shift, θ, to -90 degrees to simplify the expression. Also note that β is the index of modulation, which for PM, is the same as the maximum phase deviation, Δ.

Therefore, the phase separation between $s_0(t)$ and $s_1(t)$ is 2Δ. The data sequence, $d_i(t)$, consists of +1 and -1. The output of the lowpass filter is then given by

$$s_o(t) = \frac{A}{2}d_i(t)\sin\beta \tag{5.40}$$

Equation (5.40) is an approximation since $d_i(t)$ cannot pass undistorted through the LPF–it contains discontinuities which lead to high frequency components. The output of the filter will be a smoothed version of the data signal. The largest carrier occurs when β approaches zero, but as this happens the signal disappears. In fact, as β increases from zero, the carrier term in Eq. (5.39) decreases and the modulation term increases until, at $\beta = \pi/2$, the carrier disappears.

As we just noted, in the suppressed carrier case, $\beta=\pi/2$, and the carrier term of Eq. (5.39) goes to zero. In this case, we must resort to more complex forms of carrier recovery. One such class of systems consists of non-linear loops. We concentrate on the squaring loop. For $\beta=\pi/2$, the signal can be written as in Eq. (5.33), which we repeat as Eq. (5.41).

$$s_i(t) = Ad_i(t)\cos(2\pi f_c t) \tag{5.41}$$

Suppose we begin by squaring this signal. The result is

$$s_i^2(t) = A^2\cos^2(2\pi f_c t)$$

$$= \frac{A^2}{4} + \frac{A^2}{4}\cos(4\pi f_c t) \tag{5.42}$$

A bandpass filter or PLL can be used to extract the pure carrier term at a frequency of $2f_c$. This term has the exact same phase as the transmitted signal of Eq. (5.41). Frequency divider circuitry can then be used to generate a signal at the original carrier frequency. This process is illustrated in Fig. 5.23.

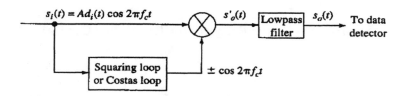

Figure 5.23 - Squaring Loop for Carrier Recovery

The squaring loop introduces a phase ambiguity of π radians or 180°. This is true since the double-frequency recovered carrier has a phase ambiguity of 2π radians. In other words, you can add 2π to the phase without changing the waveform. When the frequency is divided by two, the phase is also divided by two changing this 2π radians to π radians. We therefore have not completely recovered the carrier–just narrowed it down to two possibilities which differ in sign. This is unacceptable since an incorrect "guess" would result in the detector output yielding the exact complement of the binary data signal.

This ambiguity can be resolved by sending a known training sequence, also called a *preamble*. If the first few bits of a message are known, this can be used to resolve the phase ambiguity in the receiver. The squaring loop introduces additional noise since the non-linear operation of squaring increases the overall noise level.

A form of non-linear carrier recovery loop is the Costas loop which is shown in Fig. 5.24.

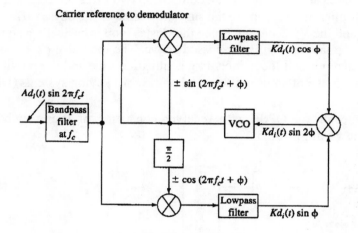

Figure 5.24 - Costas Loop

Note that the VCO operates at the carrier frequency, f_c. The loop is analogous to the squarer followed by a PLL. The output of the upper lowpass filter is given by

$$\frac{1}{2}Ad_i(t)\cos\phi$$

This output is therefore proportional to the cosine of the phase difference, ϕ. If the VCO frequency is not matched to the input frequency, this phase difference includes a linearly increasing term.

The output of the lower lowpass filter is given by

$$\frac{1}{2}Ad_i(t)\sin\phi$$

This output is therefore proportional to the cosine of the phase difference. When these two terms are multiplied together, the result is the error term, e(t).

$$e(t) = \frac{1}{4}A^2 d_i^2(t)\sin(-\phi)\cos(-\phi)$$

$$= \frac{1}{4}A^2 \sin(2\phi)$$

The loop attempts to drive this error to zero resulting in a phase difference, ϕ, of zero. The carrier can then be recovered from the output of the VCO.

The Costas loop can be thought of an comprising two PLLs fed by a single VCO. The inputs to the multiplier in this loop are the sine and cosine of the phase difference between the input carrier and the VCO output. Since the Costas loop is basically a squaring loop, it suffers from the π phase ambiguity discussed above. The loop also experiences practical problems associated with all loops including acquisition time and the possibility of false lock. For this reason, we often use differential phase shift keying systems (DPSK).

QPSK

The coherent detector for QPSK is a receiver consisting of four matched filters, or correlators, as shown in Fig. 5.25.

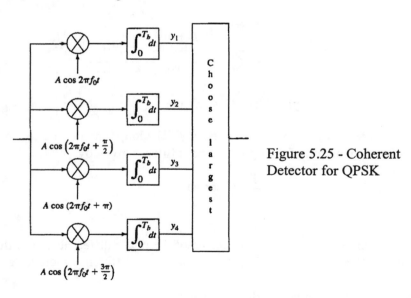

Figure 5.25 - Coherent Detector for QPSK

Note that since

$$\cos\left(2\pi f_c t + \frac{\pi}{2}\right) = -\sin(2\pi f_c t) = -\cos\left(2\pi f_c t + \frac{3\pi}{2}\right)$$

the detector could have been simplified to contain only two correlators. In that case, a two-step comparison would be required–first to find the largest output magnitude and second to test the sign of that output.

DPSK

Figure 5.26 shows a differential PSK demodulator.

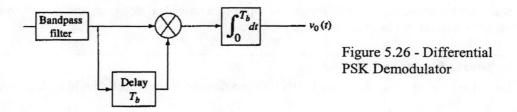

Figure 5.26 - Differential PSK Demodulator

If the signals being compared by the multiplier are identical, the input to the integrator is the square of the signal segment. Thus, provided that no change occurs between the two adjacent bit intervals, the output of the integrator is given by Eq. (5.43)

$$A^2 \int_0^{T_b} \cos(2\pi f_o t + \theta_0)\cos(2\pi f_o t + \theta_0)dt$$

$$\approx \frac{A^2 T_b}{2}$$

(5.43)

If a change does occur between the two adjacent bit intervals, the output of the integrator is given by Eq. (5.44).

$$A^2 \int_0^{T_b} \cos(2\pi f_o t + \theta_0)\cos(2\pi f_o t + \theta_1) dt$$

(5.44)

$$\approx \frac{A^2 T_b}{2} \cos(\theta_0 - \theta_1)$$

If the phase difference between the two signals is 180° (suppressed carrier), the integrator output in Eq. (5.44) becomes $-A^2 T_b/2$. Note that the integrator could be replaced by a low-pass filter with no change in the results.

Thus, the output of the integrator is a bipolar baseband signal, with the positive value obtaining if no change occurs between one interval and the next and the negative value obtaining if a change does occur. If the original data were coded using NRZ-S prior to phase modulation, the output of the demodulator would be the NRZ-L baseband signal. If the NRZ-M format is used, the output would be the inverse of the original baseband signal. The resulting baseband signal then forms the input to a baseband detector, and decisions are made to recover the original data signal.

Software Simulation Example

Load Tina and open the file called **DPSK Demodulator**. Your screen should look like this:

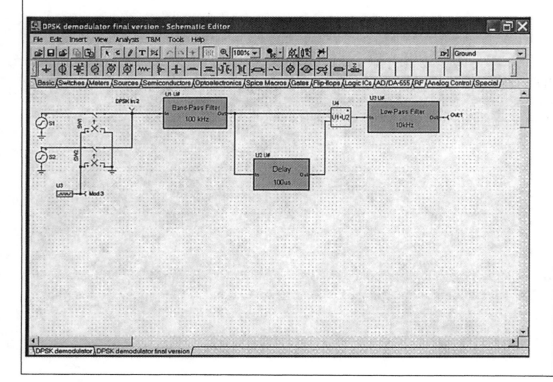

We have simulated the system of Figure 5.26. *U3* is a 2.5 kHz pulse train with a duty cycle of 50% (if you double click on this component, you will get the parameter dialogue box). If we think of the bit period as being 100 msec, this waveform represents a binary sequence of the form 001100110011.. If this type of sequence is coded differentially (using NRZ-M or NRZ-S encoding), the result is an alternating train of ones and zeros.

    Run the transient analysis simulation. You will get three waveforms as shown below.

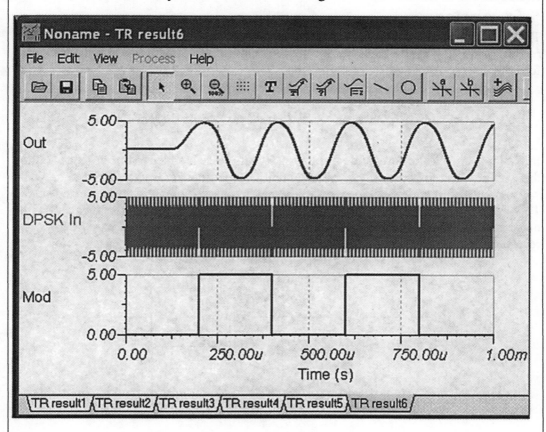

The top waveform is the output while the bottom is the pulse train input. Note that the output contains a time delay. Also observe that there is a time shift. Ideally, the positive peaks of the output should occur during the second half of each constant portion of the input pulse train. The negative peaks should occur during the first half of each of these constant portions. The offset is due to the time delay in the filters.

    You can get some insight into what is happening by varying the frequency of the input pulse train. For example, if you lower that frequency to 500 Hz or less, the entire observable part of the input should be a constant. In this case, the output should be a positive constant. You will clearly see the effects of the transient by running this special case.

5.5 PERFORMANCE

We have shown that *analog* PM is very similar to *analog* FM. Therefore the performance is also comparable. This is not the case with *digital* PM. We will show that PSK is capable of better performance than FSK. However, we pay a price in difficulty of receiver design..

5.5.1. Analog

The performance of analog PM receivers is virtually identical to that of analog FM receivers. We adapt part of that analysis from Chapter 4, and refer you to the earlier chapter for other parts.

Let us first assume that a noise-free PM waveform is transmitted.

$$\lambda_{pm}(t) = A\cos 2\pi[f_c t + k_p s(t)] \tag{5.45}$$

In Eq. (5.45), $s(t)$ is the baseband information signal. Noise is added to this PM signal along the path between the transmitter and receiver. The additive noise is assumed to be white with a power of N_o watts/Hz. The received signal, $r(t)$, is a combination of noise and signal. This enters the receiver antenna as shown in Fig. 5.27. Although the figure indicates that the noise is added prior to the receiver, this model can also be used for the case of front end receiver noise.

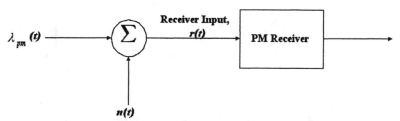

Figure 5.27 - PM Plus Noise Into Receiver

PM Receivers

A PM waveform carries its information in the form of the angle. The amplitude of the PM wave is constant. Thus, the signal information is contained in the zero crossings of the wave. The PM waveform can be clipped at a low level (i.e., effectively changed from a smooth waveform with continuous derivative to a square wave) without loss of signal information. While additive noise can significantly affect the amplitude of the PM wave, it perturbs the zero crossings to a lesser degree. Receivers therefore often clip, or limit, the amplitude of the received waveform prior to frequency detection. This provides a constant amplitude waveform as input to the discriminator. One effect of clipping is to introduce higher harmonic terms. A post-detection lowpass filter is used to reject these higher harmonics.

A block diagram of the simplified PM receiver is shown in Fig. 5.28. This receiver does not include the heterodyning and *if* strip since these do not change the signal to noise ratio.

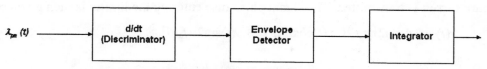

Figure 5.28 - Simplified PM Receiver

Signal to Noise Ratio

The analysis of noise in PM systems is facilitated by a phasor presentation. Figure 5.29 illustrates a phasor diagram of the received PM signal plus additive noise.

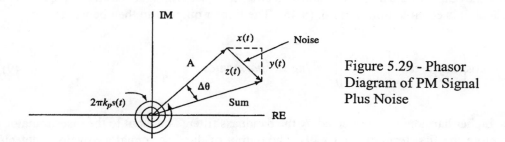

Figure 5.29 - Phasor Diagram of PM Signal Plus Noise

The signal phasor has length A and angle $2\pi k_p s(t)$. The phasor spins around according to $2\pi k_p s(t)$, while its amplitude remains constant. The noise phasor can be added to the signal if we expand the noise in quadrature components.

$$n(t) = x(t)\cos 2\pi f_c t - y(t)\sin 2\pi f_c t \tag{5.46}$$

$x(t)$ and $y(t)$ are low-frequency random processes (see Appendix C). It is the bandpass filter in the receiver which allows us to use this bandlimited quadrature representation for the noise.

The angle of the resultant phasor contains all of the useful information. The length of the vector can be disregarded, since the limiter eliminates amplitude variations.

As long as the magnitude of the noise vector does not exceed A, the maximum angular difference between the resultant and the signal vector, $\Delta\theta$, is 90°. The limiting value of 90° would only occur when the noise amplitude reaches A, and the vectors $z(t)$ and A are almost colinear and 180° out of phase. As the noise gets smaller, the angular variations around the signal vector are limited to decreasing amounts less than 90°.

The angle of the signal vector is $2\pi k_p s(t)$. For wideband PM, this angle can be considerably larger than 360°. Thus, the proportional degradation caused by the noise can be made arbitrarily small by increasing k_p.

As a first step in quantifying these observations, we shall analyze the special case of a large carrier to signal ratio. That is, assume A is much larger than $s(t)$. Thus, in the limit, a pure carrier is transmitted. The signal plus noise entering the limiter is then given by

$$r(t) = A\cos 2\pi f_c t + x(t)\cos 2\pi f_c t - y(t)\sin 2\pi f_c t \tag{5.47}$$

To trace this through the limiter, we combine terms and then eliminate the amplitude variations. We rewrite $r(t)$ as

$$r(t) = \sqrt{[A+x(t)]^2+y^2(t)}\cos\left(2\pi f_c t+\tan^{-1}\left[\frac{y(t)}{A+x(t)}\right]\right) \tag{5.48}$$

Since the output of the limiter has constant amplitude, we eliminate the amplitude term (i.e., Assume it is equal to unity) in Eq. (5.48). The limiter output can then be written as

$$\cos\left(2\pi f_c t-\tan^{-1}\left[\frac{y(t)}{A+x(t)}\right]\right) + higher\ harmonics \tag{5.49}$$

The higher harmonics are rejected by the bandpass filter. The input to the discriminator is therefore the first term in Eq. (5.49). The output of the discriminator-envelope detector combination is given by Eq. (5.50). We have taken the derivative of the inverse tangent and we have also eliminated the constant term arising from the carrier frequency.

$$\frac{d\theta}{dt} = \frac{[A + x(t)]dy/dt - y(t)dx/dt}{y^2(t) + [A + x(t)]^2} \tag{5.50}$$

The signal of Eq. (5.50) enters the final integrator of the receiver. To find the power at the output of this filter, we must examine the frequency content of Eq. (5.50). We know the power spectral densities of $x(t)$ and $y(t)$, and differentiation operations can be modeled as linear systems with $H(f)=j2\pi f$. However, before doing the mathematics we can considerably simplify Eq. (5.50) under the assumption of high carrier to noise ratio. That is, we assume that A is much larger than $x(t)$ and $y(t)$. Equation (5.50) therefore becomes

$$\frac{d\theta}{dt} = \frac{dy/dt}{A} \tag{5.51}$$

This is the noise entering the final integrator (assume that integrator cuts off above a frequency of f_m). That is, you can think of it as an ideal integrator followed by a lowpass filter. The output noise is then $y(t)/A$ for frequencies below the cutoff of the lowpass filter. Our goal is to find the output noise power, so we investigate the power spectral density of the noise. Note that the signal is not present since we assumed that the received waveform was a pure carrier plus noise. The power spectral density of $y(t)$ is N_o for frequencies between -BW/2 and +BW/2, where BW is the bandwidth of the receiver input filter. Even though the bandwidth (BW) can be many times f_m for wideband PM, the final lowpass filter will cut off any components above f_m. Therefore, the power spectral density of the output noise is as shown in Fig. 5.30.

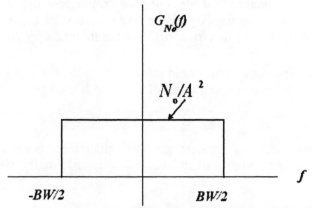

Figure 5.30 - Spectral Density of Output Noise

The output noise power is found by integrating the power spectral density over the frequency range of the lowpass filter.

$$P_N = 2\int_0^{f_m} G_{N_o}(f)df = \frac{2N_o f_m}{A^2} \tag{5.52}$$

Since the signal waveform at the output of the PM receiver is $2\pi k_p s(t)$, the output signal power is $4\pi^2 k_p^2 P_s$, where P_s is the power of the information signal, $s(t)$. The output signal to noise ratio for high CNR is then given by

$$SNR_o = \frac{4\pi^2 k_p^2 P_s}{P_N} = \frac{2A^2 k_p^2 \pi^2 P_s}{N_o f_m} \tag{5.53}$$

We can give some intuitive interpretation to Eq. (5.53) by considering the special case of a sinusoidal information signal. That is, assume $s(t) = a\cos 2\pi f_m t$. Then $P_s = a^2/2$, where we are defining SNR_i as the power of the received carrier divided by the received noise power in the PM bandwidth of $2N_o f_m$. Then

$$SNR_o = \frac{A^2 \pi^2}{N_o}\left(\frac{ak_p^2}{f_m}\right) = 2\beta^3 \pi SNR_i \tag{5.54}$$

Equation (5.54) clearly shows the effect of wideband PM. As β increases, the bandwidth increases but so does the output signal to noise ratio. The equation makes it look as if the improvement follows the third power of β. This is misleading since the input noise power to the receiver is linearly proportional to β. Therefore, if we referenced output signal to noise ratio to a signal to noise ratio that uses noise in a fixed bandwidth, the relationship would follow the square of β. This leads to a trade-off decision in the design of systems.

Threshold Effect

PM receivers experience the same type of threshold effect as FM. We therefore refer you to the previous chapter where this is discussed in detail.

5.5.2. Digital

We look first at the coherent matched filter detector. Then since there is no incoherent detector for PSK, we look at performance of differential systems. Finally, we explore performance of MPSK.

Matched Filter Detector

The performance of the matched filter (coherent) detector is completely specified by the three parameters ρ, N_o, and E. The correlation coefficient is ρ and E is the average energy of the signals used to transmit a one and a zero. The bit error rate of the binary matched filter detector is

$$
\begin{aligned}
P_e &= \frac{1}{2} erfc \left(\sqrt{\frac{E(1-\rho)}{2N_o}} \right) \\
&= Q \left(\sqrt{\frac{E(1-\rho)}{N_o}} \right)
\end{aligned}
$$

(5.55)

In phase shift keying, the two signals can be written as

$$
\begin{aligned}
s_0(t) &= A\cos(2\pi f_c t + \theta_0) \\
s_1(t) &= A\cos(2\pi f_c t + \theta_1)
\end{aligned}
$$

(5.56)

The average energy is then $A^2 T_b/2$, and the correlation coefficient is given by

$$\rho = \frac{1}{A^2T_b/2}\int_0^{T_b} A\cos(2\pi f_c t + \theta_0)A\cos(2\pi f_c t + \theta_1)dt$$

$$= \frac{1}{T_b}\int_0^{T_b}[\cos(\theta_1-\theta_0)+\cos(4\pi f_c t+\theta_1+\theta_0)dt$$

$$= \cos(\theta_1-\theta_0)+\frac{\sin(4\pi f_c T_b+\theta_1-\theta_0)-\sin(\theta_1-\theta_0)}{4\pi f_c} \approx \cos(\theta_0 - \theta_1)$$

$$(5.57)$$

The second term in the last line of Eq. (5.57) has been omitted since we assume that f_c is large enough to make it negligible relative to the first term. If this were not justified, you would have to carry that term throughout the analysis.

The complementary error function decreases as its argument increases. Therefore, the probability of error is monotonic increasing in ρ, and we can minimize the error probability by making $\rho = -1$. This value obtains if the phase difference between the two signals is 180°, This result is not surprising since the signals are then as different as possible–one is the negative of each other. We call this *antipodal* PSK. The resulting error probability is given in Eq. (5.58).

$$P_e = \frac{1}{2}erfc\left(\sqrt{\frac{A^2T_b}{2N_o}}\right)$$

$$(5.58)$$

$$= Q\left(\sqrt{\frac{A^2T_b}{N_o}}\right)$$

This error probability is plotted in Fig. 5.31. We repeat the FSK curve of Fig. 4.48. Note that the only difference between FSK (with orthogonal tone spacing) and antipodal PSK is that in the FSK case, $\rho = 0$ while in PSK, $\rho = -1$. All other parameters remain the same. If you view Eq. (5.55), you note that when $\rho = 0$, E/N_o would have to be two times as large as when $\rho = -1$ to achieve the same performance. Thus, the two curves in Fig. 5.31 are separated horizontally by 3 dB.

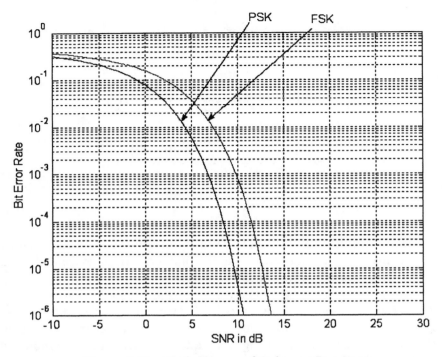

Figure 5.31 - Performance of Coherent Detectors

Differentially Coherent Detector Performance

Since the differential detector of Fig. 5.32 multiplies the signal by a delayed version of itself, it is not surprising to find that we are dealing with non-linear operations applied to the noise. The results follow those of the other non-linear detectors, such as the incoherent ASK and FSK detectors.

Let us assume that the signals in the two consecutive bit intervals are identical (i.e., a binary

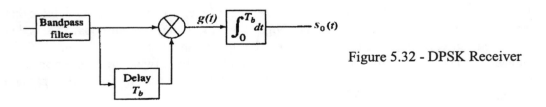

Figure 5.32 - DPSK Receiver

one is sent in NRZ-M or a zero in NRZ-S). Then the input to the integrator, $g(t)$, is given by

$$g(t) = [A\cos 2\pi f_c t + n_{nb}(t)][A\cos 2\pi f_c t_d + n_{nb}(t_d)] \tag{5.59}$$

where n_{nb} is the narrowband noise, and $t_d = t - T_b$, the time delayed by one bit period. We can expand $g(t)$ into four terms by using the quadrature expansion of the narrowband noise.

$$g(t) = [A + x(t)]\cos(2\pi f_c t)[A + x(t_d)]\cos(2\pi f_c t_d)$$

$$- [A + x(t)]\cos(2\pi f_c t)y(t)\sin(2\pi f_c t_d)$$

$$- y(t)\sin(2\pi f_c t)[A + x(t_d)\cos(2\pi f_c t_d)] \qquad (5.60)$$

$$+ y(t)y(t_d)\sin(2\pi f_c t)\sin(2\pi f_c t_d)$$

Now let us assume that the bit period is a multiple of the period of the carrier (we are assuming continuous phase modulated waveforms). This is necessary for the delayed version to align with the non-delayed version. With this assumption,

$$\cos(2\pi f_c t) = \cos(2\pi f_c t_d)$$
$$and \qquad (5.61)$$
$$\sin(2\pi f_c t) = \sin(2\pi f_c t_d)$$

We make this substitution in Eq. (5.60) and expand the sinusoidal products using trigonometric identities. We then identify the low-frequency terms that go through the filter (Actually the integrator acts like a lowpass filter since its transfer function decreases with increasing frequency). The filter output is then given by

$$s_o(t) = \frac{[A + x(t)][A + x(t_d)] + y(t)y(t_d)}{2} \qquad (5.62)$$

Recall that the noise-free output of the multiplier is an NRZ-L signal going between $+A^2$ and $-A^2$. Since we assumed that the signal in two adjacent intervals was identical, the noise-free output is $+A^2 T_b$, and an error is made if the detector output is negative. Thus,

$$P_e = PR\ \{s_o(T_b) < 0\ \} \qquad (5.63)$$

Since we are comparing to zero, we can ignore the faculty of ½ in Eq. (5.62). We now need to find the probability density of the filter output. The two noise components are sampled at two different times. In order to simplify the approach, we define four different variables using the average and differences of the noise samples.

$$w_1 = A + \frac{x(t)+x(t_d)}{2}$$

$$w_2 = \frac{x(t)-x(t_d)}{2}$$

$$w_3 = \frac{y(t)+y(t_d)}{2} \qquad (5.64)$$

$$w_4 = \frac{y(t)-y(t_d)}{2}$$

Each of these four variables is Gaussian distributed. The first has a mean of A, and the other three have a mean of zero. Using these definitions, we can rewrite the filter output as

$$s_o(t) = \left(w_1^2 - w_2^2\right) + \left(w_3^2 - w_4^2\right) \qquad (5.65)$$

The probability of error is the probability that this output is less than zero. This is given by

$$P_e = PR\left\{ \sqrt{w_1^2 + w_2^2} < \sqrt{w_3^2 + w_4^2} \right\} \qquad (5.66)$$

We don't need the square roots in Eq. (5.66). We include them to make the following observation. The first square root in this equation is a random variable which is Ricean distributed. The second square root is a random variable which is Raleigh distributed. At this point, the approach is identical to that used for incoherent detection of ASK or FSK. The result is

$$P_e = \frac{1}{2}\exp\left(\frac{-E}{N_o} \right) \qquad (5.67)$$

This is plotted in Fig. 5.33. We have repeated the result for coherent suppressed carrier BPSK on the same set of axes. Note that the probability of error is higher for the DPSK case than it is for the coherent BPSK case. This is expected since additional information about the received signal is required for the coherent case. In general, the more you know about a signal, the better the job you can do detecting it. Also please note that, although the curves appear to be almost parallel, they are not. They come from two totally different equations. Indeed, as the signal to noise ratio approaches zero (minus infinity in dB), the curves approach each other at an error rate of ½ so they could not possibly be parallel.

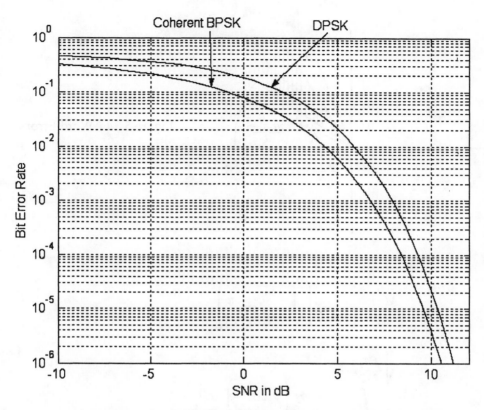

Figure 5.33 - Performance of Differentially Coherent PSK

MPSK

In MPSK, multiple bits are combined to form symbols. For example, in QPSK, bits are combined in pairs, and one of four possible sinusoidal bursts is transmitted to send each symbol. The phases are normally equally distributed around a circle. Thus, for example, in QPSK the phases would be 0°, ±90°, and 180°. In general, in MPSK, the phase separations are 360°/M, where M is the number of possible symbols. A symbol error (also known as word error) occurs if the detected phase is closer to an incorrect symbol than a correct symbol. Figure 5.34 illustrates this for a representative example. If the received phase is outside of the shaded area, a word error is made.

Calculation of the probability of the phase falling outside the shaded area is beyond the scope of this text. The equation is given by

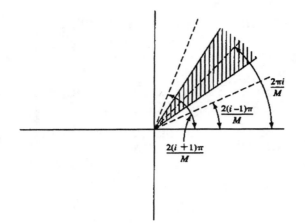

Figure 5.34 - Word Errors

$$P_M = \frac{M-1}{M} - \frac{1}{2}erf\left(\sqrt{\frac{A^2 T_s}{2N_o}}\sin\frac{\pi}{M}\right)$$

$$\tag{5.68}$$

$$- \frac{1}{\sqrt{\pi}} \int_0^{\sqrt{\frac{A^2 T_s}{2N_o}}\sin\frac{\pi}{M}} e^{-y^2} erf\left(y\ \cot\frac{\pi}{M}\right)\ dy$$

In Eq. (5.68), T_s is the symbol period. As a check on this result, suppose we let $M{=}2$. This represents the binary case. The equation reduces to

$$P_M = \frac{1}{2}\ erfc\left(\sqrt{\frac{A^2 T_s}{2N_o}}\right) \tag{5.69}$$

which is the correct result for binary PSK.

For relatively large signal to noise ratios (i.e., small word error probabilities and $M{>}2$, Eq. (5.68) can be approximated by

$$P_M \approx erfc\left(\sqrt{\frac{A^2 T_s}{2N_o}}\ \sin\left(\frac{\pi}{M}\right)\right) \tag{5.70}$$

The equations present the symbol error probability. Since M-ary communication involves combining information bits into symbols or words, we are usually more interested in the bit error probability. For example, suppose we are communicating using QPSK. We begin by combining pairs of bits into one or four possible symbols. If one of these symbols experiences a word error during transmission, this can translate back to either one or two bit errors. While we can get an average bit error rate by examining the symbol assignments, a simpler result is possible with a few assumptions.

First, we assume that if a word error is made in MPSK, the most likely event is to mistake one symbol for an adjacent symbol. Looking at Fig. 5.34 it is more likely to mistake the correct symbol for one separated by 360°/M than it is to mistake it for a symbol separated by more than this minimum angle. If the original encoding is done using a Gray code, adjacent symbols differ in only one bit position. Thus, for example, in 8-PSK, the symbols going sequentially around the circle would represent 000, 001, 011, 111, 101, 100, 110, 010. With this assignment, a circular movement by 45° translates to one bit error. With these assumptions, one symbol error translates to one bit error. The bit error rate is then approximately equal to

$$P_e = \frac{P_M}{\log_2 M} \tag{5.71}$$

To modify the signal to noise ratio, we need simply recognize that the symbol time is related to the bit period by

$$T_s = T_b \log_2 M \tag{5.72}$$

We can now combine all of these equations to present a graph of bit error rate is a function of bit signal to noise ratio. This shown in Fig. 5.35.

Figure 5.35 shows the bit error probability as a function of signal to noise ratio per bit. The signal to noise ratio per bit is $A^2 T_s/2\log_2 M$. That is, it is the signal to noise ratio per symbol divided by the number of bits in each symbol.

Note that as M increases, the error rate also increases. This makes sense since the various signal angles get closer and closer together with increasing M. This makes the signals more similar to each other. Further note that the curves for M=2 and M=4 overlap. The symbol error rate for QPSK is approximately twice that for BPSK. Since the bit rate for QPSK is assumed to be one-half of the symbol error rate (making the assumption that one symbol error translates to one bit error), the bit error rate curves overlap.

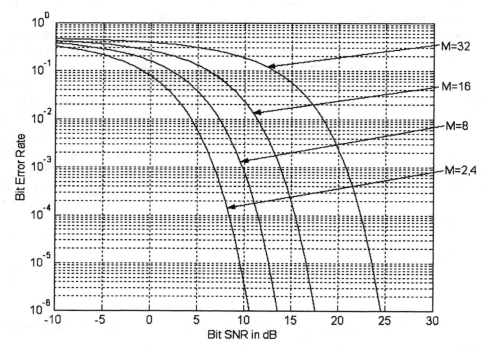

Figure 5.35 - Bit Error Rate for MPSK

5.6 Applications

Because PSK has the best performance of the three basic types of modulation (amplitude, frequency and phase), it is used wherever system design specifications are "tight", and where the design permits the extra effort required to demodulate phase-modulated waveforms. In this section, we examine MODEMS. You will find many other contemporary applications of phase modulation by researching the current literature.

5.6.1 MODEMS

The majority of low-speed MODEMs use FSK, as described in the previous chapter. These include the Bell 103 series, and the MODEMs which use the CCITT V.21, V.22 and V.23 standards. As transmission speeds increase, binary FSK can no longer be used in the limited bandwidth available in a voice channel. More bandwidth efficient techniques are needed. For medium transmission speeds in the range of 2-5 kbps, MPSK modems are popular.

Modems operating at 2400 bits per second typically use QPSK. The frequency is set at 1800 Hz, and the band of frequencies occupied by the baseband signal extends from 600 Hz to 3 kHz. That is, the band extends 1200 Hz above and below the carrier since bits are combined in pairs. The CCITT V.26 standard uses quadrature differential PSK to achieve 2400 bits per second transmission.

Modems operating at 4800 bits per second use 8-PSK. The signal constellations for the 2400 bps and 4800 bps modems are shown in Fig. 5.39.

Modems operating at 4800 bits per second use 8-PSK. The signal constellations for the 2400 bps and 4800 bps modems are shown in Fig. 5.39.

The Bell 208 series, and MODEMs using the CCITT V.27bis standard use 8DPSK, That is, they use eight different phases with differential coding. As we have seen in the previous section, as the points in the constellation get closer and closer together (i.e., we add more points), the error probability increases. Modems operating above 4800 bps typically use hybrid signaling techniques to achieve acceptable error

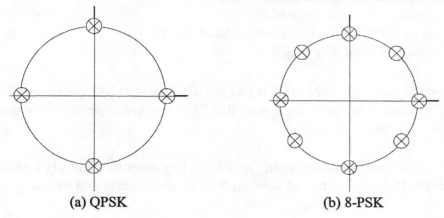

(a) QPSK   (b) 8-PSK

rates. We discuss this in the next chapter.

Figure 5.39 - Signal Constellations for QPSK and 8-PSK

Problems

5.1 A phase modulated waveform is described by

$$\lambda(t) = 50\cos(2\pi\times10^6 t + 0.001\cos2\pi\times500t)$$

(a) What is $f_i(t)$, the instantaneous frequency of $\lambda(t)$?
(b) What is the approximate bandwidth of $\lambda(t)$?
(c) Is $\lambda(t)$ a narrowband or wideband angle-modulated signal?
(d)What is $s(t)$, the information signal?

5.2 Repeat the analysis of Problem 4.25 if PM is used instead of FM. (Give answers in terms of k_p.) Does a range exist for k_p such that PM gives better performance than FM? If so, give that range.

5.3 An information signal is as shown in Fig. P5.3. The carrier frequency is 1 MHz, k_p=20. Sketch the PM waveform and label significant frequencies and values.

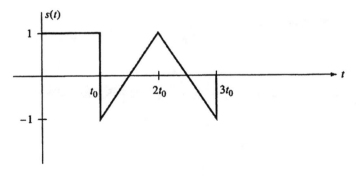

Figure P5.3

5.4 Find the power of a suppressed carrier BPSK signal that is bandlimited to a frequency between $f_c-\Delta f < f < f_c+\Delta f$. The cutoff frequencies are related to the bit rate by:
(a) $\Delta f = R_b/2$
(b) $\Delta f = R_b$
(c) $\Delta f = 2Rb$

5.5 A bipolar baseband signal is shaped by passing it through a lowpass filter with cutoff frequency $1/T_b$ (i.e., R_b). The filtered baseband signal phase modulates a carrier with maximum phase deviation $\Delta\theta = 180°$. The bit period is 1 ms. Find the bandwidth of the resulting phase-modulated signal.

5.6 Verify the entries in the table showing $s(t)$ for various $a_1(t)$ and $a_2(t)$ in Fig. 5.8.

5.7 Find the bandwidth of the OQPSK signal resulting from the signals of Fig. 5.11. Your answer should be given in terms of T_s, the sampling period.

5.8 Plot the instantaneous phase of an MSK waveform resulting from:
(a) A train of static data (all ones 's or all zeros).
(b) An alternating train of ones and zeros.

5.9 (a) Design a coherent detector for a binary transmission system that transmits the following two signals.

$$s_0(t) = A\cos 2\pi f_c t$$
$$s_1(t) = A\cos(2\pi f_c t + 90°)$$

(b) What factors affect the choice of integration time?
(c) Find the detector output under each hypothesis.

5.10 (a) Design a coherent detector for a binary transmission system that transmits the following two signals.

$$s_0(t) = A\cos 2\pi f_c t$$
$$s_1(t) = A\cos(2\pi f_c t + 180°)$$

Use only one local oscillator and correlator.
(b) Now let the phase of the local oscillator vary from its correct setting by $\Delta\theta$. Find the bit error rate as a function of $\Delta\theta$.
(c) What values of $\Delta\theta$ would cause decoding errors in the absence of noise?
(d) Show that the detector can be used to detect DPSK for any fixed value of $\Delta\theta$.

5.11 Demonstrate the performance of the DPSK detector of Fig. 5.26 for the data input

$$1011011100$$

That is, discuss the form of the transmitted DPSK signal, and find the detector output.

5.12 In the DPSK demodulator of Fig. 5.26 the delay path contains an error that causes the delay to vary between $0.9T$ and $1.1T$. Explore the effects of this "timing jitter" upon the demodulator and its performance.

5.13 A binary PSK system transmits the following two signals:

$$s_0(t) = 0.01\cos2\pi\times1000t$$
$$s_1(t) = 0.01\cos(2\pi\times1000t + \theta_1)$$

(a) Plot the probability of error of a coherent detector as a function of θ_1.
(b) Repeat part (a), assuming that the local oscillators are mismatched in phase by 45°.
(c) Assume that $\theta_1 = 180°$. Plot the probability of false alarm, P_{FA} (the probability of mistaking a transmitted zero as a received one), as a function of $\Delta\theta$, the phase mismatch.

5.14 A discriminator detector is used to detect the following two signals:

$$s_0(t) = 0.05\cos(2\pi f_c t + 90°)$$
$$s_1(t) = 0.05\cos(2\pi f_c t - 90°)$$

(a) Find the probability of error.
(b) Find the probability of error for a matched filter detector, and compare it to your answer to part (a).

5.15 Design a coherent DPSK system to transmit data at a rate of 10 kbps in additive white Gaussian noise (AWGN) with $N_o = 10^{-6}$ watt/Hz. The system must achieve a bit error rate less than 10^{-5}. Find the bandwidth of the system, and specify the minimum amplitude of the signal at the receiver.

5.16 You are given the QPSK system with the following transmitted signals:

$$s_1(t) = 10\cos(1000\pi t + 30^\circ)$$
$$s_2(t) = 10\cos(1000\pi t + 120^\circ)$$
$$s_3(t) = 10\cos(1000\pi t + 210^\circ)$$
$$s_4(t) = 10\cos(1000\pi t + 300^\circ)$$

The symbol period is 0.1 sec, and the additive noise has power $N_o = 0.01$ watt/Hz. Assume that all four signals are equally likely.

Design a matched filter detector, and find the probability of error and the probability of correct transmission.

5.17 Design a coherent detector for a binary transmission system that transmits the following two signals:

$$s_0(t) = A\cos 2\pi f_c t$$
$$s_1(t) = A\cos(2\pi f_c t + 90^\circ)$$

The signal to noise ratio is 16 dB. What is the maximum bit transmission rate if the bit error rate is to be less than 10^{-3}?

5.18 Design a PSK system to transmit data at a rate of 10 kbps in AWGN with $N_o = 10^{-6}$ watt/Hz. The channel has a bandwidth of 2 kHz. Evaluate the performance of the system.

5.19 PSK is used to send 100,000 bits per second of binary information. The additive noise has a power per BHertz of $N_o = 5 \times 10^{-6}$ watt/Hz. The carrier frequency is 1 MHz. The design specifications call for a maximum bit error rate of 10^{-4}.
(a) Design the optimum detector.
(b) What is the minimum amplitude of the received signal that is needed to meet the error rate specification?
(c) If the transmission is changed from PSK to FSK (orthogonal tone spacing), what changes must you make to your answer of part (b) to meet the specifications?

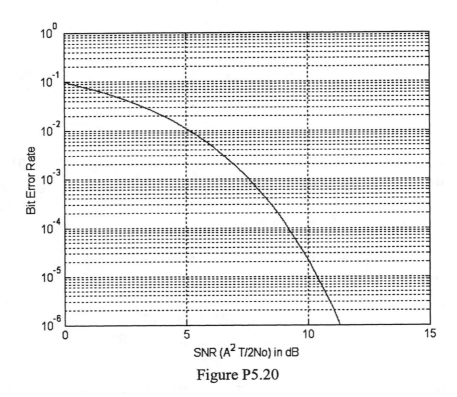

Figure P5.20

5.20 Figure P5.20 shows a curve for probability of error for a modulated digital communication system (you don't need to know whether it is FSK, PSK, ASK, or a variation of these to answer this question). Assume $A=10^{-2}$ and $R_b=1$ kbps..

(a) What is the maximum value of noise power per Hz (N_o) if the probability of error must be less than 10^{-3}?

(b) The noise power per Hertz now increases to 4 times the value found in part (a). What is the probability of error for this increased noise power?

(c) Using all the values of part (b), suppose the detector changes from coherent to incoherent. What is the new probability of bit error?

5.21 Binary information is transmitted using the following two signals:

$$s_1(t) = 6\cos(1200\pi t + 90^\circ)$$

$$s_0(t) = 6\cos(1200\pi t + 270^\circ)$$

Bits are transmitted at 400 bits per second. Noise adds to the signal and a matched filter detector is used for the receiver.

(a) Design the detector and reduce it to its simplest form.

(b) What is the correlation coefficient, ρ?

(c) Find the maximum value of noise power per Hertz such that the probability of error is less than 10^{-3}.

5.22 Figure P5.22 shows curves for probability of bit error as a function of $A^2T/2N_0$ in dB. Curves are shown for coherent detection of PSK and ASK. Three points are labeled on the curves. Find the numerical value of the x-axis markings (i.e., find the value of A, B, and C).

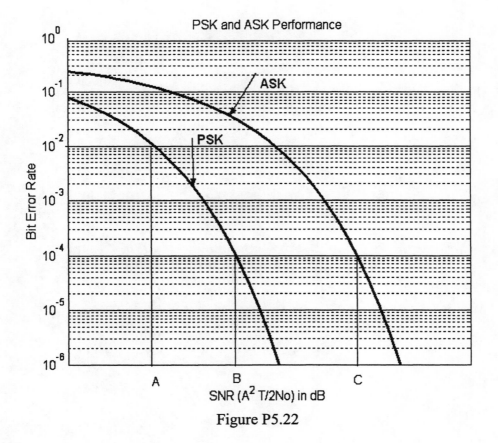

Figure P5.22

CHAPTER 6

HYBRID SYSTEMS

What we will cover and why you should care

This is the shortest chapter in the textbook. Since it combines materials from two earlier chapters, the decision was made to house it in its own chapter rather than include it in one of the earlier ones.

In the previous three chapters, we varied one of the parameters of a sinusoid to allow it to carry the information signal. In the current chapter, we explore variation of more than one parameter. We will find that this approach has the potential of allowing transmission of more information using less bandwidth.

Necessary Background

A familiarity with Fourier transform theory is necessary to understand the frequency translation process common to all modulation techniques (See Appendix A). Phase detectors require accurate tracking of carrier phase, so you need to be familiar with feedback control systems, stability, and convergence. The performance analysis requires an understanding of probability and random processes (See Appendix C).

Since this chapter combines the modulation techniques discussed in chapters 3 and 5, it would be very helpful if those two chapters were studied prior to this chapter.

6.1 Analog

We begin by investigating ways of combining multiple modulation techniques applied to a continuous sinusoidal carrier. In particular, we look at ways of simultaneously performing amplitude and phase modulation. Other combinations are often useful in specialized applications. For example, *multi-carrier modulation* (MCM) splits a signal (often data, but it could be analog) into components, and each component modulates a separate carrier signal. Although MCM was originally used by the military in analog communication systems, it has potential application to digital television. We refer you to the references for details of the specialized applications.

6.1.1. QAM

We now investigate a hybrid modulation technique that permits a form of multiplexing of signals. We have repeatedly stated that signals can be separated provided they do not overlap in time or in frequency. Double sideband AM maintains frequency separation in order to keep channels from interfering with each other. However, it uses twice the bandwidth of SSB. The fact that DSB uses twice as much bandwidth as necessary hints at the fact that it might be possible to send two DSB AM signals which overlap in both time and frequency *yet still be able to separate them* at the receiver. In fact, *quadrature amplitude modulation (QAM)* accomplishes this. We will see specific examples of this in Section 6.2 within the context of digital communication. For now, we introduce the topic for analog signals.

Suppose we have two information signals, $s_1(t)$ and $s_2(t)$, each of which is limited to frequencies below f_m. We now modulate two carriers of exactly the same frequency with these two signals. However, the carriers are in phase quadrature, that is, 90° out of phase with each other. The sum of the two AM waveforms is then given by Eq. (6.1).

$$s_{m1}(t) = s_1(t)\cos 2\pi f_c t$$

$$s_{m2}(t) = s_2(t)\sin 2\pi f_c t \tag{6.1}$$

$$s_{m1}(t) + s_{m2}(t) = s_1(t)\cos 2\pi f_c t + s_2(t)\sin 2\pi f_c t$$

Even though the two AM waveforms overlap in both frequency and time, they can be separated using the receiver shown in Fig. 6.1.

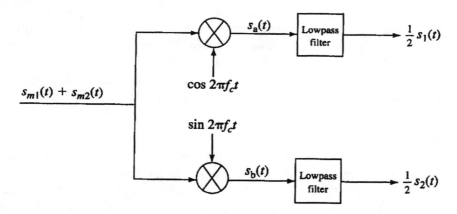

Figure 6.1 - QAM Receiver

The signal at the input of the upper lowpass filter, $s_a(t)$, is

$$s_a(t) = s_1(t)\cos^2 2\pi f_c t + s_2(t)\sin 2\pi f_c t \cos 2\pi f_c t$$

$$= \frac{1}{2}[s_1(t) + s_1(t)\cos 4\pi f_c t + s_2(t)\sin 4\pi f_c t] \tag{6.2}$$

To derive the second line of Eq. (6.2), we applied trigonometric identities to the products of sines and cosines in the first line. The output of the lowpass filter is then $s_1(t)/2$.

The signal at the input of the second lowpass filter, $s_b(t)$, is

$$s_b(t) = s_1(t)\cos2\pi f_c t\sin2\pi f_c t + s_2(t)\sin^2 2\pi f_c t$$

$$= \frac{1}{2}[s_1(t)\sin4\pi f_c t + s_2(t) - s_2(t)\cos4\pi f_c t]$$

(6.3)

The signal at the output of the bottom filter is then $s_2(t)/2$, and separation is accomplished. Of course this scheme requires perfect phase control at the receiver to avoid having one signal interfere with the other. Since phase carries some of the information, incoherent detection cannot be used. We must be able to reconstruct the carrier precisely.

Example 6.1

A QAM scheme of the type shown in Fig. 6.1 is used to simultaneously transmit two waveforms in a channel in the frequency range, $f_c \pm f_m$. The oscillators in the receiver are in error by $\Delta\theta$. Assuming that the information signals are sinusoids of equal amplitude, find the maximum value of $\Delta\theta$ such that the interference (crosstalk) is limited to -20 dB.

Solution:

We rederive Eqs. (6.1) and (6.2) for the receiver shown in Fig. 6.2.

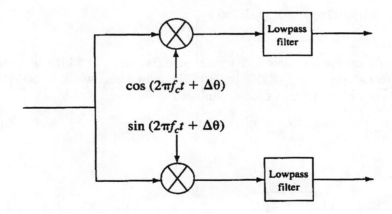

Figure 6.2 - QAM Receiver with Phase Mismatch

The output of the upper filter is now

$$\frac{1}{2}[s_1(t)\cos\Delta\theta + s_2(t)\sin\Delta\theta]$$

The output of the lower filter is

$$\frac{1}{2}[-s_1(t)\sin\Delta\theta - s_2(t)\cos\Delta\theta]$$

The ratio of the amplitude of the undesired term to that of the desired signal is $\sin\Delta\theta/\cos\Delta\theta$. Note that as the phase error approaches zero, this ratio also approaches zero as we would hope. In order for the interference to be 20 dB below the desired signal, this amplitude ratio must be 0.1. Therefore,

$$\frac{\sin\Delta\theta}{\cos\Delta\theta} \ = \ \tan\Delta\theta \ = \ 0.1$$

and the phase mismatch must be less than $\tan^{-1}(0.1) = 5.7°$.

6.1.2. AM Stereo

The concept of AM stereo is to send two independent audio signals within the 10-kHz bandwidth allocated by the FCC to each commercial AM broadcast station. Additionally, the FCC requires compatibility with existing monaural receivers. Compatibility means that a monaural receiver must recover the sum of these two signals.

Quadrature Amplitude Modulation represents one technique to send two signals simultaneously. If the two signals are designated as $s_L(t)$ and $s_R(t)$, the composite signal can be written as

$$q(t) \ = \ s_L(t)\cos 2\pi f_c t \ + \ s_R(t)\sin 2\pi f_c t \tag{6.4}$$

If both $s_L(t)$ and $s_R(t)$ are audio signals with a maximum frequency of 5 kHz, $q(t)$ occupies the band of frequencies between f_c-5 kHz to f_c+5 kHz, a total bandwidth of 10 kHz. The composite signal can be rewritten as a single sinusoid as follows:

$$q(t) \ = \ A(t)\cos[2\pi f_c t \ + \ \theta(t)] \tag{6.5}$$

where

$$A(t) \ = \ \sqrt{s_L^2(t) \ + \ s_R^2(t)}$$

$$\theta(t) \ = \ -\tan^{-1}\left(\frac{s_R(t)}{s_L(t)}\right) \tag{6.6}$$

The envelope detector in a monaural AM receiver would produce $A(t)$. This is a distorted version of the sum of the two channels, and does not meet the compatibility requirement. This is true since

$$\sqrt{s_L^2(t) \ + \ s_R^2(t)} \ \neq \ s_L(t) \ + \ s_R(t) \tag{6.7}$$

The cross product term is missing. To keep the regulatory bodies off our backs, we therefore need to investigate modifications of this system. The modification does not prove difficult, and we will delay presenting it until we continue with the analysis of this system as if it met the requirements for AM stereo.

Figure 6.3 presents a block diagram of the stereo modulator and demodulator. The phase lock loop in the demodulator is used to recover the carrier. In fact, the output of the phase lock loop is $\cos(2\pi f_c t - 45°)$. The various time functions as marked on the figure are given in Eq. (6.8).

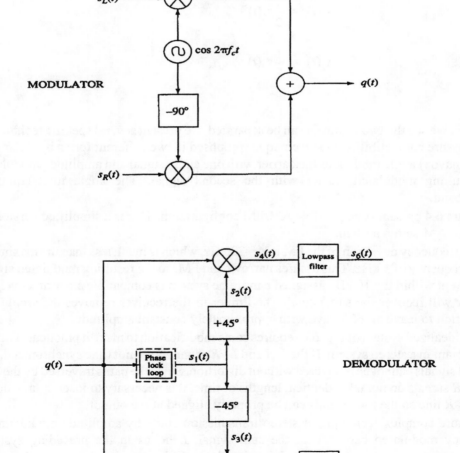

Figure 6.3 - Quadrature Modulation Stereo System

$$s_1(t) = \cos(2\pi f_c t - 45^o)$$

$$s_2(t) = \cos 2\pi f_c t$$

$$s_3(t) = \sin 2\pi f_c t$$

$$s_4(t) = s_L(t)\cos^2 2\pi f_c t + s_R(t)\sin 2\pi f_c t\cos 2\pi f_c t \qquad (6.8)$$

$$s_5(t) = s_L(t)\sin 2\pi f_c t\cos 2\pi f_c t + s_R(t)\sin^2 2\pi f_c t$$

$$s_6(t) = \frac{1}{2}s_L(t)$$

$$s_7(t) = \frac{1}{2}s_R(t)$$

Now that we see the two channels can be separated, we present several specific techniques which assure compatibility. One technique (proposed in two different forms by Belar and by Magnavox) angle modulates the carrier with one audio signal and amplitude modulates the resulting modulated carrier with the second signal. The angle modulation is narrowband.

Figure 6.4 presents one possible AM/FM configuration. This is a simplified version of the Belar AM stereo system.

The frequency deviation of the FM is Δf=320 Hz, which is much less than the maximum audio frequency of 5 kHz. This assures narrowband FM, so the resulting modulated signal can be kept within the 10 kHz assigned band. The system is compatible since an envelope detector will recover the sum signal. The limiter in the receiver removes the amplitude modulation to leave an FM wave with approximately constant amplitude.

The idealized system of Fig. 6.4 requires some modification to make it practical. Timing is important in a stereo system. If the L-R and L+R channels are not time synchronized, the original signals cannot be recovered without distortion. Since the paths traversed by the L-R and L+R signals do not take identical lengths of time, it is necessary to insert a time delay in the L-R line so the two signals can be properly aligned at the output.

A more complex technique of stereo transmission starts by amplitude modulating a frequency modulated carrier with the sum signal, L+R, as in the preceding system. However, the carrier is not angle modulated with the difference signal. Instead, the angle modulation is performed in such a way that the left channel information is carried on the lower sideband and the right channel on the upper sideband. To do so requires a relatively complex system involving a 90° phase shift between L+R and L-R signals and also

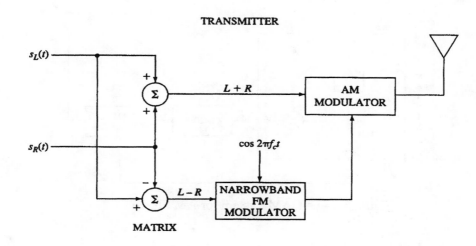

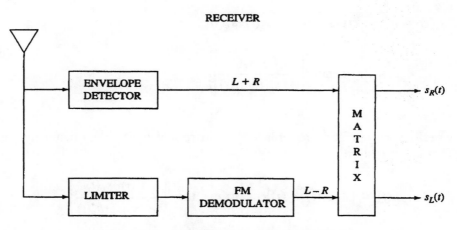

Figure 6.4 - AM/FM System for AM Stereo

employing automatic gain control (*agc*) circuitry.

One advantage of this system is that it is possible to produce a stereo effect with two monaural receivers. If one of the receivers is tuned slightly above the carrier and the other slightly below, a stereo effect is produced.

Yet a third technique is a refinement of the QAM stereo system discussed at the beginning of this section. Recall that the problem with that system was one of compatibility. A monaural receiver would recover the square root of the sum of the squares of the left and right signals. This is a distorted version of the sum waveform.

A *compatible* QAM (C-QAM) system (a simplified version of the Motorola configuration) is shown in Fig. 6.5. The system begins with a QAM signal where the two signal components are the sum and difference waveforms. The signal at $s_1(t)$ is

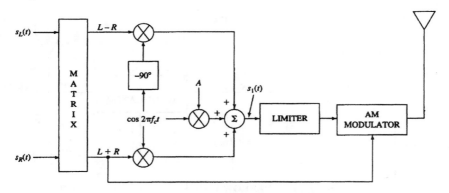

Figure 6.5 - C=QAM System for AM Stereo

$$s_1(t) = \sqrt{[a+s_L(t)+s_R(t)]^2 + [s_L(t)-s_R(t)]^2}$$

$$\times \cos\left(2\pi f_c t - \tan^{-1}\left[\frac{s_L(t)-s_R(t)}{A+s_L(t)+s_R(t)} \right] \right) \tag{6.9}$$

We now perform an operation to replace the square root multiplying factor by the sum signal. That is, we restore

$$s_2(t) = [A+s_L(t)+s_R(t)]\cos 2\pi f_c t - \tan^{-1}\left(\frac{s_L(t)-s_R(t)}{A+s_L(t)+s_R(t)} \right) \tag{6.10}$$

This is done by limiting the amplitude of the QAM waveform [Eq. (6.4)] to yield a constant amplitude, and then by amplitude modulating with the sum signal.

The system of Fig. 6.5 produces a signal which is compatible, since an envelope detector recovers the sum signal. The problem is that the stereo receiver must perform a complex function. Given the waveform of Eq. (6.10) the receiver must first replace the $L+R$ multiplying term by the square root multiplying term of Eq. (6.9). The information to do this restoration of the QAM waveform is present because the sum signal exists as the envelope, and the difference signal can be found by detecting the phase and combining this phase with the sum signal waveform. The problem is one of performing these operations with relatively simple circuitry. The receiver uses a PLL to detect the phase and then remodulates the received waveform with this phase.

6.2 Digital

We now turn our attention to hybrid techniques as applied to digital signals. In particular, we will concentrate on systems which combine ASK with PSK. Although it is possible to combine FSK with other techniques (*discrete multitone, DMT* is one form of this), it is less common than the combination we discuss below.

6.2.1. Design Tradeoffs

When we examined MASK and MPSK in Chapters 3 and 5, we found that we suffer a performance degradation as M increases. That is, if the overall extremes (amplitude and bandwidth) of a signal do not change, as we pack more and more signals into the space, they must get "closer" to each other. In orthogonal MFSK, we did not suffer the performance degradation, but the bandwidth went up as M increased. In fact, we can relate bandwidth to signal to noise ratio for the various digital signaling techniques described in the earlier chapters. In one class of practical design situation, the bit error rate is specified and we must optimize the design for that given rate. For example, a particular communication system may specify a bit error rate of 10^{-7}, and the engineer is asked to design for the maximum transmission rate within a given bandwidth and at a specified signal-to-noise ratio.

A parameter which sums up the scenario given above is the bit transmission rate per unit of system bandwidth as a function of signal to noise ratio. In order to calculate the transmission rate per Hz of bandwidth, we need to specify a bit error rate. As one example, Fig. 6.6 shows the bit rate per unit of bandwidth as function of the signal to noise ratio for a bit error rate of 10^{-4}. The points on the graph were derived from results of the previous chapters. As an example, the points for MPSK can be seen in Fig. 5.35 with the bit error rate set to 10^{-4}. Note that the given bit error rate is typical of a voice transmission system. Lower error rates are often required in data transmission systems.

The curves in the figure represent various tradeoff situations the engineer must deal with. In general, for a given signal to noise ratio, the higher we operate on the graph, the happier we are since we are squeezing more bits per second in a given bandwidth without a performance degradation.

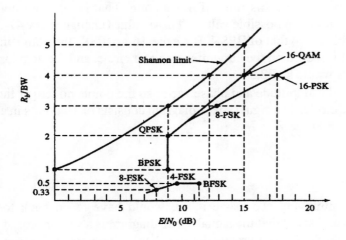

Figure 6.6 - Bit Rate/BW

It is useful to obtain an upper limit for performance. This comes from Shannon's theorem. We therefore start with the channel capacity expression we derived in Section 1.7.

$$C = BW \log_2\left(1 + \frac{S}{N}\right)$$ (6.11)

We can substitute energy per bit divided by bit period for the signal power. The noise power is N_o times the bandwidth (we are assuming white noise). Eq. (6.11) reduces to

$$C = BW \log_2\left(1 + \frac{E_b R_b}{N_o BW}\right)$$ (6.12)

Substituting R_b for C, we find

$$\frac{R_b}{BW} = \log_2\left(1 + \frac{E}{N_o}\frac{R_b}{BW}\right)$$ 6.13)

Equation (6.13) relates R_b/BW to E/N_o. This equation is plotted as the top line in Fig. 6.6.

   The challenge of hybrid systems is to design techniques that come closest to the Shannon limit in performance. QAM is one such system.

6.2.2. QAM

Looking back to the QPSK signal space diagram of Fig. 5.39, we see that the points are equally-space on the circumference of a circle. If we combined bits in groups of three, there would be eight possible values and eight points on the circumference in the signal space of 8-ary PSK or 8-PSK. Improved performance results when these points are separated as widely as possible[1]. *Quadrature amplitude modulation* (QAM) is one approach toward a different point distribution within the signal space plane.

   We present *16-QAM* as an example of this form of modulation. That is, we consider combining bits in groups of 4 yielding 16 possible values. These values change every $4T_b$, so we can expect a bandwidth that is 1/4 that of BPSK. If we use 16-ary PSK, the points in the signal space would be equally spaced on the circumference of a circle, and the angular spacing between adjacent points would be 22.5°.

   In the case of 16-QAM, both the amplitude and phase varies so the points no longer lie on the circumference of a single circle. The signal space diagram consists of 16 points in a uniform square array, as shown in Fig. 6.7.

---

[1] We ask you to accept this intuitively. To prove it, we would have to go back to orthonormal series expansions and show that the signal space diagram is a vector plot of the series coefficients. We would then have to relate distance between points to distance between two time functions.

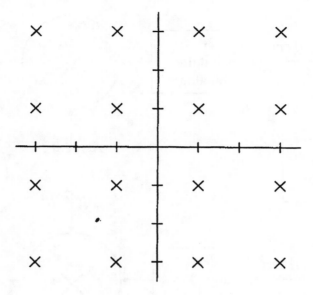

Figure 6.7 - 16-QAM Signal Space Diagram

The individual signals are of the form

$$s_i(t) = A_i \cos(2\pi f_c t + \theta_i) \tag{6.14}$$

You can think of this as simultaneously amplitude modulating a sine wave and a cosine wave and then adding the two together. The index, i, takes on values from 0 to 15. We can write this as

$$s_i(t) = RE\left\{A_i e^{j\theta_i} e^{j2\pi f_c t}\right\} \tag{6.15}$$

The signal space diagram, or *constellation*, is a plot of the first part of the brackets in Eq. (6.15) in the complex plane.

This system combines four input bits to produce one signal burst. Both the phase and the amplitude of the sinusoidal burst are modulated. A block diagram of the modulator is shown in Fig. 6.8.

The odd numbered bits in the input data stream are combined in pairs to form one of four levels which modulate the sine term. The even numbered bits are similarly combined to modulate the cosine term. The modulated sine and cosine terms are combined. The

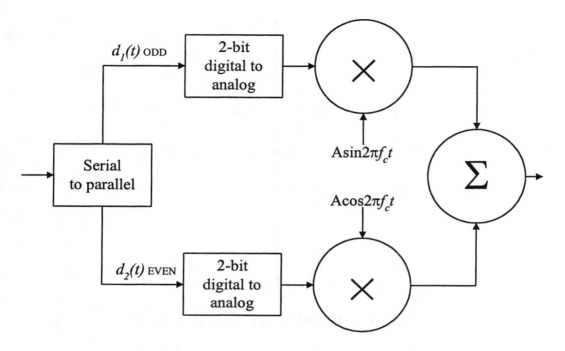

Figure 6.8 - 16-QAM Modulator

verification that this system creates the signal space diagram of Fig. 6.7 is left to the problems at the end of this chapter.

Figure 6.9 shows a demodulator for 16-QAM. This system should be compared to that of the modulator of Fig. 6.8 to verify that the original signal is recovered at the out put

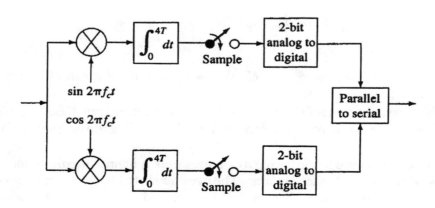

Figure 6.9 - 16 QAM Demodulator

While the 16-QAM system may appear complex as compared to simple binary systems, it possesses the potential for providing better performance for a given bandwidth.

Variations of 16-QAM

The signal space diagram for 16-QAM need not be rectangular as in Fig. 6.7. If you study this figure, you will see that the maximum signal amplitude occurs at the four outer corner points. In some applications, it is this maximum amplitude this is limited by practical considerations. In other applications, it is the average energy per bit that forms the limitation. If you make this (the limited maximum amplitude situation) into a geometry problem, you might state it as follows: *Given a maximum distance from the origin, arrange 16 points in a complex plane to maximize the minimum spacing between the points.* One alternate configuration is to maintain the same four amplitude values as the rectangular array, but rotate some of the points. This leads to the configuration of Fig. 6.10.

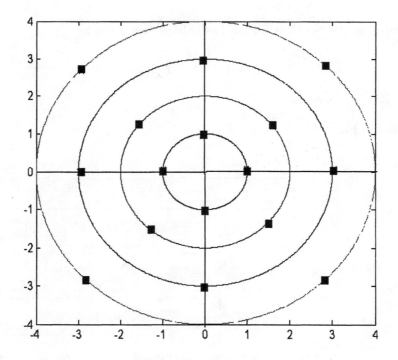

Figure 6.10 - 16-QAM Alternate Signal Space Diagram

The rectangular array has the advantage that it is easier to implement. The 2-bit A/D converters of Fig. 6.9 would have to be modified to produce the configuration of Fig. 6.10. This is true since, although there are still only four complex amplitudes, there are now twice as many values along the x and y axes. That is, if you project each point in Fig. 6.10 onto the x axis, you will find four values on each side of the zero point. If you do the same with the rectangular array of Fig. 6.7, you will only have two values on each side of the zero point. We therefore cannot think of Fig. 6.10 as a superposition of two ASK waveforms each being 2-bit.

Performance

Since the rectangular array version of 16-QAM is composed of two ASK waveforms with carriers at 90 degrees to each other. The performance can be evaluated by borrowing the results of MASK. We simply sketch the approach and give the result here.

We can show that the orthogonality leads to statistical independence. Therefore the probability of a symbol being correct is the product of the probability that each individual MASK transmission is correct. If we denote the probability of one of the ASK legs being in error as P_{ASK}, then the probability of a 16-QAM symbol being correct is $(1-P_{ASK})^2$.

We can find P_{ASK} using the MASK equations of Section 3.5. The probability of error for each of the separate legs is given by

$$P_{ASK} = 0.75 erfc\left(\sqrt{0.4SNR}\right)$$

The probability of symbol error for the 16-QAM is then given by

$$P_M = 1 - \left(1 - P_{ASK}\right)^2$$

This is plotted in Fig. 6.11.

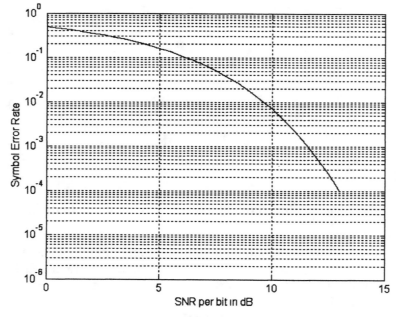

Figure 6.11 - Symbol Error for 16-QAM

6.3 MODEMS

$\mathcal{S}$ection 4.5.2 presented the concept of modems and discussed the standards as they relate to FSK. We now extend this discussion to higher speed modems using QAM. The CCITT (Comité Consultatif International Téléphonique et Télégraphique) sets international communications standards. CCITT is now known as ITU (International Telecommunications Union, which was the parent organization). CCITT standard V.29 is a 16-QAM standard used for communication up to 9600 bps. It is primarily intended for use with leased lines (so distortion is predictable and can be equalized effectively). It is the protocol used by fax modems.

To increase transmission speed beyond 9600 bps, we need to increase the number of symbols beyond 16. CCITT standard V.33 extends the speed to 14.4 kbps by using 128-QAM, while standard V.34 allows up to 28.8 kbps by using 512-QAM. Both these standard provide for automatic reduction of the bit rate is the channel signal to noise ratio is not sufficiently high. Since the signal space diagrams contain so many more points than in 16-QAM, it should not be surprising that the error rate increases. To balance the effects of this increased error rate, the V.33 and V.34 standards include *trellis coded modulation* [TCM]. In this form of transmission, the various symbols are not independent of each other. By introducing memory, errors can be reduced. We do not discuss TCM in this text.

Problems

6.1 Verify the three FSK points shown in Fig. 6.6. Assume coherent detection.

6.2 Derive the function of time for each of the 16 possible signals resulting from the 16-QAM modulator of Fig. 6.8. Plot these as points in a signal space, and compare your answer with Fig. 6.7.

6.3 Derive an expression for the signals at the inputs to the two analog-to-digital converters in the 16-QAM demodulator of Fig. 6.8.

6.4 You are given a PSK system in which the two signals are

$$s_1(t) = 10\cos2\pi \times 10^6 t$$
$$s_0(t) = 10\sin2\pi \times 10^6 t$$

Bits are transmitted at a rate of 1 million per second. The additive noise has a power $N_o = 10^{-4}$ watt/Hz.

(a) Design a matched filter detector for this transmission system, and simplify the system as much as possible.

(b) Find the probability of bit error for the detector of part (a).

(c) Repeat parts (a) and (b) for the following signals:

$$s_1(t) = 10\cos2\pi \times 10^6 t$$
$$s_0(t) = 20\sin2\pi \times 10^6 t$$

(Note that the amplitudes do not match each other in part (c).)

6.5 Verify all of the points indicated in Fig. 6.6.

6.6. You decide to design a binary communication system that combines both PSK and ASK. The signals used to send bits are given by

$$s_1(t) = 2\cos(2\pi \times 10^6 t)$$

$$s_0(t) = 4\cos(2\pi \times 10^6 t + 180^\circ)$$

NOTE that the amplitudes are not equal. Bits are transmitted at 1000 bps, and noise of power (N_o) equal to 5×10^{-3} watts/Hz adds during transmission.

(a) Design the optimum detector

(b) Find the probability of error for your detector.

APPENDIX A

FOURIER ANALYSIS

A.1 Fourier Series

A function can be approximately represented over a given interval by a linear combination of members of an *orthogonal set* of functions. If the set of functions is denoted as $g_n(t)$, this statement can be written as in Eq. (A.1).

$$s(t) \approx \sum_{n=-\infty}^{\infty} c_n g_n(t) \qquad \text{(A.1)}$$

An orthogonal set of functions is a set with the property that a particular operation performed between any two distinct members of the set yields zero. You have learned that vectors are orthogonal if they are at right angles. The *dot product* of any two distinct right-angle vectors is zero. This means that one vector has nothing in common with another. The projection of one vector on another is zero. A function can be considered as an infinite dimensional vector (think of forming a sequence by sampling the function), so the concepts from vector spaces have direct analogy to function spaces. There are many possible orthogonal sets of functions, just as there are many possible orthogonal sets of three dimensional vectors (e.g. consider any rotation of the three rectangular unit vectors).

One such set of orthogonal functions is the set of harmonically related sines and cosines. That is, the functions

$\sin 2\pi f_o t$, $\sin 4\pi f_o t$, $\sin 6\pi f_o t$.....

$\cos 2\pi f_o t$, $\cos 4\pi f_o t$, $\cos 6\pi f_o t$

form an orthogonal set for any choice of f_o. These functions are orthogonal over the interval between any starting point, t_o, and $t_o + 1/f_o$. That is,

$$\int_{t_o}^{t_o + \frac{1}{f_o}} g_n(t) g_m(t) dt = 0 \quad \text{for all} \quad n \neq m \qquad \text{(A.2)}$$

where $g_n(t)$ is any member of the set of functions, and $g_m(t)$ is any other member. This can be verified by a simple integration using the "cosine of sum" and "cosine of difference" trigonometric identities.

471

In the first sentence of this section, we used the word "approximately". We are implying that this relationship cannot always be made an equality. An orthogonal set of time functions is said to be a *complete set* if the approximation of Eq. (A.1) can be made into an equality (with the word equality being interpreted in some special sense) through proper choice of the c_n weighting factors, and for *s(t)* being any member of a certain class of functions. The three rectangular unit vectors form a complete orthogonal set in three-dimensional space, while unit vectors in the x and y direction, by themselves, form an orthogonal set which is not complete.

We state without proof that the set of harmonic time functions,

$$\cos 2\pi n f_o t, \sin 2\pi n f_o t$$

(where n can take on any integer value between zero and infinity) is an orthogonal *complete* set in the space of time functions defined in the interval between t_o and $t_o + 1/f_o$. Therefore a time function[1] can be expressed, in the interval between t_o and $t_o + 1/f_o$ by a linear combination of sines and cosines. In this case, the word "equality" is interpreted not as a pointwise equality, but in the sense that the *distance* between *s(t)* and the series representation, approaches zero as more and more terms are included in the sum. Distance is defined as

$$\int_{t_o}^{t_o+\frac{1}{f_o}} \left| s(t) - \sum_{n=0}^{\infty} c_n g_n(t) \right|^2 dt \tag{A.3}$$

This is what we will mean when we talk of equality of two time functions. This type of equality will be sufficient for all of our applications.

For convenience, we define the period of the function, T, as

$$T = \frac{1}{f_o} \tag{A.4}$$

Any time function, *s(t)*, can then be written as

$$s(t) = a_0 \cos(0) + \sum_{n=1}^{\infty} [a_n \cos 2\pi n f_o t + b_n \sin 2\pi n f_o t] \tag{A.5}$$

---

[1]In the case of Fourier series, the class of time functions is restricted to be that class which has a finite number of discontinuities and a finite number of maxima and minima in any one period. Also, the integral of the magnitude of the function over one period must exist (i.e., not be infinite).

Equation (A.5) applies in the time interval,

$$t_o < t < t_o + T$$

An expansion of this type is known as a *Fourier series*. We note that the first term in Eq. (A.5) is simply "a_o" since cos(0)=1. The proper choice of the constants, a_n and b_n is indicated by Eq. (A.6).

$$a_0 = \frac{1}{T} \int_{t_o}^{t_o+T} dt$$

$$a_n = \frac{2}{T} \int_{t_o}^{t_o+T} \cos 2\pi n f_o t \ dt \qquad (A.6)$$

$$b_n = \frac{2}{T} \int_{t_o}^{t_o+T} s(t) \sin 2\pi n f_o t \ dt$$

Note that a_0 is the *average* of the time function, $s(t)$. It is reasonable to expect this term to appear by itself in Eq. (A.5) since the average value of the sines or cosines is zero. In any equality the time average of the left side must equal the time average of the right side.

A more compact form of the Fourier Series described above is obtained if one considers the orthogonal, complete set of complex harmonic exponentials, that is, the set made up of the time functions

$$\exp(j2\pi n f_o t)$$

where n is any integer, positive or negative. This set is orthogonal over a period of $1/f_o$ sec. Recall that the complex exponential can be viewed as (actually, it is defined as) a vector of length "1" and angle, "$n2\pi f_o t$" in the complex two-dimensional plane. Thus,

$$\exp(j2\pi n f_o t) = \cos 2\pi n f_o t + j\sin 2\pi n f_o t \qquad (A.7)$$

As before, the series expansion applies in the time interval between t_o and $t_o + 1/f_o$. Therefore, any time function, $s(t)$, can be expressed as a linear combination of these exponentials in the interval between t_o and $t_o + T$ ($T = 1/f_o$ as before).

$$s(t) = \sum_{n=-\infty}^{\infty} c_n e^{j2\pi n f_o t} \qquad (A.8)$$

The c_n are given by

$$c_n = \frac{1}{T} \int_{t_o}^{t_o + T} s(t) e^{-j2\pi n f_o t} \, dt \qquad\qquad (A.9)$$

This is verified by multiplying both sides of Eq. (A.8) by $\exp(-j2\pi f_o t)$ and integrating both sides.

The basic results are summed up in Eqs. (A.5) and (A.8). Any time function can be expressed as a weighted sum of sines and cosines or a weighted sum of complex exponentials in an interval. The rules for finding the weighting factors are given in Eqs. (A.6) and (A.9).

The right side of Eq. (A.5) represents a periodic function outside of the interval $t_o < t < t_o + T$. In fact, its period is T. Therefore, if $s(t)$ happened to be periodic with period T, even though the equality in Eq. (A.5) was written to apply only within the interval $t_o < t < t_o + T$, *this equality does indeed apply for all time!* (Please give this statement some thought!)

In other words, if $s(t)$ is periodic, and we write a Fourier series that applies over one complete period, the series is equivalent to $s(t)$ for all time.

A.2 The Fourier Transform

The vast majority of interesting signals extend for all time and are nonperiodic. One would certainly not go through any great effort to transmit a periodic waveform since all of the information is contained in one period. Instead, one could either transmit the signal over a single period, or transmit the values of the Fourier series coefficients in the form of a list. The question therefore arises as to whether or not we can write a Fourier series for a nonperiodic signal.

A nonperiodic signal can be viewed as a limiting case of a periodic signal, where the period of the signal, T, approaches infinity. If the period approaches infinity, the fundamental frequency, f_o, approaches zero. The harmonics get closer and closer together, and in the limit, the Fourier series summation representation of $s(t)$ becomes an integral. In this manner, we could develop the Fourier integral (transform) theory.

To avoid the limiting processes required to go from Fourier series to Fourier integral, we will take an axiomatic approach. That is, we will *define* the Fourier transform, and then show that this definition is extremely useful. There need be no loss in motivation by approaching the transform in this "pull out of a hat" manner since its extreme versatility will become rapidly obvious.

What is a *transform*? Recall that a common everyday *function* is a set of rules which substitutes one number for another number. That is, $s(t)$ is a set of rules which assigns a number, $s(t)$, in the *range* to any number, t, in the *domain*. You can think of a function as a box which spits out a number whenever you stick in a number. In a similar manner, a *transform* is a set of rules which substitutes one function for another function. It can be thought of as a box which spits out a function whenever you stick in a function.

We define one particular transform as follows:

$$S(f) = \int_{-\infty}^{\infty} s(t)e^{-j2\pi ft}\, dt \qquad (A.10)$$

Since t is a dummy variable in integration, the result of the integral evaluation (after the limits are plugged in) is not a function of t, but only a function of f. We have therefore given a rule that assigns to every function of t (with some broad restrictions required to make the integral of Eq. (A.10) converge) a function of f.

The extremely significant Fourier transform theorem states that, given the Fourier transform of a time function, the original time function can always be uniquely recovered. The transform is unique. Either $s(t)$ or its transform, $S(f)$, uniquely characterizes a function. This is crucial! Were it not true, the transform would be useless.

An example of a useless transform (the *Roden transform*) follows:

To every function, $s(t)$, assign the function

$$R(f) = f^2 + 1.3$$

This transform defines a function of f for every function of t. The reason it has not become famous is that, among other factors, it is not unique. Given that the Roden transform of a time function is $f^2 + 1.3$, you have not got a prayer of finding the $s(t)$ which led to that transform.

Actually, the Fourier transform theorem goes one step further than stating uniqueness. It gives the rule for recovering $s(t)$ from its Fourier transform. The rule exhibits itself as an integral, and is almost of the same form as the original transform rule. That is, given $S(f)$, one can recover $s(t)$ by evaluating the following integral,

$$s(t) = \int_{-\infty}^{\infty} S(f)e^{j2\pi ft}\, df \qquad (A.11)$$

Equation (A.11) is sometimes referred to as the *inverse transform of $S(f)$*. It follows that this is also unique.

There are infinitely many unique transforms[2]. Why then has the Fourier transform achieved such widespread fame and use. Certainly it must possess properties which make it far more useful than other transforms.

Indeed, we shall presently discover that the Fourier transform is useful in a way which is analogous to the common logarithm. (Remember them from high school?) In order to multiply two numbers together, we can find the logarithm of each of the numbers, add the logs, and then find the number corresponding to the resulting logarithm. One goes through all of this trouble in order to avoid multiplication (a frightening prospect to school students).

$$a \quad \times \quad b \quad = \quad c$$

$$\Downarrow \qquad \Downarrow \quad \Downarrow \qquad \qquad \Uparrow$$

$$\log(a) \quad + \quad \log(b) \quad = \quad \log(c)$$

An operation which must often be performed between two time functions is *convolution*. It is enough to scare even those few who are not frightened by multiplication. In Section A.4, we will show that if the Fourier transform of each of the two time functions is first found, a much simpler operation can be performed upon the transforms which corresponds to convolution of the time functions. That operation, which corresponds to convolution of the two time functions, is multiplication of the two transforms. (Multiplication is no longer difficult once one graduates from high school.) Thus, we will multiply the two transforms together, and then find the time function corresponding to the resulting transform.

Notation

We will usually use the same letter for the time function and its corresponding transform, the upper case version being used for the transform. That is, if we have a time function called $g(t)$, we shall call its transform, $G(f)$. In cases where this is not possible, we find it necessary to adopt some alternative notational forms to associate a time function with its transform. The script capital $\mathfrak{F}$ and $\mathfrak{F}^{-1}$ are often used to denote taking the transform, or the inverse transform respectively. Thus, if $S(f)$ is the transform of $s(t)$, we can write

$$\mathcal{F}[s(t)] = S(f)$$

$$\mathcal{F}^{-1}[S(f)] = s(t)$$

---

[2]As two examples, consider either time scaling or multiplication by a constant. That is, define $S_1(f)=s(2f)$ or $S_2(f)=2s(f)$. Then suppose $s(t) = \sin t$, $S_1(f) = \sin 2f$ and $S_2(f)=2\sin f$.

A double-ended arrow is also often used to relate a time function to its transform, known as a transform pair. Thus we would write

$$s(t) \leftrightarrow S(f)$$

$$S(f) \leftrightarrow s(t)$$

A.3 Singularity Functions

We must introduce a new kind of function before proceeding to applications of Fourier theory. The new function arises whenever we analyze periodic functions. This new entity is part of the class of functions known as *singularities*. These can be thought of as derivatives of the unit step function. We begin by finding the Fourier transform of a gating function. We simply use the table in Appendix D to arrive at the result of Fig. A.1.

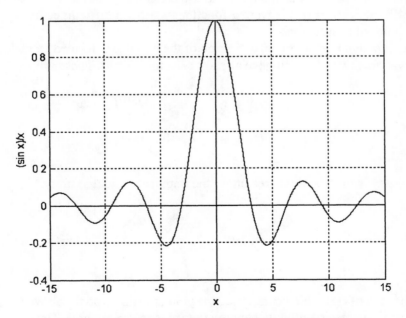

Figure A.1 - Fourier Transform of Gating Function

Functions of the type illustrated in Fig. A.1 are common in communication studies. To avoid having to repeatedly rewrite this type of function, we define the *sinc* function as

$$sinc(x) = \frac{\sin x}{x} \qquad (A.12)$$

Suppose we now wish to find the Fourier transform of a constant, *s(t)=A*, for all *t*. Plugging *s(t)=A* into the defining integral for the Fourier transform yields,

$$S(f) = \int_{-\infty}^{\infty} Ae^{-j2\pi ft} \, dt \qquad\qquad (A.13)$$

This integral does not converge. If we viewed the constant time function as a pulse whose width approaches infinity, we see the Fourier transform approaches infinity at the origin. Thus, it can only be timidly suggested that, in the limit, the height of the transform goes to infinity and the width to zero. This sounds like a pretty ridiculous function. Indeed, it is not a function at all since it is not defined at $f=0$. If we insist on saying anything about the Fourier transform of a constant, we must now restructure our way of thinking.

This restructuring begins by defining a new function which is not really a function at all. This new "function" is given the name, *impulse*. It is actually part of a class of operations known as *distributions*. We will see that when this new type of function operates on a function the result is a number. This puts the distribution somewhere between a function (which operates upon numbers to produce numbers) and a transform (which operates upon functions to produce functions).

We define this new entity as *impulse*, and use the Greek letter, *delta* (δ) to denote it. While we will write $\delta(t)$ as if this were a function, we avoid difficulties by *defining* its behavior in all possible situations.

The usual, though non-rigorous, definition of the impulse is formed by making three simple observations, two of which have already been mentioned. These are

$$\delta(t) = 0, \quad t \neq 0$$

$$\delta(t) \rightarrow \infty, \quad t = 0$$

The third property is that the total area under the impulse is equal to unity.

$$\int_{-\infty}^{\infty} \delta(t) \, dt = 1 \qquad\qquad (A.14)$$

Since all of the area of $\delta(t)$ is concentrated at one point, the limits on the above integral can be moved toward the origin without changing the value of the integral. Thus,

$$\int_{a}^{b} \delta(t) \, dt = 1 \qquad\qquad (A.15)$$

as long as $a<0$ and $b>0$.

One can also observe that the integral of $\delta(t)$ is $U(t)$, the unit step function. That is,

$$\int_{-\infty}^{t} \delta(\tau) \, d\tau = \begin{cases} 1, & t>0 \\ 0, & t<0 \end{cases} = U(t) \qquad (A.16)$$

We called the definition comprised of the above three observations "non-rigorous". Some elementary study of *singularity functions* shows that these properties do not uniquely define the impulse function. That is, there are other functions in addition to the impulse which satisfy the above three conditions. However, these three conditions can be used to indicate (not prove) a fourth. This fourth property serves as a unique definition of the delta function, and will represent the only property of the impulse which we ever make use of. We shall integrate the product of an arbitrary time function with $\delta(t)$.

$$\int_{-\infty}^{\infty} s(t)\delta(t) \, dt = \int_{-\infty}^{\infty} s(0)\delta(t) \, dt \qquad (A.17)$$

In Eq. (A.17) we have claimed that we could replace $s(t)$ by a constant function equal to $s(0)$ without changing the value of the integral. This requires some justification.

Suppose that we have two new functions, $g_1(t)$ and $g_2(t)$, and we consider the product of each of these with a third function, h(t). Suppose further that $h(t)$ is zero for part of the time axis. As long as $g_1(t) = g_2(t)$ at all values of t for which $h(t)$ is non-zero, then $g_1(t)h(t) = g_2(t)h(t)$. At those value of t which $h(t)=0$, the values of $g_1(t)$ and $g_2(t)$ have no effect on the product. One possible example is illustrated in Fig. A.2.

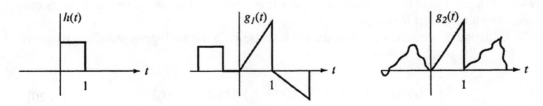

Figure A.2 - Example of $h(t)$, $g_1(t)$ and $g_2(t)$

For the functions illustrated, we see that

$$g_1(t)h(t) = g_2(t)h(t) \qquad (A.18)$$

Returning to Eq. (A.17), we note that $\delta(t)$ is zero for all $t \neq 0$. Therefore, the product of $\delta(t)$ with any time function only depends on the value of the time function at $t = 0$. Figure A.3 illustrates several possible functions which will have the same product with $\delta(t)$ as does $s(t)$.

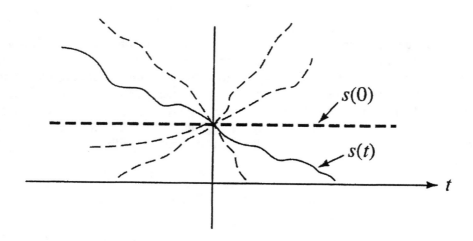

Figure A.3 - Functions Having Same Product With $\delta(t)$

Out of the infinity of possibilities, the constant time function is a wise choice since we can now factor it out of the integral to get

$$\int_{-\infty}^{\infty} s(t)\delta(t) \ dt = s(0) \int_{-\infty}^{\infty} \delta(t) \ dt = s(0) \tag{A.19}$$

This is a significant result, and we will refer to it as the *sampling property* of the impulse. Note that a great deal of information about $s(t)$ has been lost since the result depends only upon the value of $s(t)$ at one point.

A change of variables yields a shifted impulse with the analogous sampling property.

$$\int_{-\infty}^{\infty} s(t)\delta(t - t_o) \ dt = \int_{-\infty}^{\infty} s(k + t_o)\delta(k) \ dk = s(t_o) \tag{A.20}$$

Equations (A.19) and (A.20) are the only things that one must know about the impulse. Indeed, either of these can be treated as the definition of the impulse.

It is now a simple matter to find the Fourier transform of the impulse.

$$\delta(t) \leftrightarrow \int_{-\infty}^{\infty} \delta(t)e^{-j2\pi ft}dt = e^{-j2\pi ft}\Big|_{t=0} = e^0 = 1 \tag{A.21}$$

This is indeed a very nice Fourier transform for a function to have. One can guess that it will prove significant since it is the unity, or identity, multiplicative element. That is, anything multiplied by one is left unchanged.

Let us now return to the evaluation of the transform of a constant, $s(t) = A$. We observed earlier that the defining integral does not converge.

$$A \leftrightarrow \int_{-\infty}^{\infty} Ae^{-j2\pi ft} \, dt \tag{A.22}$$

For $f \neq 0$, this integral is bounded by $A/\pi f$. For $f = 0$, the integral blows up.

Since the integral defining the Fourier transform and that used to evaluate the inverse transform are quite similar, one might guess that the transform of a constant is an impulse. That is, since an impulse transforms to a constant, a constant should transform to an impulse. In the hope that this guess is valid, let us find the inverse transform of an impulse.

$$\delta(f) \leftrightarrow \int_{-\infty}^{\infty} \delta(f)e^{j2\pi ft}df = 1 \tag{A.23}$$

Our guess was correct! The inverse transform of $\delta(f)$ is a constant. Therefore, by applying a scaling factor, we have

$$A \leftrightarrow A\delta(f) \tag{A.24}$$

If we take the inverse transform of a shifted impulse, we develop the additional transform pair,

$$Ae^{j2\pi f_o t} \leftrightarrow A\delta(f - f_o) \tag{A.25}$$

The "guess and check" technique deserves some comment. We stressed earlier that the uniqueness of the Fourier transform is extremely significant. That is, given $S(f)$, $s(t)$ can be uniquely recovered. Therefore, the guess technique is a perfectly rigorous one to use in order to find the Fourier transform of a time function. If we can somehow guess at an $S(f)$

that yields $s(t)$ when plugged into the inversion integral, we have found the one and only transform of $s(t)$. As in the above example, this technique is very useful if the transform integral cannot be easily evaluated, while the inverse transform integral evaluation is a simple one.

Example

Find the Fourier transform of $s(t) = \cos 2\pi f_o t$.

Solution

We make use of Euler's identity to express the cosine function in the following form:

$$\cos 2\pi f_o t = \frac{1}{2}e^{j2\pi f_o t} + \frac{1}{2}e^{-j2\pi f_o t}$$

The Fourier transform of the cosine is then the sum of the transforms of the two exponentials, which we found in Eq. (A.25). Therefore,

$$\cos 2\pi f_o t \leftrightarrow \frac{1}{2}\delta(f - f_o) + \frac{1}{2}\delta(f + f_o)$$

We can now reveal a deception we have been guilty of. Although the Fourier transform is specified by a strictly mathematical definition, and f is just an independent functional variable, we have slowly tried to brainwash you into thinking of this variable as *frequency*. Indeed, the choice of f for the symbol of this independent variable brings the word frequency to mind. We have seen several Fourier transforms that are not identically zero for negative values of f. In fact, we shall see in Section A.5 that the transform *cannot* be zero for negative f in the case of real time functions. Since the definition of frequency (repetition rate) has no meaning for negative values, we could never be completely correct in calling f a frequency variable.

Suppose we view the positive-f axis only. For this region, the transform of $\cos 2\pi f_o t$ is non-zero only at the point $f = f_o$. The only time in your earlier education when you have experienced the definition of frequency is that in which the time function is a pure sinusoid. Since for the pure sinusoid, the positive f-axis appears to have meaning when interpreted as frequency, we shall consider ourselves justified in calling f a frequency variable.

A.4 Convolution

We are now ready to investigate the scary operation referred to at the end of Section A.2. The *convolution* of two time functions, $r(t)$ and $s(t)$, is defined by the following integral operation,

$$r(t) * s(t) = \int_{-\infty}^{\infty} r(\tau)s(t - \tau)d\tau = \int_{-\infty}^{\infty} s(\tau)r(t - \tau)d\tau \qquad \text{(A.26)}$$

the asterisk notation is conventional and is read, "$r(t)$ convolved with $s(t)$." The second integral in Eq. (A.26) results from a change of variables, and it proves that convolution is *commutative*. That is, $r(t)*s(t)=s(t)*r(t)$.

Note that the convolution of two functions of "t" is, itself, a function of t since τ is a dummy variable of integration. The integral of Eq. (A.26) is, in general, very difficult to evaluate in closed form.

We now investigate the operation of convolving an arbitrary time function with $\delta(t)$.

$$\delta(t)*s(t) = \int_{-\infty}^{\infty} \delta(\tau)s(t - \tau)d\tau = s(t - 0) = s(t) \tag{A.27}$$

This shows that any function convolved with an impulse remains unchanged.

If we convolve $s(t)$ with the shifted impulse, $\delta(t-t_o)$, we find

$$\delta(t - t_o)*s(t) = \int_{-\infty}^{\infty} \delta(\tau - t_o)s(t - \tau)d\tau = s(t - t_o) \tag{A.28}$$

In summation, convolution of $s(t)$ with an impulse function does not change the function form of $s(t)$. It may only cause a time shift in $s(t)$ if the impulse does not occur at $t=0$.

As promised, the *convolution theorem* states that the Fourier transform of a time function which is the convolution of two time functions is equal to the product of the two corresponding Fourier transforms. That is, if

$$r(t) \leftrightarrow R(f)$$

$$s(t) \leftrightarrow S(f)$$

then

$$r(t)*s(t) \leftrightarrow R(f)S(f) \tag{A.29}$$

The proof is straightforward. We simply evaluate the Fourier transform of the convolution.

$$\mathcal{F}[r(t)*s(t)] = \int_{-\infty}^{\infty} e^{-j2\pi ft} \left[\int_{-\infty}^{\infty} r(\tau)s(t - \tau)d\tau \right] dt$$

$$(A.30)$$

$$= \int_{-\infty}^{\infty} r(\tau) \left[\int_{-\infty}^{\infty} e^{-j2\pi ft}s(t - \tau)dt \right] d\tau$$

We now make a change of variables in the inner integral by letting $t - \tau = k$. We then have

$$\mathcal{F}[r(t)*s(t)] = \int_{-\infty}^{\infty} r(\tau)e^{-j2\pi f\tau} \left[\int_{-\infty}^{\infty} s(k)e^{-j2\pi fk}dk \right] d\tau \qquad (A.31)$$

The integral in the square brackets is simply $S(f)$. Since $S(f)$ is not a function of τ, it can be pulled out of the outer integral. This yields the desired result,

$$\mathcal{F}[r(t)*s(t)] = S(f) \int_{-\infty}^{\infty} r(\tau)e^{-j2\pi f\tau}d\tau = S(f)R(f) \qquad (A.32)$$

and the theorem is proved.

Convolution is an operation performed between two functions. They need not be functions of the independent variable t. We could just as easily have convolved two Fourier transforms together to get a third function of f.

$$H(f) = R(f)*S(f) = \int_{-\infty}^{\infty} R(k)S(f - k)dk \qquad (A.33)$$

Since the integral defining the Fourier transform and that yielding the inverse transform are quite similar, one might guess that convolution of two transforms corresponds to multiplication of the two corresponding time functions. Indeed, one can prove, in an analogous way to the above proof, that

$$R(f)*S(f) \leftrightarrow r(t)s(t) \qquad \text{(A.34)}$$

To prove this, simply calculate the inverse Fourier Transform of $R(f)*S(f)$. Equation (A.32) is called the *time convolution theorem* and Eq. (A.34), the *frequency convolution theorem*.

A.4.1 Parseval's Theorem

There is little similarity between the waveshape of a function and that of its Fourier transform. However, certain relationships do exist between the energy of a time function and the energy of its transform. Here, we use *energy* to denote the integral of the square of the function. This term is used since it would represent the amount of energy, in watt-seconds, dissipated in a 1Ω resistor if the time signal represented the voltage across or the current through the resistor. This relationship proves useful if we know the *transform* of a time function and wish to know the energy of the time function. We do not need to go through the effort of evaluating the inverse transform.

Parseval's Theorem is such a relationship. It is derived from the frequency convolution theorem. Starting with that theorem, we have

$$r(t)s(t) \leftrightarrow R(f) * S(f)$$

$$\mathscr{F}[r(t)s(t)] = \int_{-\infty}^{\infty} r(t)s(t)e^{-j2\pi ft} \, dt \qquad \text{(A.35)}$$

$$= \int_{-\infty}^{\infty} R(k)S(f - k) \, dk$$

Since the above equality holds for all values of f, we can let $f=0$. For this value of f, Eq. (A.35) becomes,

$$\int_{-\infty}^{\infty} r(t)s(t)dt = \int_{-\infty}^{\infty} R(k)S(-k) \, dk \qquad \text{(A.36)}$$

Equation (A.36) is one form of Parseval's formula. It can be made to relate to energy by further taking the special case of

$$s(t) = r^*(t) \qquad \text{(A.37)}$$

The Fourier transform of the conjugate, $\Im[r*(t)]$, is given by the conjugate of the transform, reflected around the vertical axis, $R*(-f)$. You should take the time now to prove this last statement.

Using this result in Eq. (A.36) we find

$$\int_{-\infty}^{\infty} |r^2(t)| \ dt = \int_{-\infty}^{\infty} |R^2(f)| \ df \tag{A.38}$$

We have used the fact that the product of a function with its complex conjugate is equal to the magnitude of the function, squared[3]. (Convince yourself that the magnitude squared is the same as the magnitude of the square of a complex number).

Equation (A.38) shows that the energy of the time function is equal to the energy of its Fourier transform.

A.5 Properties of the Fourier Transform

We now illustrate some of the more important properties of the Fourier transform. One can certainly go through technical life without making use of any of these properties, but to do so would involve considerable repetition and extra work. The properties allow us to derive something *once* and then to used the result for a variety of applications. They also allow us to predict the behavior of various systems.

A.5.1 Real/Imaginary-Even/Odd

The following table summarizes properties of the Fourier transform based upon observations made upon the time function.

	Time Function	Fourier Transform
A	Real	Real part even, imaginary part odd
B	Real and Even	Real and even
C	Real and odd	Imaginary and odd
D	Imaginary	Real part odd; imaginary part even
E	Imaginary and Even	Imaginary and even
F	Imaginary and Odd	Real and odd

Proof:

The Fourier transform defining integral can be expanded using Euler's identity as follows:

---

[5]The time signals we deal with in the real communications world are real functions of time. However, as in basic circuit analysis, complex mathematical functions are often used to represent sinusoids. A complex number is used for the magnitude and phase angle of a sinusoid. Therefore, although complex signals do not exist in real life, they are often used in "paper" solutions of problems.

$$S(f) = \int_{-\infty}^{\infty} s(t)e^{-j2\pi ft} \, dt$$

$$= \int_{-\infty}^{\infty} s(t)\cos 2\pi ft \, dt - j\int_{-\infty}^{\infty} s(t)\sin 2\pi ft \, dt \qquad (A.39)$$

$$= R + jX$$

R is an even function of f since when f is replaced with $-f$, the function does not change. Similarly, X is an odd function of f.

If $s(t)$ is first assumed to be real, R becomes the real part of the transform and X is the imaginary part. Thus, *property A* is proved.

If in addition to being real, $s(t)$ is even, then $X=0$. This is true since the integrand in X is odd (the product of even and odd) and integrates to zero. Thus, *property B* is proved.

If $s(t)$ is now real and odd, the same argument applies, but $R = 0$. This proves *property C*.

Now we let $s(t)$ be imaginary. X now becomes the imaginary part of the transform, and R is the real part. From this simple observation, *properties D, E, and F* are verified.

A.5.2 Time Shift

The Fourier transform of a shifted time function is equal to the product of the transform of the original time function with a complex exponential.

$$s(t - t_o) \leftrightarrow e^{-j2\pi ft_o}S(f) \qquad (A.40)$$

Proof

The proof follows directly from evaluation of the transform of $s(t-t_o)$.

$$\mathcal{F}[s(t - t_o)] = \int_{-\infty}^{\infty} s(t - t_o)e^{-j2\pi ft} \, dt = \int_{-\infty}^{\infty} s(\tau)e^{-j2\pi(\tau+t_o)} \, d\tau \qquad (A.41)$$

The second integral follows from a change of variables, letting $\tau=(t-t_o)$. We now pull the part that does not depend on τ to the front of the integral, and note that the remaining part is a Fourier transform of $s(t)$. Finally, we get

$$S(f - f_o) \leftrightarrow e^{j2\pi f_o t} s(t) \tag{A.42}$$

A.5.3 Frequency Shift

The time function corresponding to a shifted Fourier transform is equal to the product of the time function of the unshifted transform with a complex exponential.

Proof

The proof follows directly from evaluation of the inverse transform of $S(f\text{-}f_o)$.

$$\int_{-\infty}^{\infty} S(f - f_o) e^{j2\pi ft} df = \int_{-\infty}^{\infty} S(k) e^{j2\pi t(k + f_o)} dk \tag{A.43}$$

In the second integral, we have made a change of variables letting $k = f - f_o$. We now pull the part of the integrand that does not depend on k in front of the integral, and recognize that the remaining integral is the inverse Fourier transform of $s(t)$. This yields,

$$S(f - f_o) \leftrightarrow e^{j2\pi f_o t} s(t) \tag{A.44}$$

A.5.4 Linearity

Linearity is undoubtedly the most important property of the Fourier transform.

The Fourier transform of a linear combination of time function is a linear combination of the corresponding Fourier transforms.

$$as_1(t) + bs_2(t) \leftrightarrow aS_1(f) + bS_2(f) \tag{A.45}$$

where a and b are any constants.

Proof

The proof follows directly from the definition of the Fourier transform, and from the fact the integration is a linear operation.

A.5.5 Modulation Theorem

The modulation theorem is very closely related to the frequency shift theorem. We treat it as a separate entity since it forms the basis of the entire study of amplitude modulation.

The result of multiplying a time function by a pure sinusoid is to shift the original transform both up and down by the frequency of the sinusoid (and to cut the amplitude in half).

We start by assuming that $s(t)$ is given, with its associated Fourier transform. $s(t)$ is then multiplied by a cosine waveform to yield

$$s(t)\cos 2\pi f_o t$$

where the frequency of the cosine is f_o. The Fourier transform of this waveform is given by

$$\mathscr{F}[s(t)\cos 2\pi f_o t] = \frac{1}{2}S(f-f_o) + \frac{1}{2}S(f+f_o) \tag{A.46}$$

Proof
The proof of the modulation theorem follows directly from the frequency shift theorem. We split $\cos 2\pi f_o t$ into two exponential components and then apply the frequency shift theorem to each component.

A.6 Periodic Functions

We found that the Fourier transform of the cosine function is composed of two impulses occurring at the frequency of the cosine, and at the negative of this frequency. We will now show that the Fourier transform of any periodic function of time is a discrete function of frequency. That is, the transform is non-zero only at discrete points along the f-axis. The proof follows from Fourier series expansions and the linearity of the Fourier transform.

Suppose we find the Fourier transform of a time function, $s(t)$, which is periodic with period T. We can express the function in terms of the complex Fourier series representation.

$$s(t) = \sum_{n=-\infty}^{\infty} c_n e^{jn2\pi f_o t}$$
$$where \tag{A.47}$$
$$f_o = \frac{1}{T}$$

We previously established the transform pair,

$$Ae^{j2\pi f_o t} \leftrightarrow A\delta(f - f_o) \tag{A.48}$$

From this transform pair and the linearity property of the transform, we have

$$\mathcal{F}[s(t)] = \sum_{n=-\infty}^{\infty} c_n \mathcal{F}[e^{jn2\pi f_o t}] \tag{A.49}$$

Example
Find the Fourier transform of the periodic function made up of unit impulses, as shown in Fig. A.4.

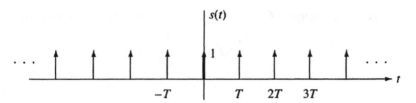

Figure A.4 - Periodic $s(t)$ for Example

The function is given by

$$s(t) = \sum_{n=-\infty}^{\infty} \delta(t - nT)$$

Solution
The Fourier transform is given by Eq. (A.49).

$$S(f) = \sum_{n=-\infty}^{\infty} c_n \delta(f - nf_o)$$

where

$$f_o = \frac{1}{T}$$

$$c_n = \frac{1}{T} \int_{-\frac{T}{2}}^{\frac{T}{2}} s(t) e^{-jn2\pi f_o t}\ dt$$

Within the range of integration, the only contribution of $s(t)$ is that due to the impulse at the origin. Therefore,

$$c_n = \frac{1}{T} \int_{-\frac{T}{2}}^{\frac{T}{2}} \delta(t) e^{-jn2\pi f_o t} \, dt = \frac{1}{T}$$

Finally, the Fourier transform of the pulse train is given by

$$S(f) = \frac{1}{T} \sum_{n=-\infty}^{\infty} \delta(f - nf_o)$$

where

$$f_o = \frac{1}{T}$$

The function of the previous example has an interesting Fourier series expansion. All of the coefficients are equal. Each frequency component possesses the same amplitude as every other component. This is analogous to the observation that the Fourier transform of a single impulse is a constant. This similarity leads us to examine the relationship between the Fourier transform of a periodic function and the Fourier transform of one period of this function.

Suppose that $s(t)$ represents a single period of the periodic function, $s_p(t)$. We can then express the periodic function as a sum of shifted versions of $s(t)$.

$$s_p(t) = \sum_{n=-\infty}^{\infty} s(t - nT) \tag{A.50}$$

Since convolution with an impulse simply shifts the original function, Eq. (A.50) can be rewritten as Eq. (A.50).

$$s_p(t) = s(t) * \sum_{n=-\infty}^{\infty} \delta(t - nT) \tag{A.51}$$

Convolution in the time domain is equivalent to multiplication of the Fourier transforms. The Fourier transform of the train of impulses was found in the example. Transforming Eq. (A.51) then yields

$$S_p(t) = S(f) \sum_{n=-\infty}^{\infty} \frac{1}{T} \delta(f - nf_o) \tag{A.52}$$

Equation (A.52) shows that the Fourier transform of the periodic function is simply a sampled and scaled version of the transform of a single period of the waveform.

APPENDIX B

LINEAR SYSTEMS

B.1 The System Function

We begin by defining some common terms. A *system* is a set of rules that associates an *output* time function to every *input* time function. This is shown in block diagram form as Fig. B.1.

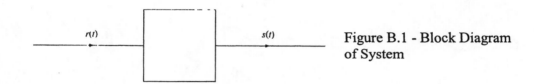

Figure B.1 - Block Diagram of System

The input, or *source* signal, is *r(t)*; *s(t)* is the output, or *response* signal due to this input. The actual physical structure of the system determines the exact relationship between *r(t)* and *s(t)*.

A single-ended arrow is used as a shorthand method of relating an input to its resulting output. That is,

$$r(t) \rightarrow s(t)$$

is read, "an input, *r(t)*, causes an output, *s(t)*."

For example, suppose the system under study is an electric circuit. Then *r(t)* could be an input voltage or current signal, and *s(t)* could be a voltage or current measured anywhere in the circuit. We would not modify the block diagram representation of Fig. B.1 even though the circuit schematic would have *two* wires for each voltage. The single lines in the figure represent signal flow.

In the special case of a two-terminal electrical network, *r(t)* could be a sinusoidal input voltage across two terminals, and *s(t)* could be the current flowing into one of the terminals due to the impressed voltage. In this case, the relationship between *r(t)* and *s(t)* is the *complex impedance* between the two terminals of the network.

Any system can be described by specifying the response, *s(t)*, associated with *every* possible input, *r(t)*. This is an obviously exhaustive process. We would certainly hope to find a much simpler way of characterizing the system.

Before introducing alternate techniques of describing systems, some additional basic definitions are needed.

A system is said to obey *superposition* if the output due to a sum of inputs is the sum of the corresponding individual outputs. That is, given that the response (output) due to an excitation (input) of $r_1(t)$ is $s_1(t)$, and that the response due to $r_2(t)$ is $s_2(t)$, then the response due to $r_1(t)+r_2(t)$ is $s_1(t)+s_2(t)$.

Restating this, a system that obeys superposition has the property that if

$$r_1(t) \rightarrow s_1(t)$$

and

$$r_2(t) \rightarrow s_2(t)$$

then

$$r_1(t) + r_2(t) \rightarrow s_1(t) + s_2(t)$$

Some thought should convince you that in order for a circuit to obey superposition, the source-free, or transient response (response due to initial conditions) must be zero [let $r_2(t)=0$ to prove this]. In practice, one often replaces a circuit having non-zero initial conditions with one containing zero initial conditions. Additional sources are added to simulate the initial condition contributions.

A concept closely related to superposition is *linearity*. Assume again that $r_1(t) \rightarrow s_1(t)$ and $r_2(t) \rightarrow s_2(t)$. The system is said to be linear if the following relationship holds true for all values of the constants, a and b.

$$ar_1(t) + br_2(t) \rightarrow as_1(t) + bs_2(t)$$

In the remainder of this text, we will use the words *linearity* and *superposition* interchangeably.

A system is said to be *time invariant* if the response due to an input is not dependent upon the actual time of occurrence of the input. That is, a time shift in input signal causes an equal time shift in the output waveform. In symbolic form, if

$$r(t) \rightarrow s(t)$$

then

$$r(t - t_o) \rightarrow s(t - t_o)$$

for all real t_o.

A sufficient condition for an electrical network to be time invariant is that its component values do not change with time (assuming unchanging initial conditions). That is, resistance, capacitances, and inductances remain constant.

Returning to the task of characterizing a system, we shall see that for a time invariant linear system, a very simple description is possible. That is, instead of requiring that we know the response due to *every* possible input, it will turn out that we need only know the output for one *test* input.

Convolution of any function with an impulse yields the original function. That is,

$$r(t) = r(t) * \delta(t)$$

$$= \int_{-\infty}^{\infty} r(\tau)\delta(t - \tau)d\tau \qquad \text{(B.1)}$$

Although one must always use extra caution in working with impulses, let us assume that the integral can be considered as a limiting case of a sum, as given in Eq. (B.2).

$$r(t) = \lim_{\Delta\tau \to 0} \sum_{n=-\infty}^{\infty} r(n\Delta\tau)\delta(t - n\Delta\tau)\Delta\tau \qquad \text{(B.2)}$$

Equation (B.2) represents a weighted sum of delayed impulses. Suppose now that this weighted sum forms the input to a linear time invariant system. The output would then be a weighted sum of delayed outputs due to a single impulse.

Suppose now that we know the system's output due to a single impulse. Let us denote that output as $h(t)$, the *impulse response*. The output due to the input of Eq. (B.2) is then given by

$$s(t) = \lim_{\Delta\tau \to 0} \sum_{n=-\infty}^{\infty} r(n\Delta\tau)h(t - n\Delta\tau)\Delta\tau \qquad \text{(B.3)}$$

If we now take the limit, this becomes an integral.

$$s(t) = \int_{-\infty}^{\infty} r(\tau)h(t - \tau)d\tau = r(t) * h(t) \qquad \text{(B.4)}$$

Equation (B.4) states that the output due to *any* input is found by convolving that input with the system's response to an impulse. All we need to know about the system is its impulse response. Equation (B.4) is known as the *superposition integral equation*.

The Fourier transform of the impulse is unity. Therefore, in an intuitive sense, $\delta(t)$ contains all frequencies to an equal degree. This observation hints at the impulse's suitability as a test function for system behavior. On the negative side, it is not possible to produce a perfect impulse in real life. We can only approximate it with a large amplitude, very narrow pulse.

Taking the Fourier transform of Eq. (B.4) yields,

$$S(f) = R(f)H(f)$$

$$H(f) = \frac{S(f)}{R(f)} \tag{B.5}$$

The Fourier transform of the impulse response is the ratio of the output Fourier transform to the input Fourier transform. It is given the name, *transfer function* or *system function*, and it completely characterizes a linear time invariant system.

B.2 Filters

The word "filter" traditionally referred to the removal of the undesired parts of something. In linear system theory, it was probably originally applied to systems which eliminate undesired frequency components from a time waveform. The term has evolved to include systems which simply weight the various frequency components of a signal.

Many of the communication systems we discuss contain *ideal distortionless filters*. We therefore begin our study by defining *distortion*.

A distorted time signal is one whose basic shape has been altered. $r(t)$ can be multiplied by a constant and shifted in time without changing the basic shape of the waveform.

In mathematical terms, we consider $Ar(t-t_o)$ to be an undistorted version of $r(t)$, where A and t_o are any real constants. Of course, A cannot equal zero. The Fourier transform of $Ar(t-t_o)$ is found from the time shift property.

$$Ar(t - t_o) \leftrightarrow Ae^{-j2\pi ft_o}R(f) \tag{B.6}$$

We can consider this as the output of a linear system with input $r(t)$ and system function,

$$H(f) = Ae^{-j2\pi ft_o} \tag{B.7}$$

This is illustrated in Fig. B.2. Since $H(f)$ is complex, we have plotted its magnitude and phase. The real and imaginary parts would have also sufficed, but would not have been as instructive.

Let us view Fig. B.2 intuitively. It seems reasonable that the magnitude function turned out to be a constant. This indicates that all frequencies of $r(t)$ are multiplied by the same factor. But why did the phase turn out to be a linear function of frequency? Why aren't all frequencies shifted by the "same amount"? The answer is clear from a simple example.

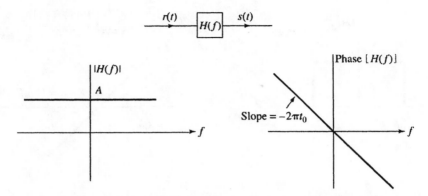

Figure B.2 - Characteristic of Distortionless System

Suppose we wish to shift a 1 Hz sinusoid, $r(t) = \cos 2\pi t$, by one second. This represents a phase shift of 2π radians, or 360°. If we now wish to shift a signal of twice the frequency, $r(t) = \cos 4\pi t$, by the same one second, we would have to shift the phase by 4π radians, or 720°. If we shifted the second signal by only 360°, it would only be delayed by 0.5 seconds instead of the required one second.

Now consider a general signal composed of many frequency components. If we delay all components by the same angular phase, we would not be delaying them by the same amount of time and the signal would be severely distorted. In order to delay by the same amount of time, the phase shift must be proportional to frequency.

B.2.1 Ideal Lowpass Filter

An *ideal lowpass filter* is a linear system which acts like an ideal distortionless system provided that the input signal contains no frequency components above the *cutoff* frequency of the filter. Frequency components above this cutoff are completely blocked from appearing at the output. The cutoff frequency is the maximum frequency passed by the filter, and we denote this as f_m. The system function is then given by

$$H(f) = \begin{cases} Ae^{-j2\pi f t_o}, & |f| < f_m \\ 0, & |f| > f_m \end{cases}$$ (B.8)

The transfer function of the ideal lowpass filter is shown in Fig. B.3. Note that since $h(t)$ is real, the magnitude of $H(f)$ is even and the phase is odd. The impulse response of the ideal lowpass filter is found by computing the inverse Fourier transform of $H(f)$.

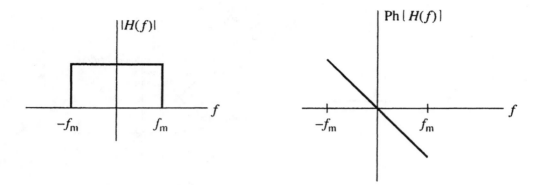

Figure B.3 - Ideal Lowpass Filter Characteristic

$$h(t) = \int_{-f_m}^{f_m} A e^{-j2\pi f t_o} e^{j2\pi ft} \, df$$

$$= \frac{A \sin 2\pi f_m(t - t_o)}{\pi(t - t_o)}$$

(B.9)

This impulse response is shown in Fig. B.4.

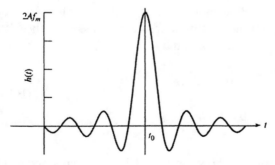

Figure B.4 - Impulse Response of
Ideal Lowpass Filter

The amount of delay, t_o, is proportional to the slope of the phase characteristic. The cutoff frequency is proportional to the peak of $h(t)$, and inversely proportional to the spacing between zero-axis crossings of the function. That is, as f_m increases, the peak of $h(t)$ increases, and the width of the shaped pulse decreases–the response gets taller and skinnier.

B.2.2 Ideal Bandpass Filter

Rather than pass frequencies between zero and f_m as in the case of the lowpass filter, the ideal bandpass filter passes frequencies between two non-zero frequencies, f_L and f_H. The filter acts like an ideal distortionless system provided that the input signal contains no frequency components outside of the filter *passband*. The system function of the ideal bandpass filter is given by

$$H(f) = \begin{cases} Ae^{-j2\pi f t_o}, & f_L < |f| < f_H \\ \\ 0, & otherwise \end{cases} \tag{B.10}$$

This system function is illustrated in Fig. B.5.

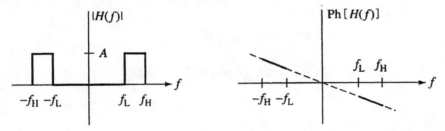

Figure B.5 - Ideal Bandpass Filter Characteristic

The impulse response of the bandpass filter can be found by evaluating the inverse Fourier transform of $H(f)$. Alternatively, we can save a lot of work by deriving this from the lowpass filter impulse response and the frequency shifting theorem. If we denote the lowpass system function as $H_{lp}(f)$, the bandpass function can be expressed as in Eq. (B.11).

$$H(f) = H_{lp}\left(f - \frac{f_L + f_H}{2}\right) + H_{lp}\left(f + \frac{f_L + f_H}{2}\right) \tag{B.11}$$

Figure B.6 shows the relationship between the functions.

We have illustrated the system functions as if they were real functions of frequency. That is, for purposes of the derivation, we are assuming $t_o = 0$. This approach is justified by the time invariance of the system. When we are finished, we can simply insert a time shift and the associated phase factor. Alternatively, you can view Fig. B.6 as a plot of the magnitudes of the functions, and carry the exponential phase term through every step of the derivation.

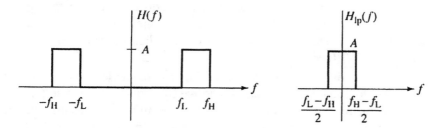

Figure B.6 - Bandpass and Lowpass Characteristics

If we define the midpoint of the passband (average of f_L and f_H) as f_{av},

$$f_{av} = \frac{f_L + f_H}{2} \qquad\qquad (B.12)$$

then the impulse response is then given by

$$h(t) = h_{lp}(t)e^{j2\pi f_{av}t} + h_{lp}(t)e^{-j2\pi f_{av}t}$$
$$= 2h_{lp}(t)\cos 2\pi f_{av}t = 2h_{lp}(t)\cos[\pi(f_L + f_H)t] \qquad\qquad (B.13)$$

From Eq. (B.9), we have that $h_{lp}(t)$ is given by

$$h_{lp}(t) = \frac{A\sin\pi(f_H - f_L)t}{\pi t} \qquad\qquad (B.14)$$

Combining Eqs. (B.13) and (B.14), and reinserting the time shift (phase factor), we find the impulse response of the ideal bandpass filter.

$$h(t) = \frac{2A\sin[\pi(f_H - f_L)(t - t_o)]\cos[\pi(f_L + f_H)(t - t_o)]}{\pi(t - t_o)} \qquad\qquad (B.15)$$

The impulse response is illustrated in Fig. B.7.

The outline of this waveform resembles the impulse response of the lowpass filter. Note that as the two limiting frequencies become large compared to the difference between them, the impulse response starts resembling a shaded-in version of the lowpass impulse response and its mirror image. This happens when the center frequency of the bandpass filter becomes large compared to the width of its passband.

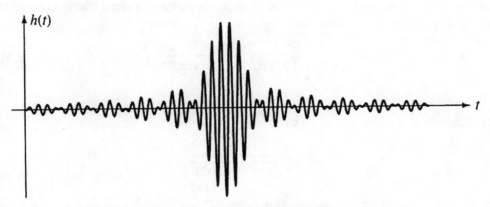

Figure B.7 - Impulse Response of Ideal Bandpass Filter

B.3 Practical Filters

We now present circuits which approximate the ideal lowpass and bandpass filters. Throughout this section, we assume that the closer $H(f)$ approaches the system function of an ideal filter, the more the filter will behave in an ideal manner in our applications. This fact is not at all obvious. A "small" change in $H(f)$ can lead to relatively large changes in $h(t)$. One can examine the consequences and categorize the effects of deviation from the constant amplitude characteristic or from the linear phase characteristic of the ideal system function.

B.3.1 Lowpass Filter

The simplest passive approximation to a lowpass filter is the single energy-storage device circuit. An example is the RC circuit of Fig. B.8.

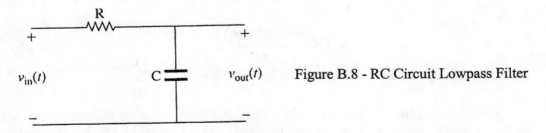

Figure B.8 - RC Circuit Lowpass Filter

If the output is taken across the capacitor, this circuit approximates a lowpass filter. This is true since, as the frequency increases, the capacitor behaves as a short circuit. The transfer function is given by

$$H(f) = \frac{1/j2\pi fC}{R + 1/j2\pi fC} = \frac{1}{1 + j2\pi fRC} \tag{B.16}$$

The magnitude and phase are given by

$$|H(f)| = \frac{1}{\sqrt{1 + (2\pi fRC)^2}} \tag{B.17}$$

$$\theta(f) = -\tan^{-1}(2\pi fRC)$$

If we set RC to $1/2\pi$, the magnitude of the transfer function drops to $1/\sqrt{2}$ at a frequency of 1 Hz. This is the 3-dB cutoff frequency of the filter ($20\log(1/\sqrt{2})$ is approximately -3 dB)[1].

Figure B.9 shows the magnitude and phase of the RC circuit transfer function. In Fig. B.9(a) we use a logarithmic frequency axis while in Fig. B.9(b) we use a linear frequency axis. Superimposed on each set of curves is the equivalent gain curve for an ideal lowpass filter with a cutoff frequency of 1 Hz. In particular, if we view the linear frequency plot, there is a dramatic difference between the RC approximation and the ideal lowpass filter characteristic. Keep in mind that in a logarithmic gain curve, a drop of 20 dB is a decrease by a factor of 10.

These curves, and the following response waveforms, were developed using MicroCap, a computer simulation program. Similar curves would result using other SPICE-based computer simulation programs, or if we skip the simulation, and plot the functions in Eq. (B.16) using MATLAB.

$$h(t) = \frac{\sin 2\pi(t - t_o)}{\pi t} \tag{B.18}$$

We could continue to analyze the RC filter distortion using the techniques derived earlier in this chapter. We choose instead to contrast the output of the RC circuit to that of an ideal lowpass filter for several representative inputs.

---

[1]The decibel (dB) is 20 times the log of the amplitude ratio.

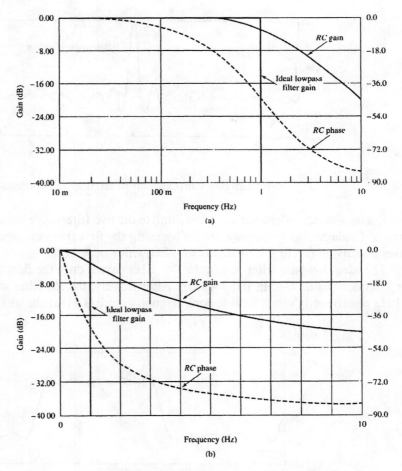

Figure B.9 - Characteristic of RC Circuit

Let us first view the impulse response of the two systems. The impulse response of the ideal lowpass filter is

$$h(t) \;=\; \frac{\sin 2\pi(t-t_o)}{\pi(t-t_o)} \tag{B.19}$$

The impulse response of the RC circuit, with $RC=2\pi$, is given by

$$h(t) \;=\; e^{-2\pi t}\; U(t)$$

These two impulse responses are shown in Fig. B.10, where we have arbitrarily chosen the delay of the ideal filter to be 10 seconds so the distinct plots can be easily seen.

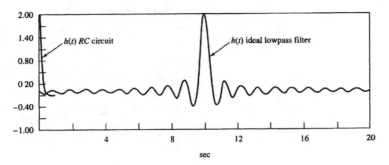

Figure B.10 - Comparison of Impulse Responses

Let us now consider a square wave input to the two filters. We have simulated a square wave of fundamental frequency 1/4 Hz by using the first five non-zero terms in a Fourier series expansion (up to the harmonic at a frequency of 9/4 Hz).

The ideal lowpass filter with cutoff of 1 Hz passes only the first two non-zero terms (i.e., frequencies of 1/4 Hz and 3/4 Hz). Alternatively, the *RC* filter with 3-dB frequency at 1 Hz significantly distorts these components. Figure B.11(a) shows the input waveform,

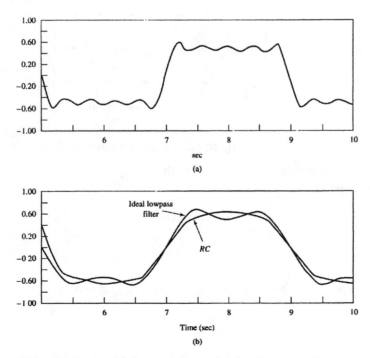

Figure B.11 - Comparison of Square Wave Responses

and Fig. B.11(b) shows both the ideal lowpass filter and the *RC* output function. Not only does the *RC* filter distort the waveform in the passband, but it also admits significant signal energy beyond the cutoff frequency.

In these examples, we have seen several types of distortion. At this point, we would probably agree that an *RC* network is not a very good lowpass filter except in limited applications. This leads us to explore more complex forms of practical filters.

There are several types of approximations to the ideal lowpass filter, each exhibiting unique characteristics. *Butterworth filters* produce no ripple in the passband and attenuate unwanted frequencies outside of this band. They are known as *maximally flat* filters since they are designed to force the maximum number of derivatives of $H(f)$ (at $f=0$) to be zero. *Chebyshev filters* attenuate unwanted frequencies more effectively than Butterworth filters, but they exhibit *ripple* in the passband. Other important classical filters include the elliptic, parabolic, Bessel, Papoulis, and Gaussian.

We limit the current discussion to Butterworth filters, and refer you to the references for more detail on the broad topic of filter design. The amplitude characteristic of the ideal lowpass filter can be approximated by the function,

$$|H_n(f)| = \frac{1}{\sqrt{1 + (2\pi f)^{2n}}} \qquad \text{(B.20)}$$

This function is sketched, for several values of n, in Fig. B.12. We have only illustrated the positive half of the f-axis since the function is even. We have chosen $f_m=1/2\pi$ (1 radian/sec) for the illustration, but a simple scaling process can be used to design a filter for any cutoff frequency. Note that as n gets larger, the amplitude characteristic approaches that of the ideal lowpass filter.

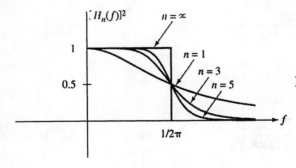

Figure B.12 - Butterworth Gain Functions

If $h(t)$ is real (as it must be for a real system), the real part of $H(f)$ is even, while the imaginary part is odd. Therefore,

$$H(f) = H^*(-f) \qquad \text{(B.21a)}$$

and

$$|H(f)|^2 = H(f)H^*(f) \tag{B.21b}$$

This observation, coupled with Eq. (B.20), is sufficient to design Butterworth filters.

B.3.2 Bandpass Filter

The simplest passive approximation to a bandpass filter is the double energy-storage device circuit. An example of the *RLC* circuit is shown in Fig. B.13.

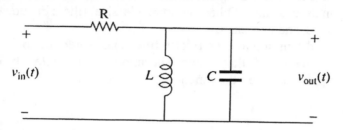

Figure B.13 - RLC Bandpass Circuit

If the output is taken across the parallel *LC* combination, this circuit approximates a bandpass filter. This is true since, as the frequency approaches zero the inductor behaves as a short circuit, and as the frequency approaches infinity, the capacitor behaves as a short circuit. The circuit response therefore approaches zero at both extremes, and peaks somewhere between the extremes. The transfer function is given by Eq. (B.22).

$$H(f) = \frac{j2\pi fL}{R - (2\pi f)^2 RLC + j2\pi fL} \tag{B.22}$$

The magnitude of this is

$$|H(f)| = \frac{1}{\sqrt{R^2[1/2\pi fL) - 2\pi fC]^2 + 1}} \tag{B.23}$$

The magnitude peaks at $2\pi f = \sqrt{1/LC}$. This point is known as the *resonant frequency* of the filter. The ratio of complex impedance to R is related to the Q of the circuit. Figure B.14 shows a computer simulation of the circuit characteristics, where we have selected $R=L=C=1$.

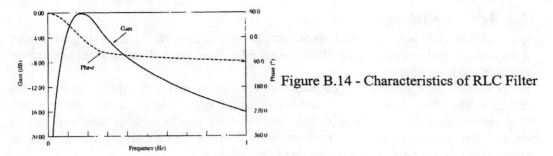

Figure B.14 - Characteristics of RLC Filter

The impulse response of the *RLC* circuit is given by the inverse Fourier transform of *H(f)*. Therefore, for *R=L=C=*1,

$$h(t) = 1.15e^{-t/2}\sin(1.15t) \qquad (B.24)$$

This should be compared to the impulse response of the ideal bandpass filter derived in Eq.(B.15) and repeated below.

$$h(t) = \frac{2A\sin[\pi(f_H - f_L)(t - t_o)]\cos[\pi(f_L + f_H)(t - t_o)]}{\pi(t - t_o)}$$

Figure B.15 shows the impulse response of the *RLC* circuit and the impulse response of the ideal bandpass filter, where we have chosen $f_H = 0.1$ Hz and $f_L = 0.25$ Hz, the 3-dB points of the *RLC* circuit response. Note that the *Q* of this filter is extremely low since the ratio of bandwidth to center frequency is close to unity.

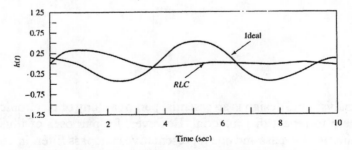

Figure B.15 - Comparison of Impulse Responses

As in the case of the lowpass filter, improvements are possible by including additional components. We shall reserve presentation of the bandpass Butterworth filter to the next section, where we show the active version of this filter.

B.4 Active Filters

In Section B.4, we examined some simple realizations of filters using inductors, capacitors, and resistors. Such filters are called *passive* since all component parts either absorb or store energy.

A filter is called *active* if it contains devices which deliver energy to the rest of the circuit. Active filters do not absorb part of the desired signal energy, as do passive filters. They are versatile and simple to design, and arbitrary causal transfer functions can be realized. For some applications, such as audio filtering, the passive filter requires an impractically large number of inductors and capacitors.

The basic building block of active filters is the *operational amplifier (op-amp)*. The op-amp has characteristics approaching those of an ideal, infinite gain, amplifier: infinite input resistance, zero output resistance and infinite voltage gain. Practical op-amps suffer from limited bandwidths.

The basic configuration of a single op-amp is shown in Fig. B.16. We indicate input and feedback impedances, Z_{in} and Z_f respectively. These could be single components (e.g., resistors, capacitors or inductors), or combinations of components.

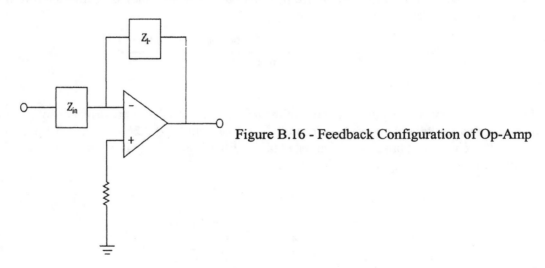

Figure B.16 - Feedback Configuration of Op-Amp

Active-filter analysis and design is an essential part of a study of electronics. We shall not take the time here to repeat that material. However, for purposes of illustration, we present one representative lowpass and one representative bandpass filter. In each case, we include the computer simulated characteristics. The lowpass filter is shown in Fig. B.17 and the bandpass filter in Fig. B.18.

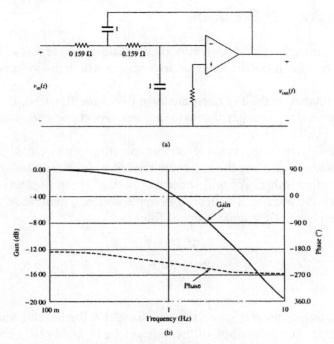

Figure B.17 - Active Lowpass Filter

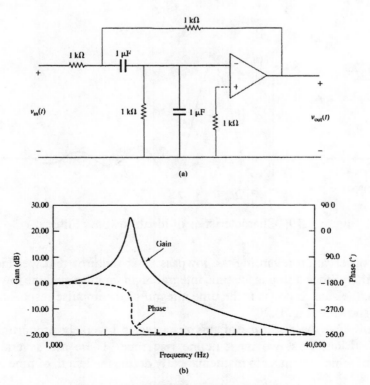

Figure B.18 - Active Bandpass Filter

B.5 Time Bandwidth Product

In designing a communication system, an important consideration is that of the *bandwidth* of the system. The bandwidth is the range of frequencies which the system is capable of handling.

Bandwidth is related to the Fourier transform of a time function. It is not directly definable in terms of the time function, unless we use intuitive statements about how quickly the function changes value.

Physical quantities of importance in communication system design include the minimum width of a time pulse and the minimum time in which the output of a system can jump from one level to another. We will show that both of these physical quantities are related to bandwidth. We start with a specific example and then generalize the result.

The impulse response of the ideal lowpass filter is

$$h(t) = \frac{\sin 2\pi f_m(t - t_o)}{\pi(t - t_o)} \tag{B.25}$$

This $h(t)$, and the corresponding $H(f)$ are shown in Fig. B.19. We use this transform pair to make two observations. First, the width of the largest lobe of $h(t)$ is $1/f_m$. This width is inversely proportional to the bandwidth of the signal. In fact, since the resulting bandwidth (difference between lowest and highest frequency) is f_m, the product of the pulse width with the bandwidth is unity.

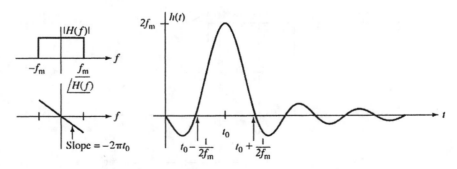

Figure B.19 - Characteristics of Ideal Lowpass Filter

The second observation regarding the lowpass filter requires that we find the step response of this filter. Since a step is the time integral of an impulse, and the lowpass filter is a linear system, the step response is the time integral of the impulse response. The step response, $a(t)$, is shown in Fig. B.20.

We now show that the *rise time* of this response is inversely proportional to the bandwidth of the filter. First we must define rise time. There are several common definitions, each of which attempts to mathematically define the length of time it takes the

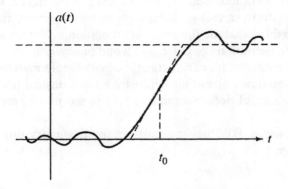

Figure B.20 - Step Response of Lowpass Filter

output to respond to a change, or jump, in the input. In practice, it is difficult to define the exact time at which the output has finished responding to the input jump. We present one particular definition which is well suited to our application.

The rise time is defined as the time required for a signal to go from the initial to the final value along a ramp with constant slope equal to the maximum slope of the function. This is shown as a dashed line in Fig. B.15. The maximum slope of $a(t)$ is the maximum value of the derivative, $h(t)$. This maximum is given by $2f_m$. The rise time of the step response is then

$$t_r = \frac{1}{2f_m} \tag{B.26}$$

Since the bandwidth of the filter is f_m, we see that the rise time and bandwidth are inversely related, and their product is equal to 0.5.

Although we have only illustrated the inverse relationship between rise time and bandwidth (or pulse width and bandwidth) for the ideal lowpass filter, the observation applies in general. That is, rise time is inversely proportional to bandwidth for any system. The product will not necessary be 0.5, but the product will be a constant.

We now verify this observation for a particular definition of bandwidth and pulse width. Suppose that a time function and its Fourier transform are as shown in Fig. B.21.

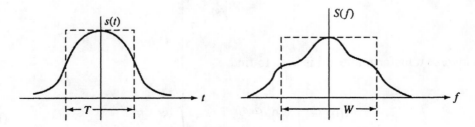

Figure B.21 - Definition of Pulse Width and Bandwidth

We emphasize that the actual shape is not intended to be that shown in the figure. Indeed, if the pulse is time-limited, the transform cannot go identically to zero over any range of frequencies. It is also unrealistic to think that the functions are monotonic. We present the pictures only to help understand the definitions of pulse width and bandwidth.

We define the pulse width, T, as the width of a rectangle whose height matches $s(0)$, and area is the same as that under the time pulse. This is illustrated as a dashed line in the figure. Note that this is not a meaningful definition unless $s(0)$ is the maximum of the waveform.

Equivalently, we define bandwidth, BW, using a pulse in the frequency domain as illustrated in the figure. We then have,

$$T = \frac{\int_{-\infty}^{\infty} s(t)dt}{s(0)}$$

(B.27)

$$BW = \frac{\int_{-\infty}^{\infty} S(f)df}{S(0)}$$

The product of these two is

$$TBW = \frac{\int_{-\infty}^{\infty} s(t)dt \int_{-\infty}^{\infty} S(f)df}{s(0)S(0)}$$

(B.28)

We now use the Fourier transform integral to find

$$S(0) = \int_{-\infty}^{\infty} s(t)e^{-j2\pi ft}dt \bigg|_{f=0} = \int_{-\infty}^{\infty} s(t)dt$$

(B.29)

The inverse transform integral is used to find

$$s(0) = \int_{-\infty}^{\infty} S(f)e^{j2\pi ft}df \bigg|_{t=0} = \int_{-\infty}^{\infty} S(f)df$$

(B.30)

Substituting Eqs. (B.29) and (B.30) into Eq. (B.28), we find

$$TBW = 1 \qquad\qquad (B.31)$$

The product of the pulse width with the bandwidth is unity–the two parameters are inversely related.

   This shows that the faster we desire a signal to change from one level to another, the more space on the frequency axis we must allow. This proves significant in digital communications where the bit transmission rate is limited by the bandwidth of the channel.

APPENDIX C

PROBABILITY AND NOISE

C.1 Basic Elements of Probability Theory

Probability theory can be approached either using theoretical mathematics or through empirical reasoning. The *mathematical approach* embeds probability theory within a study of *abstract set theory*. In contrast, the *empirical approach* satisfies one's intuition. In our basic study of communications, we will find the empirical approach to be sufficient, though advanced study and referring to current literature will require extending these concepts using principles of set theory.

Before we define what is meant by probability, we must extend our vocabulary by defining some important terms.

An *experiment* is a set of rules governing an operation which is performed.

An *outcome* is the result realized after performing the experiment one time.

An *event* is a combination of outcomes.

Consider the experiment defined by flipping a single die (half of a pair of dice) and observing which of the six faces is at the top when the die comes to rest. (Please notice how precise we are being. If you simply say "flip a die", you could mean that you observe the *time* at which it hits the floor.) There are six possible outcomes, these being any one of the six surfaces of the die facing upward after the performance of the experiment.

There are many possible events (64 to be precise). One event would be that of "an even number of dots showing". This event is a combination of the three outcomes: two dots, four dots, and six dots. Another event is "one dot". This latter event is known as an *elementary event* since it is the same as one of the outcomes. Of the 64 possible events, six represent elementary events. You should be able to list the 64 events. Try it! If you come up with only 62 or 63, you are probably missing the combination of all outcomes and/or the combination of no outcomes.

C.1.1 Probability

We now define what is meant by the probability of an event. Suppose that an experiment is performed N times, where N is very large. Furthermore, suppose that in n of these N experiments, the outcome belongs to an event, A (e.g., Consider flipping a die 1000 times and in 495 of these flips, the outcome is "even". Then $N=1000$ and $n=495$). If N is large enough, the probability of event A is given by the ratio n/N. That is, the probability is the fraction of times that the event occurs. Formally, we define the probability of event A as

$$Pr\{A\} = \lim_{N \to \infty} \frac{n_A}{N} \tag{C.1}$$

In Eq. C.1), n_A is the number of times that the event "A" occurs in N performances of the experiment. This definition is intuitively satisfying. For example, if a coin were flipped many times, the ratio of the number of heads to the total number of flips would approach ½. We therefore define the probability of a head to be ½. This simple example shows why N must approach infinity. Suppose, for example, you flipped a coin 3 times and in two of these flips, the outcome were heads. You would certainly not be correct in assuming the probability of heads to be 2/3!

Suppose that we now consider two different events, A and B, with probabilities

$$Pr\{A\} = \lim_{N \to \infty} \frac{n_A}{N} \quad and \quad Pr\{B\} = \lim_{N \to \infty} \frac{n_b}{N} \tag{C.2}$$

If A and B could not possibly occur at the same time, we call them *disjoint*. The events, "an even number of dots" and "two dots" are not disjoint in the die-throwing example, while the events "an even number of dots" and "an odd number of dots" are disjoint.

The probability of event A *or* event B is the ratio of the number of times A or B occurs divided by N. If A and B are disjoint, this is seen to be

$$Pr\{A \ or \ B\} = \lim_{N \to \infty} \frac{n_A + n_B}{N} = Pr\{A\} + Pr\{B\} \tag{C.3}$$

Equation C.3) expresses the "additivity" concept. That is, if two events are disjoint, the probability of their "sum" is the sum of their probabilities.

Since each of the outcomes (elementary events) is disjoint from every other outcome, and each event is a sum of outcomes, we see that it would be sufficient to assign probabilities only to the elementary events. We could derive the probability of any event from these given probabilities. For example, in the die-flipping experiment, the probability of an even outcome is the sum of the probabilities of "2 dots", "4 dots" and "6 dots".

Example

Consider the experiment of flipping a coin twice and observing which side is facing up when the coin comes to rest. List the outcomes, the events, and their respective probabilities.

Solution

The outcomes of this experiment are (letting H denote heads and T, tails)

    HH, HT, TH, and TT

We shall assume that somebody has used intuitive reasoning or has performed this experiment enough times to establish that the probability of each of the four outcomes is 1/C. There are 16 events, or combinations or these outcomes. These are given by

    {HH}, {HT}, {TH}, {TT}
    {HH,HT}, {HH,TH}, {HH,TT}, {HT,TH}, {HT,TT}, {TH,TT}

{HH,HT,TH}, {HH,HT,TT}, {HH,TH,TT}, {HT,TH,TT}
{HH,HT,TH,TT}, and {∅}

Note that the comma within the curly brackets is read, "or". Thus, the events {HH,HT} and {HT,HH} are identical and we list this only once. For completeness we have included the zero event, denoted {∅}. This is the event made up of none of the outcomes, and is called the *null* event. We also include the event consisting of all of the outcomes, the so-called *certain* event.

Using the additivity rule, the probability of each of these events is the sum of the probabilities of the outcomes comprising each event. Therefore,

$Pr\{HH\} = Pr\{HT\} = Pr\{TH\} = Pr\{TT\} = 1/4$

$Pr\{HH,HT\} = Pr\{HH,TH\} = Pr\{HH,TT\} = Pr\{HT,TH\}= Pr\{HT,TT\} =$
$= Pr\{TH,TT\} = \frac{1}{2}$

$Pr\{HH,HT,TH\}=Pr\{HH,HT,TT\}=Pr\{HH,TH,TT\}=Pr\{HT,TH,TT\} = 3/4$

$Pr\{HH,HT,TH,TT\} = 1$

$Pr\{∅\} = 0$

The last two probabilities indicate that the event made up of all four outcomes is the *certain* event. It has probability "1" of occurring since each time the experiment is performed, the outcome must belong to this event. Similarly, the null event has probability zero of occurring since each time the experiment is performed the outcome does not belong to the zero event.

C.1.2 Conditional Probabilities

We would like to be able to tell if one random variable has any effect on another. In the die experiment, if we knew the time at which the die hit the floor, would it tell us anything about which face was showing? In a more practical case, if we knew the frequency of a random noise signal, would this tell us anything about its amplitude? These questions lead naturally into a discussion of *conditional probabilities*.

Let us examine two events, A and B. The probability of event A *given that event B has occurred* is defined by

$$Pr\{A/B\} = \frac{Pr\{A\ AND\ B\}}{Pr\{B\}} \tag{C.4}$$

For example, if A represented two dots appearing in the die experiment and B represented an even number of dots, the probability of A given B would be the probability of two dots assuming that we know the outcome is either two, four, or six dots. Thus, the conditioning statement has reduced the scope of possible outcomes from six to three. We would intuitively expect the answer to be 1/3. Now using Eq. (C.4), the probability of A AND B is the probability of getting two AND an even number of dots simultaneously (In set theory, this is known as the *intersection*). This is simply the probability of two dots, or 1/6. The probability of B is the probability of two, four or six dots, which is ½. The ratio is 1/3 as expected.

Similarly, we could have defined event *A* as "an even number of dots" and event *B* as "an odd number of dots". The event "*A AND B*" would therefore be the zero event, and Pr{*A/B*} would be zero. This is reasonable since the probability of an even outcome assuming that an odd outcome occurred is clearly zero.

Two events, *A* and *B*, are said to be *independent* if

$$Pr\{A/B\} = Pr\{A\} \tag{C.5}$$

That is, if *A* and *B* are independent, the probability of *A* given that *B* occurred is simply the probability of *A*. Knowing that *B* has occurred tells nothing about *A*. Plugging this into Eq. (C.4) shows that independence implies

$$Pr\{A \text{ and } B\} = \frac{Pr\{A \text{ and } B\}}{Pr\{B\}} = Pr\{A\}Pr\{B\} \tag{C.6}$$

You have probably used this fact before in simple experiments. For example, we assumed that the probability of flipping a coin and having it land with heads facing up was ½. The probability of flipping the coin twice and getting two heads is ½ x ½ = 1/4. This is true since the events are independent of each other.

Example

A coin is flipped twice. Four different events are defined.

  A is the event of getting a head on the first flip.

  B is the event of getting a tail on the second flip.

  C is the event of a match between the two flips.

  D is the elementary event of a head on both flips.

(a) Find Pr{A}, Pr{B}, Pr{C}, Pr{D}, Pr{A/B}, and Pr{C/D}.

(b) Are A and B independent? Are C and D independent?

Solution

(a) The events are defined by the following combination of outcomes.

  A = {HH, HT}

  B = {HT, TT}

  C = {HH, TT}

  D = {HH}

Therefore,

  Pr{A} = Pr{B} = Pr{C} = ½

  Pr{D} = 1/4

(b) In order to find Pr{A/B} and Pr{C/D} we use Eq. (C.4).

$$Pr\{A/B\} = \frac{Pr\{A\ AND\ B\}}{Pr\{B\}}$$

$$Pr\{C/D\} = \frac{Pr\{C\ AND\ D\}}{Pr\{D\}}$$

The event {A AND B} is {HT}. The event {C AND D} is {HH}. Therefore,

$$Pr\{A/B\} = \frac{1/4}{1/2} = 0.5$$

$$Pr\{C/D\} = \frac{1/4}{1/4} = 1$$

Since Pr{A/B} = Pr{A}, the event of a head on the first flip is independent of that of a tail on the second flip. Since Pr{C/D} ≠ Pr{C}, the event of a match and that of two heads are not independent.

C.1.3 Random Variables

We would like to perform several forms of analysis on the probabilities. As such, it is not too satisfying to have symbols such as "heads", "tails", and "two dots" floating around. It would be preferable to work with numbers. We therefore associate a real number with each possible outcome of an experiment. In the single flip of the coin experiment, we could associate the number "0" with "tails" and "1" with "heads". We could just as well have associated "π" with "heads" and "207" with "tails".

The mapping (function) which assigns a number to each outcome is called a *random variable*.

Once the random variable is assigned, we can perform many forms of analysis. We could, for example, plot the various outcome probabilities as a function of the random variable. An extension of that type of plot is the *distribution function, F(x)*. If the random variable is denoted by[1] "X", then the distribution function, $F(x_o)$, is defined by

$$F(x_o) = Pr\{X < x_o\} \tag{C.7}$$

We note that the set, $\{X < x_o\}$ defines an event, or combination of outcomes.

---

[1]We shall use upper-case letters for random variables and lower-case letters for the values that they can take on. Thus, $X=x_o$ means that the random variable X is equal to the number, x_o.

Example

Assign two different random variables to the "one flip of the die" experiment, and plot the two resulting distribution functions.

Solution:

The first assignment we will choose is the one that is naturally suggested by this particular experiment. That is, we assign the number "1" to the outcome described by the face with one dot facing up. We assign, "2" to "two dots", "3" to "three dots", and so on. We therefore see that the event $\{X \leq x_o\}$ includes the one-dot outcome if x_o is between 1 and 2. If x_o is between 2 and 3, the event includes the one-dot and two-dot outcomes. Thus, the distribution function is as shown in Fig. C.1(a).

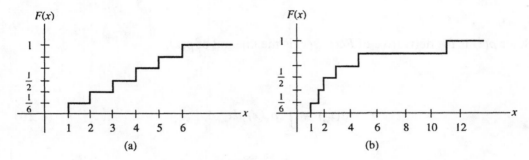

Figure C.1 - Distribution Functions

Let us now choose a different assignment of the random variable, this one representing a less natural choice.

Outcome	Random Variable
One Dot	1
Two Dots	π
Three Dots	2
Four Dots	$\sqrt{2}$
Five Dots	11
Six Dots	5

We have chosen strange numbers to illustrate that the mapping is arbitrary. The resulting distribution function is plotted as Fig. C.1(b). As an example, let us verify one point on the distribution function, the point for $x = 3$. The event $\{X \leq 3\}$ is the event made up of the three outcomes: one dot, three dots, and four dots. This is true since the value of the random variable assigned to each of these three outcomes is less than 3.

A distribution function can never decrease with increasing argument. This is true since an increase in argument can only add outcomes to the event, and the probabilities of these added outcomes cannot be negative. We also easily verify that

$$F(-\infty) = 0 \quad and \quad F(+\infty) = 1 \tag{C.8}$$

C.1.4 Probability Density Function

The *probability density function* is defined as the derivative of the distribution function. Using the symbol $p_X(x)$ for the density, we have

$$p_X(x) = \frac{dF(x)}{dx} \tag{C.9}$$

Since $p(x)$ is the derivative of $F(x)$, $F(x)$ is the integral of $p(x)$.

$$F(x_o) = \int_{-\infty}^{x_o} p_X(x) \, dx \tag{C.10}$$

The random variable can be used to define any event. For example, $\{x_1 < X \le x_2\}$ defines an event. Since the events $\{X \le x_1\}$ and $\{x_1 < X \le x_2\}$ are disjoint, the additivity principle can be used to prove that

$$Pr\{X \le x_1\} + Pr\{x_1 < X \le x_2\} = Pr\{X <+ x_2\}$$

$$or \tag{C.11}$$

$$Pr\{x_1 < X \le x_2\} = Pr\{X \le x_2\} - Pr\{X \le x_1\}$$

Combining equations (C.10) and (C.11), we have the important result,

$$Pr\{x_1 < X \le x_2\} = \int_{-\infty}^{x_2} p_X(x)dx - \int_{-\infty}^{x_1} p_X(x)dx$$

$$\tag{C.12}$$

$$= \int_{x_1}^{x_2} p_X(x)dx$$

We now see why $p_X(x)$ is called a density function. The probability that X is between any two limits is given by the area under the density function between these two limits.

Since the distribution function can never decrease with increasing argument, its slope, the density function, can never be negative[2]. Since the distribution function approaches unity as the argument approaches infinity, the integral of the density function over infinite limits must be unity.

The examples given previously (die and coin) result in density functions which contain impulses. The random variables associated with such experiments are known as *discrete random variables*. Another class of experiments gives rise to random variables with continuous density functions. This is logically called the class of *continuous random variables*. We present several frequently occurring continuous random variable density functions in Section C.2. For now, we examine the simplest of these functions, the *uniform density function*. This is shown in Figure C.2(a), where a and b are specified parameters. The height of the density must be such that the total area is unity.

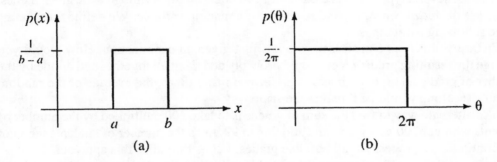

Figure C.2 - Uniform Density Function

Let's look at one practical experiment that results in a uniformly distributed random variable. Suppose you were asked to turn on a sinusoidal generator. The output of the generator would be of the form,

$$v(t) = A\cos(2\pi f_o t = \theta) \tag{C.13}$$

Since the absolute time at which you turn on the generator is random, it would be reasonable to expect that θ is uniformly distributed between 0 and 2π (this is true provided that f_o is much larger than the reciprocal of your reaction time). It thus would follow the density function shown in Fig. C.2(b).

---

[2]We could infer the same conclusion from Eq. (C.12). If the probability density function were negative over any range of values, we could integrate the curve over that range to get a negative result. This would imply that the probability of the variable being in that range is negative, but this is impossible.

Example

A random variable is uniformly distributed between 1 and 3. Find the probability that the variable is in the range between 1.5 and 2.

Solution:

The density function is as shown in Fig. C.2(a), where a is 1 and b is 3. In order for this to integrate to unity, the height of the density must be ½. The probability that the variable is between 1.5 and 2 is simply the integral under the curve between these limits. This is equal to 1/4.

C.1.5 Expected Values

Expected values, or averages, are important in communications. The average of the square of a voltage is closely related to the power associated with that voltage. The power of a noise voltage is an important measure of the level of disturbance caused by that voltage.

There are several types of expected values that come up often enough to be given names. These are the *mean*, *variance*, and *moments* of a random variable. We define these terms in the following paragraphs.

Picture yourself as a professor who has just given an examination. How would you average the resulting grades? You would probably add them all together and divide by the number of grades. If an experiment is performed many times, the average of the random variable resulting would be found in the same way.

An alternate way to find the sum of grades is to take 100 multiplied by the number of students who got 100 as a grade and add this to 99 times the number of students who got 99. Continue this process for all possible grades. Let us formalize this approach.

Let x_i, I=1, 2, ..., M, represent the possible values of the random variable, and let n_i represent the number of times the outcome associated with x_i occurs. The average of the random variable, after N performances of the experiment, is

$$X_{avg} = \frac{1}{N}\sum_i n_i x_i = \sum_i \frac{n_i}{N} x_i \qquad (C.14)$$

Since x_i ranges over all possible values of the random variable, we observe that

$$\sum_i n_i = N \qquad (C.15)$$

As N approaches infinity, n_i/N becomes the probability, $\Pr\{x_i\}$. Therefore,

$$X_{avg} = \sum_i x_i \Pr\{x_i\} \qquad (C.16)$$

This average value is known as the *mean, expected value,* or *first moment* of X, and is given the symbol $E\{x\}$, X_{avg}, m_x, or $\bar{x}$. The words "expected value" should not be taken too literally since they do not always lend themselves to an intuitive definition. As an example, suppose we assign 1 to heads and 0 to tails in the coin flip experiment. The expected value of the random variable is ½. However, no matter how many times you perform the experiment, you will never obtain an outcome with associated random variable of ½.

Now suppose that we wish to find the average value of a continuous random variable. We can use the result of Eq. (C.16) if we first round off the continuous variable to the nearest multiple of Δx. Thus, if X is between $(k-\frac{1}{2})\Delta x$ and $(k+\frac{1}{2})\Delta x$, we round it off to $k\Delta x$. The probability of X being in this range is given by the integral under the probability density function.

$$Pr\{(k - \frac{1}{2})\Delta x < x \le (k + \frac{1}{2})\Delta x\} = \int\limits_{(k - \frac{1}{2})\Delta x}^{(k + \frac{1}{2})\Delta x} p_X(x) \; dx \qquad \text{(C.17)}$$

If Δx is small, this can be approximated by $p_X(k\Delta x)\Delta x$. Therefore, Eq. (C.16) can be rewritten as

$$X_{avg} = \sum_{k = -\infty}^{\infty} k\Delta x p_X(k\Delta x)\Delta x \qquad \text{(C.18)}$$

As Δx approaches zero, this becomes

$$X_{avg} = m_x = \int\limits_{-\infty}^{\infty} x p_X(x) dx \qquad \text{(C.19)}$$

Equation (C.19) is very important. It tells us that to find the average value of x, we simply weight x by the density function and integrate the product.

The same approach can be used to find the average of any function of the random variable. For example, again suppose that you are a professor who gave an exam, but instead of entering the raw percentage score in your grade book, you enter some function of this score, such as e^x, where x is the raw score. If you now wish to average the grade book entries, you would follow the reasoning used earlier [Eqs. (C.14) through (C.19)] with the result that $x p_X(x)$ in Eq. (C.19) gets replaced by $e^x p_X(x)$.

In general, if y is defined as $y = g(x)$, the expected value of y is given by

$$Y_{avg} = [g(x)]_{avg} = \int\limits_{-\infty}^{\infty} g(x) p_X(x) dx \qquad \text{(C.20)}$$

Equation (C.20) is extremely significant and useful. It tells us that to find the expected value of a function of x, we simply integrate that function weighted by the *density of X*. It is not necessary to first find the density of the new random variable.

We often find the expected value of the random variable raised to a power. This is given the name, *moment*. Thus, the expected value of x^n is known as the *n-th moment* of the random variable, X.

If we first shift the random variable by its mean and then take a moment of the resulting shifted variable, the *central moment* results. Thus, the n-th central moment is given by the expected value of $(x-m_x)^n$.

The *second central moment* is extremely important as it is related to power. It is given the name, *variance*, and the symbol σ^2. Thus, the variance is given by

$$\sigma^2 = E\{(x - m_x)^2\} = \int (x - m_x)^2 p P_X(x)\, dx \qquad \text{(C.21)}$$

The variance gives a measure of how far we can expect the variable to deviate from its mean value. As the variance gets larger, the density function tends to "spread out". The square root of the variance, σ, is known as the *standard deviation*.

Example

X is uniformly distributed as shown in Fig. C.3. Find $E\{x\}$, $E\{x^2\}$, $E\{\cos x\}$ and $E\{(x-m_x)^2\}$.

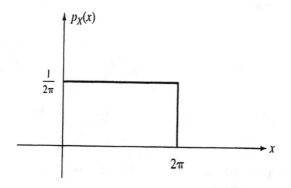

Figure C.3 - Uniform Density for Example

Solution:

We apply Eq. (C.20) to find,

$$E\{x\} = \int_{-\infty}^{\infty} x p_X(x) dx = \frac{1}{2\pi} \int_0^{2\pi} x\, dx = \pi$$

$$E\{x^2\} = \int_{-\infty}^{\infty} x^2 p_X(x) dx = \frac{1}{2\pi} \int_0^{2\pi} x^2 dx = \frac{4}{3}\pi^2$$

$$E\{\cos x\} = \int\limits_{-\infty}^{\infty} \cos x \, p_X(x)dx = \frac{1}{2\pi}\int\limits_{0}^{2\pi} \cos x \, dx = 0$$

$$E\{(x-\pi)^2\} = \int\limits_{-\infty}^{\infty}(x-\pi)^2 p_X(x)dx = \frac{1}{2\pi}\int\limits_{0}^{2\pi}(x-\pi)^2 dx = \frac{\pi^2}{3}$$

C.1.6 Functions of a Random Variable

"Everybody talks about the weather, but nobody does anything about it". We, as communication engineers, would be open to the same type of criticism if all we ever did was make statements such as "there is a 42% probability that the noise will be annoying". A significant part of communication engineering involves changing noise from one form to another in the hopes that the new form will be less annoying than the old. We must therefore study the effects of processing on random phenomena.

Consider a function of a random variable, $y=g(x)$, where X is a random variable with known density function. A representative function is shown in Figure C.4.

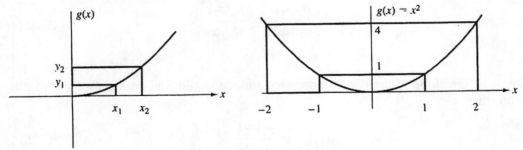

Figure C.4 - Representative $g(x)$

Since X is random, Y is also random. We are interested in finding the density function of Y.

The event, $\{x_1 < X \le x_2\}$ corresponds to[3] the event, $\{y_1 < Y \le y_2\}$, where

$$y_1 = g(x_1) \quad and \quad y_2 = g)(x_2)$$

---

[3]We are assuming that y_1 is less than y_2 if x_1 is less than x_2. That is, g(x) is monotonic increasing and has a positive derivative. If this is not the case, the inequalities would have to be reversed.

The two events are identical since they include the same outcomes. We are assuming for the moment that $g(x)$ is a single-valued function. Since the events are identical, their probabilities must also be equal.

$$Pr\{x_1 > X \geq x_2\} = Pr\{y_1 < Y \leq y_2\} \tag{C.22}$$

and in terms of the densities,

$$\int_{x_1}^{x_2} P_{xX}(x)dx = \int_{y_1}^{Y_2} p_Y(y)dy \tag{C.23}$$

If we now let x_2 get very close to x_1, in the limit Eq. (C.23) becomes

$$P_X(x_1)dx = p_Y(Yy)dy \tag{C.24}$$

and finally,

$$p_Y(y_1) = \frac{P_X(x_1)}{dy/dx_1}) = \frac{P_X(x_1)}{dy/dx} \tag{C.25}$$

If $y_1 > y_2$, the slope of the curve is negative and we would find (you should prove this result) that,

$$p_Y(y_1) = -\frac{P_X(x_1)}{dy/dx} \tag{C.26}$$

We can account for both of these cases by writing,

$$p_Y(y_1) = \frac{P_X(x_1)}{|dy/dx|} \tag{C.27}$$

Finally, writing $x_1 = g^{-1}(y_1)$, and realizing that y_1 can be a variable (i.e., replace it with y), we have

$$p_Y(y) = \frac{P_X[g^{-1}(y)]}{|dy/dx|} \tag{C.28}$$

If the function g(x) is not monotonic, the event $\{y_1 < Y < y_2\}$ can correspond to more than one interval of the variable X. For example if $g(x)=x^2$, then the event $\{1 < Y \le 4\}$ is the same as the event $\{1 < X \le 2\}$ or $\{-2 < X \le -1\}$. This is shown in Figure C.4(b). Therefore,

$$\int_1^2 p_X(x)dx + \int_{-2}^{-1} p_X(x)dx = \int_1^4 p_Y(y)dy \qquad (C.29)$$

In terms of the density functions, this would mean that $g^{-1}(y)$ has two values. Denoting these values as x_a and x_b, then

$$p_Y(y) = \left.\frac{p_X(x)}{|dy/dx|}\right|_{x=x_a} + \left.\frac{p_X(x)}{|dy/dx|}\right|_{x=x_b} \qquad (C.30)$$

Example

A random voltage, v, is put through a full-wave rectifier. The input voltage is uniformly distributed between -2 volts and +2 volts. Find the density of the output of the full-wave rectifier.

Solution:

Calling the output y, we have $y = g(v)$, where $g(v)$ and the density of V are sketched in Fig. C.5. Note that we have let the random variable be equal to the value of voltage.

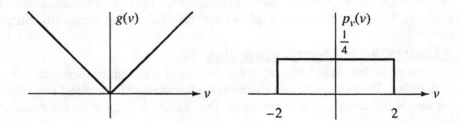

Figure C.5 - $g(v)$ and $p(v)$ for Example

At every value of V, $|dg/dv| = 1$. For $y > 0$, $g^{-1}(y) = \pm y$. For $y < 0$, $g^{-1}(y)$ is undefined. That is, there are no values of v for which $g(v)$ is negative. Equation (C.29) is then used to find

$$p_Y(y) = p_X(y) + p_X(-y) \quad y>0$$

$$p_Y(y) = 0 \quad y<0$$

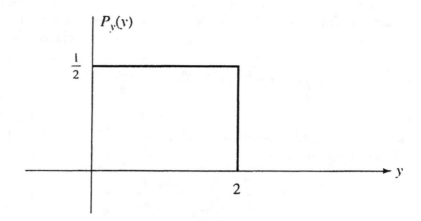

Figure C.6 - Resulting Density for Example

This result is shown in Fig. C.6.

C.2 Frequently Encountered Density Functions

We introduced the basic concepts of probability in section C.1. The uniform density function was presented. While some experiments in communications lead to this density, the majority of random variables we encounter follow densities other than uniform. The current section explores several of the most frequently encountered densities.

C.2.1 Gaussian Random Variables

The most common density confronted in the real world is called the *Gaussian (or normal) density function*. The reason it is so common is attributed to the *central limit theorem*, a theorem we shall discuss in a few moments. The density is defined by the equation,

$$p_X(x) = \frac{1}{\sqrt{2\pi}\sigma} \exp\left[\frac{-(x-m)^2}{2\sigma^2}\right] \tag{C.31}$$

where m and σ are given constants. The Gaussian density function is sketched in Fig. C.7. The parameter m dictates the center position or symmetry point of the density. Evaluating the integral of $xp_X(x)$ would show that m is the mean value of the variable. The other parameter, σ, indicates the spread of the density. Evaluating the integral of $(x-m)^2 p_X(x)$ would show that σ^2 is the variance of the variable, so σ is the standard deviation. As σ increases, the bell-shaped curve gets wider and the peak decreases. Alternatively, as σ

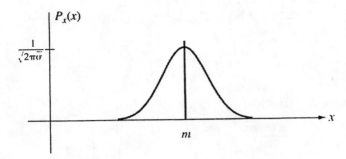

Figure C.7 - Gaussian
Probability Density

decreases, the density sharpens into a narrow pulse with a higher peak (the area must always be unity).

To evaluate probabilities that Gaussian variables are within certain ranges, we find it necessary to integrate the density. The function of Eq. (C.31) cannot be integrated in closed form, although software such as MATLAB can easily be used. The Gaussian density is sufficiently important that this integral has been computed and tabulated under the names, *error function* (*erf*) and *Q function*. The error function is defined by the following integral.

$$erf(x) = \frac{2}{\sqrt{\pi}} \int_0^x e^{-u^2} \, du \qquad (C.32)$$

It can be shown that erf(∞) = 1. Therefore,

$$\frac{2}{\sqrt{\pi}} \int_x^\infty e^{-u^2} du = \frac{2}{\sqrt{\pi}} \int_0^\infty e^{-u^2} du - \frac{2}{\sqrt{\pi}} \int_0^x e^{-u^2} du \qquad (C.33)$$

$$= erf(\infty) - erf(x) = 1 - erf(x)$$

For convenience, this last expression is tabulated under the name, *complementary error function* (erfc). Thus, the relationship between the error function and the complementary error function is given by

$$erfc(x) = 1 - erf(x) \qquad (C.34)$$

Both the error function and the complementary error function are tabulated in Appendix E. The area under a Gaussian density with any values of m and σ can be expressed in terms of error functions. For example, the probability that X is between x_1 and x_2 is given by

$$Pr\{x_1 < X \le x_2\} = \frac{1}{\sqrt{2\pi}\sigma} \int_{x_1}^{x_2} \exp\left[\frac{(x - m)^2}{2\sigma^2}\right] dx \tag{C.35}$$

We now make the following change of variables,

$$u = \frac{x - m}{\sqrt{2}\sigma}$$

to get

$$Pr\{x_1 < X \le x_2\} = \frac{1}{\sqrt{\pi}} \int_{\frac{x_1 - m)}{\sqrt{2}\sigma}}^{\frac{x_2 - m}{\sqrt{2}\sigma}} e^{-u^2} du \tag{C.36}$$

$$= \frac{1}{2} erf\left(\frac{x_2 - m}{\sqrt{2}\sigma}\right) - \frac{1}{2} erf\left(\frac{x_1 - m}{\sqrt{2}\sigma}\right)$$

We have assumed that both x_1 and x_2 are greater than m since the error function is not defined for negative arguments. Example C.7 will deal with a situation where this assumption is not valid.

A companion to the error function is the *Q-function*. It is sometimes called the complementary error function, or *co-error function*. However, the Q-function differs from the complementary error function of Eq. C.34) by a constant multiplier and a scaling factor. The Q-function is defined as follows:

$$Q(x) = \int_{x}^{\infty} \frac{1}{\sqrt{2\pi}} e^{-u^2/2} du \tag{C.37}$$

The integrand is a unit-variance zero-mean Gaussian density function. Note that
    Q(-∞) = 1
    Q(-x) = 1 - Q(x)
We can relate the Q-function to the error function by making a change of variables.

$$Q(x) = \frac{1}{\sqrt{2\pi}} \int_x^\infty e^{-u^2/2} \, du = \frac{1}{\sqrt{\pi}} \int_{x/\sqrt{2}}^\infty e^{-v^2} \, dv$$

$$= \frac{1}{2} erfc \left(\frac{x}{\sqrt{2}} \right) \tag{C.38}$$

The Q-function is tabulated in Appendix E.

Either the Q-function or the error function contains the necessary information to evaluate integrals of Gaussian density functions. Some feel that the Q-function is more satisfying and easier to work with since the integrand is a normalized Gaussian density.

Let us now find the probability that the random variable is between two limits using the Q-function.

$$Pr\{x_1 < X \le x_2\} = \frac{1}{\sqrt{2\pi}\sigma} \int_{x_1}^{x_2} \exp \left[\frac{(x - m)^2}{2\sigma^2} \right] dx \tag{C.39}$$

Using the Q-function, the required change of variables is

$$Pr\{x_1 < X \le x_2\} = \frac{1}{\sqrt{2\pi}} \int_{(\frac{x_1 - m}{\sigma})}^{\frac{x_2 - m}{\sigma}} e^{-u^2/2} \, du \tag{C.40}$$

$$= Q \left(\frac{x_1 - m}{\sigma} \right) - Q \left(\frac{x_2 - m}{\sigma} \right)$$

Although the error function has traditionally been much more common than the Q-function, there are indications that the Q-function will predominate in the future. It makes absolutely no difference which you use to solve a problem (you had better get the same answer either way). The only question you should have is which type of table is more readily available. Of course if you use the wrong table, you will get the wrong answer.

Now that we are familiar with the Gaussian density, let's return to the discussion of why it occurs so frequency in the real world. It results whenever a large number of factors contribute to an end result, as in the case of static in broadcast radio. There are two conditions that must be satisfied before the sum of many random variables starts to appear Gaussian. The first relates to the individual variances, and to their infinite sum. The sum must approach infinity as the number of variables added together, approaches infinity. The second condition is satisfied if the component densities go to zero outside some range (this is a sufficient but not a necessary condition). Since all quantities we deal with in the real world have bounded ranges, they satisfy this condition.

Example

A binary communication system is one that sends only one of two possible messages. A simple example of a binary system is one in which either *zero* or *one* volt is sent. Consider such a system in which the transmitted voltage is corrupted by additive atmospheric noise. If the receiver receives anything above 1/2 volt (i.e., the midpoint), it assumes that a *one* was sent. If it receives anything below 1/2 volt, it assumes that a *zero* was sent. Measurements show that if one volt is transmitted, the received signal level is random and has a Gaussian density with m = 1 and σ = 1/2. Find the probability that a transmitted *one* will be interpreted as a *zero* at the receiver (i.e., a *bit error*).

Solution:

The received signal level has a Gaussian density with $m = 1$ and $\sigma^2 = (1/2)^2$. Thus, if we designate the random variable as V, we have

$$p_V(v) = \sqrt{\frac{2}{\pi}} \exp\left[\frac{-(v - 1)^2}{2(0.5)^2}\right]$$

Since any value received below a level of 0.5 is called "0", the probability that a transmitted "1" will be interpreted as a "0" at the receiver is simply the probability that the random variable V is less than 0.5. This is given by the integral

$$\int_{-\infty}^{0} 0.5 p_V(v) dv = \sqrt{\frac{2}{\pi}} \int_{-\infty}^{0.5} \exp[-2(v - 1)^2] dv$$

To reduce this to a form that can be found in a table of error functions, we make the change of variable,

$$u = \sqrt{2}(v - 1)$$

to get

$$Pr(error) = \frac{1}{\sqrt{\pi}} \int_{-\infty}^{-\frac{\sqrt{2}}{2}} e^{-u^2} du$$

This is not yet in the form of an error function. However, since $\exp(-u^2)$ is an even function, we can take the mirror image of the integral limits without changing the value of the integral.

$$Pr(error) = \frac{1}{\sqrt{\pi}} \int_{\frac{\sqrt{2}}{2}}^{\infty} e^{-u^2} \, du$$

This is now seen to be related to the complementary error function.

$$Pr(error) = \frac{1}{2} erfc\left(\frac{\sqrt{2}}{2}\right) = 0.159$$

Had we desired to use the Q function, we would make the change of variables,

$$u = 2(v - 1)$$

to get

$$Pr(e) = \frac{1}{\sqrt{2\pi}} \int_{1}^{\infty} e^{-u^2/2} \, du = Q(1) = 0.159$$

Thus, on the average, one would expect 159 out of every 1000 transmitted 1's to be misinterpreted as 0's at the receiver. This is an extremely poor level of performance.

C.2.2 Rayleigh Density Function

The *Rayleigh density function* is defined by Eq. (C.41).

$$p_X(x) = \begin{cases} \dfrac{x}{K^2} \exp\left(\dfrac{-x^2}{2K^2}\right), & x > 0 \\[2mm] 0, & x < 0 \end{cases} \tag{C.41}$$

K is a given constant. Figure C.8 shows the density function for two different values of K.

The Rayleigh density function is related to the Gaussian density function. In fact, the square root of the sum of the squares of two zero-mean Gaussian distributed random variables is, itself, Rayleigh. If we transform from rectangular to polar coordinates, the radius is given by

$$r = \sqrt{x^2 + y^2}$$

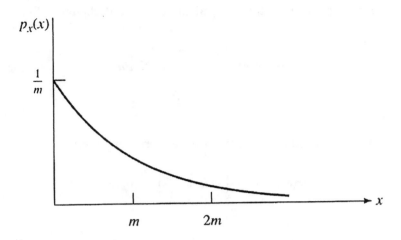

Figure C.8 - Rayleigh Density Function

If x and y are both Gaussian, in most cases (the restriction is one of *independence*, a term we have not yet defined) r will be Rayleigh. As an example, suppose that you were throwing darts at a target on a dartboard and that the horizontal and vertical components of your error were Gaussian distributed with zero mean (a fair assumption if you are not biased by wind, gravity or a muscle twitch). The *distance* from the center of the target to the dart position would then be Rayleigh distributed.

The *chi-square* distribution is closely related to the Rayleigh. If we were to consider r^2 as the variable instead of r, we would find that variable to be chi-square distributed. That is, a chi-square distribution results from summing the squares of Gaussian variables. If we sum two such variables, the result is chi-square with 2 *degrees of freedom*. In general, if

$$z = x_1^2 + x_2^2 + x_3^2 + ... + x_n^2$$

and x_i are Gaussian with $m=0$, z will be chi-square with n degrees of freedom. If the Gaussian variables all have $r = 1$, the actual form of the density is given by

$$p(z) = \begin{cases} \dfrac{(z)^{n/2-1}}{2^{n/2}(n/2 - 1)!} \, e^{-z/2}, & z > 0 \\[4mm] 0, & z < 0 \end{cases} \tag{C.42}$$

The "!" in Eq. (C.42) indicates the *factorial* operation.

C.2.3 Exponential Random Variables

We occasionally deal with random variables that are exponentially distributed. This occurs in problems where we view the pattern of waiting times (this is important in communication network traffic studies). It also shows up in examining the life of some systems, where we are interested in the mean time between failures (*MTBF*).

The exponential density is defined by Eq. (C.43).

$$p_X(x) = \begin{cases} \dfrac{1}{m}e^{-x/m}, & x > 0 \\[3mm] 0, & x < 0 \end{cases} \qquad (C.43)$$

This density is shown in Figure C.9.

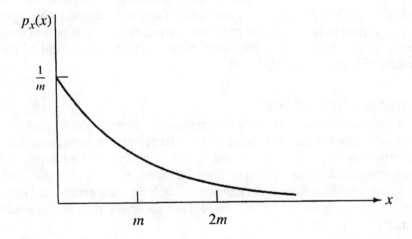

Figure C.9 - Exponential Density Function

The parameter, *m*, is the mean value of the variable, as can be verified by a simple integration (convince yourself you can do this).

C.2.4 The Ricean Density Function

We stated in Section C.2.2 that the Rayleigh density function results when we take the square root of the sum of the squares of two zero-mean Gaussian densities. When we analyze digital communication systems in the absence of a signal (when noise alone is present), we will often be dealing with variables that are Rayleigh distributed.

We will also encounter situations where a transmitted signal is embedded in noise (which adds in the channel). In these cases, our receivers will sometimes effectively take the square root of the sum of the squares of two quantities. One of these will be zero-mean Gaussian distributed, but the other results from an addition of signal and noise. The quantity we then observe is of the form shown in Eq. (C.44).

$$z = \sqrt{[s + x]^2 + y^2(t)} \qquad\qquad\qquad (C.44)$$

In this equation, x and y are zero-mean Gaussian random functions, and s is the signal. If s is zero, z follows a Rayleigh density function. If s is not zero, z follows a more complex density known as a *Ricean density*. The Ricean density is given by the equation,

$$p(z) = \frac{z}{\sigma^2} \exp\left[- \frac{1}{2\sigma^2}(z^2 + s^2(2\right] I_o\left(\frac{sz}{\sigma^2} \right) \qquad\qquad (C.45)$$

In Eq. (C.45), s is the signal value (this will normally be a specific time sample, $s(T)$), and I_o is a *modified Bessel function of zero order*. We discuss Bessel functions in Chapter 4 (when we deal with FM). For now, you can think of this as a function which you look up in a table. You simple plug in s, z, and σ^2 to find the argument of the Bessel function, and then you look in a table of Bessel functions. Note from Eq. (C.45) that when s=0, this reduces to the Rayleigh density [$I_o(0)$=0].

C.3 Random Processes

Some of the noise encountered in a communication system is *deterministic*, as in the case of some types of jamming signals in a radar system. In these cases, the noise can be described completely by a formula, graph, or equivalent technique. Other types of noise are composed of components from so many sources that we find it more convenient to analyze them as *random processes*. Each time the experiment is performed, we observe a *time function* as the outcome. This contrasts with the one-dimensional outcomes studied earlier in this chapter.

We usually assume that additive noise is a random process that is *Gaussian*. This means that if the process is sampled at any point in time, the probability density of the sample is Gaussian. This assumption proves important because if we start with a Gaussian random process and put it through any linear system, the system output will also be a Gaussian process. This fact will be used many times in examining the performance of various digital communication systems.

In addition to knowing that the noise process is Gaussian, it will be necessary to know something about the relationship between various time samples. We do this by examining the *autocorrelation function* and its Fourier transform, known as the *power spectral density*.

Until this point, we have considered single random variables. All averages and related parameters were simply numbers. Now we shall add another dimension to the study - the dimension of time. Instead of talking of numbers only, we will now be able to characterize random functions. The advantages of such a capability should be clear. Our approach to random function analysis begins with the consideration of discrete-time functions since these will prove to be a simple extension of random variables.

Imagine a single die being flipped 1000 times. Let X_i be the random variable assigned to the outcome of the ith flip. Now list the 1000 values of the random variables

$$x_1, x_2, x_3, ... x_{999}, x_{1000}$$

For example, if the random variable is assigned to be equal to the number of dots on the top face after flipping the die, a typical list might resemble that shown below

4,6,3,5,1,4,2,5,3,1,4,5,....

Suppose that all possible sequences of random variables are now listed. We would then have a collection of 6^{1000} entries, each one resembling that shown above. This collection is known as the *ensemble* of possible outcomes. This ensemble, together with associated statistical properties, forms a *random process*. In this particular example, the process is discrete-valued and discrete-time.

If we were to view one digit, say the third entry, a random variable would result. In this example, the random variable would represent that assigned to the outcome of the third flip of the die.

We can completely describe the above random process by specifying the 1000-dimensional probability density function,

$$p(x_1, x_2, x_3, ..., x_{999}, x_{1000})$$

This 1000-dimensional probability density function is used in the same way as a one-dimensional probability density function. Its integral must be unity.

$$\int_{-\infty}^{\infty} \int_{-\infty}^{\infty} \cdots \int p(x_1,x_2,x_3,...,x_{999},x_{1000})dx_1 dx_2...dx_{1000} = 1 \qquad \text{(C.46)}$$

The probability that the variables fall within any specified 1000-dimensional volume is the integral of the probability density function over that volume. The expected value of any single variable can be found by integrating the product of that variable with the probability density function. Thus, for example, the expected value of x_1 is

$$\int_{-\infty}^{\infty} \int_{-\infty}^{\infty} \cdots \int x_1 p(x_1,x_2,x_3,...,x_{999},x_{1000})dx_1 dx_2...dx_{1000} \qquad \text{(C.47)}$$

This is known as a *first-order average*.

In a similar manner, the expected value of any multidimensional function of the variables is found by integrating the product of that function with the density function. Thus, for example, the expected value of the product $x_1 x_2$ is given by

$$\int_{-\infty}^{\infty} \int_{-\infty}^{\infty} \cdots \int x_1 x_2 \; p(x_1,x_2,x_3,...,x_{999},x_{1000})dx_1 dx_2...dx_{1000} \qquad \text{(C.48)}$$

This is a *second-order average*.

We can continue this process for higher order averages. In many instances it proves sufficient to specify only the first- and second-order averages (moments). That is, one would specify

$E\{x_i\}$ and $E\{x_i x_j\}$ for all i and j

Since we are interested in continuous functions of time, we now extend this example to the case of an infinite number of random variables in each list and an infinite number of lists in the ensemble. It may be helpful to refer back to this simple example (i.e., 1000 flips of the die) from time to time.

The most general form of the random process results if our simple experiment could yield an infinite range of values as the outcome, and if the period between performances of the experiment approaches zero.

Another way to arrive at a random process is to perform a discrete experiment, but to each outcome assign a *time function* instead of a number. When the samples of the process are time functions, we call this a *stochastic process*. As an example, consider the experiment defined by picking a 6-volt *dc* generator from an infinite inventory in a warehouse. The voltage of the selected (call it the ith) generator, $v_i(t)$, is then displayed on an oscilloscope. The waveform, $v_i(t)$, is a *sample function* of the process. $v_i(t)$ will not be a perfect constant of 6 volts due to imperfections in the generator construction and also due to *rf* pickup when the wires are acting as an antenna. Since we assume an unlimited inventory of generators, there are an infinite number of possible sample functions of the process. This infinite group of samples form the ensemble. Each time we choose a generator and measure its voltage, a sample function from this infinite ensemble results.

If we were to sample the voltage at a specific time, $t = t_o$, the sample, $v(t_o)$, is a random variable. Since $v(t)$ is assumed to be continuous with time, there are an infinity of random variables associated with the process.

Having introduced the concept with the generator example, let's now speak in general terms. Let $x(t)$ represent a stochastic process. $x(t)$ is then an infinite ensemble of all possible sample functions. For every specific value of time, $t = t_o$, $x(t_o)$ is a random variable.

Suppose we now examine the first- and second-order averages of the process. Note that instead of having a discrete list of numbers as in the die example, we have continuous functions of time. The first moment is then a function of time. This mean value is given by

$$m(t) = E\{x(t)\}erbrace \tag{C.49}$$

The second moments are found by averaging the product of two different time samples. We use the symbol R_x for this moment, and call it the *autocorrelation*. The autocorrelation is then given by

$$R_x(t_1,t_2) = E\{x(t_1)x(t_2)\} \tag{C.50}$$

Both the mean and autocorrelation must be thought of as averages taken over the entire ensemble of time functions. To find $m(t_o)$, we must average all samples across the ensemble at time t_o. To find the autocorrelation, we must average the product of $x(t_1)$ with $x(t_2)$ across the ensemble. This is generally difficult, and the ensemble averages are virtually impossible to perform experimentally.

In practice, we could find the mean by measuring the voltage of a great number of generators at time t_o and average the resulting numbers. For the example of *dc* generators, we would expect this average to be independent of t_o. Indeed, most processes we consider have mean values which are independent of time.

A process with *overall statistics* that are independent of time is called a *stationary* (or strict-sense stationary) process. If only the mean and second moment are independent of time, the process is *wide-sense stationary*. Since we are primarily interested in power, and power depends upon the second moment, wide-sense stationarity will be sufficient for our analyses.

If a process, $x(t)$, is stationary, then the shifted process, $x(t-T)$, has the same statistics independent of the value of T. Clearly, $m(t_o)$ cannot depend upon t_o for a stationary process.

Viewing the autocorrelation of a stationary process, we have

$$R_x(t_1,t_2) = Ee\{x(t_1)x(t_2)\} = E\{x(t_1 - T)x(t_2 - T)\} \qquad (C.51)$$

In the last equality of Eq. (C.51), we have shifted the process by T. If we now let $T=t_1$, we find

$$R_x(t_1,t_2) = E\{x(0)x(t_2 - t_1)bnrbrace \qquad (C.52)$$

Equation (C.52) indicates that the autocorrelation of a stationary (wide-sense is sufficient) process depends only on the time spacing between the two samples, t_2-t_1. That is, the left-hand time point can be placed anywhere, and as long as the right-hand point is separated from this by t_2-t_1, the autocorrelation remains unchanged. Since the independent variable of the autocorrelation is effectively one-dimensional instead of two-dimensional, we use the argument τ, and refer to the autocorrelation of a stationary process as $R_x(\tau)$. Thus

$$R_x(\tau) = E\{x(t)x(t - \tau)\} = E\{x(t)x(t + \tau)brbrace \qquad (C.53)$$

The last equality results from adding τ to each of the arguments. This shows that autocorrelation is an even function.

If t_2 and t_1 are widely separated such that $x(t_1)$ and $x(t_2)$ are independent, the autocorrelation reduces to

$$R_x(\tau) = E\{x(t)x(t - \tau)\} = E\{x(t)\} E\{x(t - \tau)\} = E\{x(t - \tau)\} = m^2 \qquad (C.54)$$

The average value of a random time function is the *dc* value, and most communication channels will not pass *dc* (they contain bandpass filters). Therefore, most of the processes we consider have mean values equal to zero. In this case, the value of τ at which $R_x(\tau)$ goes to zero represents the time over which the process is correlated. If two samples are separated by this length of time, one sample has no effect upon the other.

Time Averages

Suppose you were asked to find the average value of the voltage in the *dc* generator example. You would have to measure the voltage of many generators (at any given time) and then compute the average. Once you were told that the process is stationary, you would probably be tempted to take one generator and average its voltage over a large time interval. You would expect to get 6 volts as the result using either technique. That is, you would reason that

$$m_v = E\{v(t)\} = \lim_{T \to \infty} \frac{1}{T} \int_{-\frac{T}{2}}^{\frac{T}{2}} v_i(t)\ dt \qquad (C.55)$$

Here, $v_i(t)$ is one sample function of the ensemble. This approach does not always result in the correct answer. Suppose, for example, that one of the generators was burned out, and you happened to choose this particular generator. You would erroneously think that the $m_v=0$.

Most of the processes we encounter have the property that any sample function contains all of the essential information about the process. Such processes are known as *ergodic*. The generator example process is ergodic as long as none of the generators is "exceptional".

A process which is ergodic must also be stationary. This is true since, once we agree that all averages can be found from a single time sample, the averages can no longer be a function of the time at which they are computed. Alternatively, a stationary process need not be ergodic (consider the burned-out generator example).

The autocorrelation of an ergodic process is given by

$$R_x(\tau) = E\{x(t)x(t + \tau)\} = \lim_{T \to \infty} \frac{1}{T} \int_{-\frac{T}{2}}^{\frac{T}{2}} x(t)x(t + \tau)\ dt \qquad (C.56)$$

The autocorrelation, $R(t)$, is a function of time. We define $G(f)$ as the Fourier transform of the autocorrelation. $G(f)$ is called the *power spectral density* for reasons that will become obvious in a moment.

$$G_x(f) = \mathscr{F}[R_x(t)] = \int\limits_{-\infty}^{\infty} R_x(t)e^{-j2\pi\pi t}\, dt \qquad \text{(C.57)}$$

The autocorrelation is then the inverse transform of the power spectral density.

$$R_x(t) = \mathscr{F}^{-1}[G_x(f)] = \int\limits_{-\infty}^{\infty} G_x(f)e^{j2\pi ft}\, df = 2\int\limits_{0}^{\infty} G_x(f)e^{j2\pi ft}\, df \qquad \text{(C.58)}$$

In the last equality, we have doubled the positive-half range of the integral. This results from the fact that the power spectral density must be real and even since the autocorrelation is even. We can now relate the average power to the power spectral density.

$$P_{av} = E\{x^2(t)\} = R_x(0) = 2\int\limits_{0}^{\infty} G_x(f)\, df \qquad \text{(C.59)}$$

Equation (C.59) is a very important result. It says that to find the power of a random time function, we integrate the power spectral density over all positive values of f (the frequency variable) and then double the result.

The other important result we need is the effect that a filter has on the power of a random signal. If a stochastic process forms the input to a filter, as shown in Fig. C.10, the output is also a stochastic process. That is, each sample function of the input process yields a sample function of the output process. We wish to find the statistics of the output process.

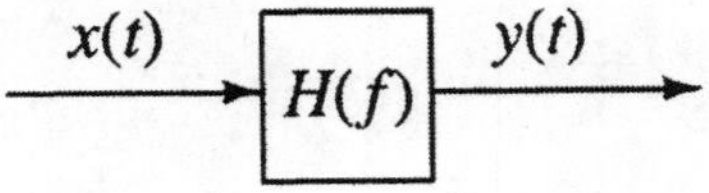

$$x(t) \longrightarrow \boxed{H(f)} \xrightarrow{\; y(t) \;}$$

Figure C.10 - Stochastic Process
as Input to Filter

We begin with the mean value. The mean is given by

$$E\{yY(t)\} = E\left\{ \int\limits_{-\infty}^{\infty} h(\tau)x(t - \tau)d\tau \right\} \qquad \text{(C.60)}$$

The average of a sum is the sum of the averages. Therefore, with some broad restrictions (finite mean value and a stable system) we can interchange the order of taking the expected value and integrating. If we assume that $x(t)$ is stationary, we have

$$E\{y(t)\} = \int_{-\infty}^{\infty} E\{h(\tau)x(t - \tau)\}\, d\tau = \int_{-\infty}^{\infty} h(\tau)E\{x(t - \tau)\}\, d\tau$$

$$\tag{C.61}$$

$$= m_x \int_{-\infty}^{\infty} h(\tau)\, d\tau = m_x H(0)$$

Most of the random processes we encounter in this text have zero average value. If the input to the filter has zero average value, the output mean is also zero.

We now evaluate the autocorrelation of the output process. We assume that the input process is stationary.

$$R_y(t_1,t_2) = E\{y(t_1)y(t_2)\}$$

$$= E\left\{ \int_{-\infty}^{\infty} h(\tau)x(t_1 - \tau)d\tau \int_{-\infty}^{\infty} h(\tau)x(t_2 - \tau)d\tau \right\} \tag{C.62}$$

We once again interchange the order of taking the expected value and integrating. We combine the two integrals (using two different symbols for the dummy variable of integration) to get

$$R_y(t_1,t_2) = \int_{-\infty}^{\infty} h(\tau_1)d\tau_1 \int_{-\infty}^{\infty} E\{x(t_1 - \tau_1)x(t_2 - \tau_2)\}h(\tau_2)d\tau_2$$

$$\tag{C.63}$$

$$= \int_{-\infty}^{\infty} h(\tau_1)d\tau_1 \int_{-\infty}^{\infty} R_x(t_1 - t_2 + \tau_2 - \tau_1)h(\tau_2)d\tau_2$$

Note that the result does not depend on the values of t_1 and t_2, but only on their difference. Therefore, the output process is wide sense stationary. The autocorrelation is a function of only one variable, the spacing between the two time points. We have used the notation τ for this spacing. Doing the same here, the result becomes,

$$R_y(\tau) = \int_{-\infty}^{\infty} \int_{-\infty}^{\infty} R_x(\tau - \tau_1 + \tau_2)h(\tau_1)h(\tau_2)d\tau_1 d\tau_2 \tag{C.64}$$

Equation (C.64) shows that the output autocorrelation is the result of convolving the input autocorrelation first with $h(t)$ and then with $h(-t)$. Therefore,

$$R_y(t) = R_x(t) * h(t) * h(-t) \tag{C.65}$$

Taking the Fourier transform of this equation, and recognizing that the Fourier transform of $h(-t)$ is the complex conjugate of the transform of $h(t)$, we have

$$G_y(f) = G_x(f)H(f)H^*(f) = G_x(f)|H(f)|^2 \tag{C.66}$$

Equation (C.66) has an intuitive interpretation. $G(f)$ is the power spectral density. It is not surprising that the output power spectral density is weighted by the square magnitude of the transfer function since this is what would happen to the power of a single sinusoid that goes through the filter.

Example

A received signal is made up of two components, signal and noise.

$$r(t) = s(t) + n(t)$$

The signal can be considered as a sample of a random process since random amplitude fluctuations are introduced by turbulence in the air. You are told that the autocorrelation of the signal process is

$$R_s(\tau) = 2er^{-|\tau|}$$

The noise is a sample function of a random process with autocorrelation

$$R_n(\tau) = e^{-2|\tau|}$$

Both processes have zero mean value, and they are independent of each other.

     Find the autocorrelation and total power of $r(t)$.

Solution:

From the definition of autocorrelation, we have

$$\begin{aligned}
R_r(\tau) &= E\{r(t)r(t+\tau)\} \\
&= E\{[s(t) + n(t)][s(t+\tau) + n(t+\tau)]\} \\
&= E\{x(t)x(t+\tau)\} + E\{s(t)n(t+\tau)\} \\
&\quad + E\{x(t+\tau)n(t)\} + E\{s(t+\tau)n(t+\tau)\}
\end{aligned}$$

Since the signal and noise are independent,

$$E\{s(t+\tau)n(t)\} = E\{s(t+\tau)\}E\{n(t)\} = 0$$

and

$$E\{s(t)n(t+\tau)\} = E\{s(t)\}E\{n(t+\tau)\} = 0$$

Finally, the autocorrelation is given by

$$R_r(\tau) = R_s(\tau) + R_n(\tau) = 2e^{-|\tau|} + e^{-2|\tau|}$$

The total power of $r(t)$ is $R_r(0)$, or 3 watts.

Example (The random telegraph Signal)

Evaluate the autocorrelation of the *random telegraph waveform* as shown in Fig. C.11.

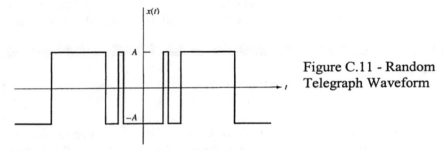

Figure C.11 - Random
Telegraph Waveform

This is a binary waveform that can only take on one of two values, +A or -C. The probabilities of each of these values are equal (i.e., ½). Assume that transitions occur randomly and that there are an average of λ transitions per second. The probability that n transitions occur in a positive time interval τ is given by a *Poisson distribution*,

$$Pr(n,\tau) = \frac{(\lambda\tau)^n}{n!} ex^{-\lambda\tau}$$

Solution:

We first find the autocorrelation of the process.

$$R_x(\tau) = E\{x(t)x(t+\tau)\}$$

The product inside the expected value signs of this equation is either $+A^2$ or $-A^2$. The plus sign obtains if there are an even number of transitions in the interval, and the minus sign obtains if there are an odd number. For a given value of τ, the probability of an even number of transitions is found by summing the Poisson probably distribution over all even values of n. Similarly, the probability of an odd number of transitions is the sum of the Poisson distribution over all odd values of n. Therefore,

$$PR(even) = e^{-\lambda\tau} \sum_{\substack{n=0 \\ n=even}}^{\infty} \frac{(\lambda\tau)^n}{n!}$$

$$PR(odd) = e^{-\lambda\tau} \sum_{\substack{n=odd}}^{\infty} \frac{(\lambda\tau)^n}{n!}$$

The autocorrelation is then

$$R_x(\tau) = A^2 \, Pr(even) - A^2 \, Pr(odd)$$

$$= A^2 e^{-\lambda\tau} \sum_{n=0}^{\infty} (-1)^n \frac{(\lambda\tau)^n}{n!}$$

$$= A^2 \, e^{-\lambda\tau} \, e^{-\lambda\tau} = A^2 \, e^{-2\lambda\tau}$$

The result applies for positive time intervals. We know that the autocorrelation must be an even function, so we can write the autocorrelation as

$$R_x(\tau) = A^2 \, e^{-0\,2\lambda|\tau|}$$

This result is shown in Fig. C.12.

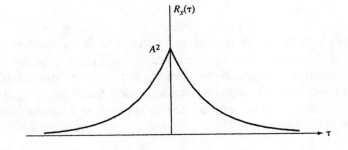

Figure C.12 - Autocorrelation of
Random Telegraph Waveform

C.4 White Noise

Suppose that $x(t)$ is a stochastic process with a constant power spectral density, as shown in Fig. C.13. This process contains all frequencies "to an equal degree". Since white light is composed of all frequencies (colors), the process described above is known as *white noise*.

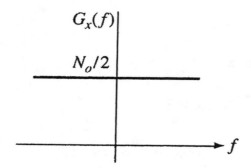

Figure C.13 - Power Spectral Density of White Noise

Suppose now that white noise forms the input to an ideal bandpass filter with a passband extending from a low frequency cutoff of f_L to a high frequency cutoff of f_H. The power of the output for the filter is given by

$$P_{out} = 2\int_0^\infty G_x(f)|H(f)|^2 \, df = 2\int_{f_L}^{f_H} G_x(f) \, df = +2\int_{f_L}^{f_H} \frac{N_o}{2} \, df = N_o(f_H - f_L) \quad \text{(C.67)}$$

The output of the bandpass filter consists of all components of the input lying within the passband of the filter. The output power can therefore be considered to be that portion of the input power in the frequency range between f_L and f_H. We see from equation (C.67) that this is proportional to the bandwidth, with the proportionality factor being N_o. Therefore, N_o is the *power per Hz* of the noise waveform. The total power in a band of frequencies is the product of N_o with the bandwidth.[4]

---

[4]The power spectral density of Figure C.13 is known as the *two-sided power spectral density*. Since the power spectral density is real and even and the equation for power contains a factor of two, we sometimes define a *one-sided power spectral density* with a value of twice that of the two-sided density. The one-sided density of white noise therefore has a height of N_o instead of $N_o/2$, and to find the power in a band of frequencies, we simply integrate (without the factor of 2).

The autocorrelation of white noise is the inverse Fourier transform of the power spectral density. Therefore,

$$R_x(\tau) = \frac{N_o}{2}\,\delta(\tau) \qquad\qquad (C.68)$$

The average power of a stochastic process is $R_x(0)$, which for this case is infinity. Therefore, white noise cannot exist in real life (thankfully, signals with infinite power do not exist–were this not the case, you probably would not be here to read my textbook). However, many types of noise encountered can be assumed to be approximately white.

Since the autocorrelation of white noise is zero for t ≠ 0, two different time samples of white noise are uncorrelated even if they are taken very close together. Thus knowing the sample value of white noise at one instant of time tells us absolutely nothing about its value an instant later[5]. From a practical standpoint, this is an unfortunate situation. It would appear to make the elimination of noise more difficult.

Up to this point, we have said nothing about the actual probability distributions of the process. We have talked only about the first and second moment. There are an infinity of random processes with the same first and second moments. (The analogy to mechanics is that given the center of gravity and moment of inertia of an object, the exact shape can be any of an infinity of possibilities). Each random variable, $x(t_o)$, has a certain probability density. By considering only the mean and second moment, we are not telling the whole story.

Because of the central limit theorem, most processes we encounter are Gaussian. Once we know that a random variable is Gaussian, the density function is completely specified by its mean and second moment.

We now examine several types of noise encountered in communication systems, and determine whether the white noise model is appropriate for these noise sources.

C.5 Narrowband Noise

Most communication systems with which we deal contain bandpass filters. Therefore, white noise appearing at the input to the system will be shaped into bandlimited noise by the filtering operation. If the bandwidth of the noise is relatively small compared to the center frequency, we refer to this as *narrowband noise*. We have no problem deriving the power spectral density and autocorrelation of this noise, and these quantities are sufficient to analyze the effect of linear systems. However, we will often be dealing with multipliers, and the frequency analysis approach is not sufficient since non-linear operations are present. In such cases, it proves useful to have a trigonometric expansion for the noise signals. The form of this expansion is

---

[5]The zero correlation is a necessary, but not sufficient condition for independence. However, for Gaussian processes, two uncorrelated samples are independent.

$$n(t) = x(t)\cos 2\pi f_o t - y(t)\sin 2\pi f_o t \tag{C.69}$$

In equation (C.69), $n(t)$ is the noise waveform and f_o is a frequency (often the center) within the band occupied by the noise. Since sine and cosine vary by 90 degrees, $x(t)$ and $y(t)$ are known as the *quadrature components* of the noise.

Equation (C.69) can be derived by starting with exponential notation,

$$n(t) = Re\left\{r(t)e^{j2\pi f_o t}\right\} \tag{C.70}$$

where $r(t)$ is a complex function with a low-frequency bandlimited Fourier transform, *Re* is the "real part of", and the exponential has the effect of shifting the frequencies of $r(t)$ by f_o. Expanding the exponential using Euler's identity and letting $x(t)$ be the real part of $r(t)$ and $y(t)$ be the imaginary part, we have

$$n(t) = Re\ \{\ [x(t) + jy(t)]\ (\cos 2\pi f_o t + j\sin 2\pi f_o t)\}$$
$$= x(t)\cos 2\pi f_o t - y(t)\sin 2\pi f_o t \tag{C.71}$$

This is the same as Eq. (C.69).

Explicit solution of Eq. (C.69) for $x(t)$ and $y(t)$ is not simple. One way to solve it is using Hilbert transforms.

Hilbert Transform

The *Hilbert transform* of a time function is obtained by shifting all frequency components by 90°. The Hilbert transform operation can therefore be represented by a linear system with $H(f)$ as shown in Fig. C.14.

Note that the phase function of a real system must be odd. The system function is then given by

$$H(f) = -j\ sgn(f) \tag{C.72}$$

The impulse response of this system is the inverse transform of $H(f)$. This is given by

$$h(t) = \frac{1}{\pi t} \tag{C.73}$$

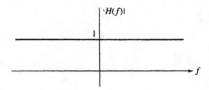

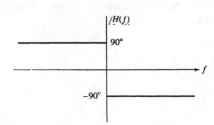

Figure C.14 - The Hilbert Transform

The Hilbert transform of $s(t)$ is then given by the convolution of $s(t)$ with $h(t)$. Let us denote the transform by $\hat{s}(t)$. Then

$$\hat{s}(t) = \frac{1}{\pi} \int_{-\infty}^{\infty} \frac{s(\tau)}{t - \tau} d\tau \qquad (C.74)$$

If we take the Hilbert transform of a Hilbert transform, the effect in the frequency domain is to multiply the signal transform by $H^2(f)$. But $H^2(f) = -1$, so we return to the original signal which a change of sign. This indicates that the inverse Hilbert transform equation is the same as the transform relationship except with a minus sign. Therefore,

$$s(t) = - \frac{1}{\pi} \int_{-\infty}^{\infty} \frac{\hat{s}(\tau)}{t - \tau} d\tau \qquad (C.75)$$

Example
Find the Hilbert transform of the following time signals.
(a) $s(t) = \cos(2\pi f_o t + \theta)$
(b)

$$s(t) = \frac{\sin t}{t} \cos 1200\pi t$$

c)

$$s(t) = \frac{\sin t}{t} \sin 200\pi t$$

Solution:

Although the Hilbert transform is defined by a convolution operation, it is almost always easier to avoid time convolution by working with Fourier transforms.

(a) The Fourier transform of $s(t)$ is given by

$$S(f) = \frac{1}{2}[\delta(f - f_o) + \delta(f + f_o)] \, e^{-j\theta f/f_o}$$

Note that the phase shift of θ radians is equivalent to a time shift of $\theta/2\pi f_o$ seconds. We now multiply this by $-j\text{sgn}(f)$ to get

$$\hat{S}(f) = \frac{1}{2}\left[- j\delta(f - f_o) + j\delta(f + f_o)\right]e^{-j\theta f/f_o}$$

The quantity in square brackets is the Fourier transform of a sine wave. Therefore,

$$\hat{s}(t) = \sin(2\pi f_o t + \theta)$$

This result is not surprising since the Hilbert transform is a 90-degree phase-shifting operation.

(b) Let us denote

$$x(t) = \frac{\sin t}{t}$$

The Fourier transform of $s(t)$ is then given by

$$S(f) = \frac{1}{2}X(f - 100) + \frac{1}{2}X(f + 100)$$

Since $X(f)$ is bandlimited to $f = \pm 1$, the first term in $S(f)$ occupies frequencies between 99 and 101 Hz while the second term occupies frequencies between -101 and -99 Hz. When $S(f)$ is multiplied by $-j\text{sgn}(f)$ we find

$$\hat{S}(f) = -\frac{1}{2}jX(f - 100) + \frac{1}{2}jX(f + 100)$$

The inverse transform yields

$$\hat{s}(t) = x(t)\sin 200\pi t = \frac{\sin t}{t}\sin 200\pi t$$

(c) We use the fact that the Hilbert transform of a Hilbert transform is the negative of the original function. Therefore, by inspection, we have

$$\hat{s}(t) = -x(t)\cos 200\pi t = -\frac{\sin t}{t}\cos 200\pi t$$

We are now ready to return to the solution of Eq. (C.69). If $x(t)$ and $y(t)$ are assumed to be bandlimited to frequencies below f_o, we can take the Hilbert transform of both sides of Eq. (C.69) to get

$$\hat{n}(t) = x(t)\sin 2\pi f_o t + y(t)\cos 2\pi f_o t \tag{C.76}$$

If Eq. (C.69) is multiplied by $\cos 2\pi f_o t$ and Eq. (C.76) is multiplied by $\sin 2\pi f_o t$, when the two expressions are added together, $y(t)$ is eliminated yielding

$$n(t)\cos 2\pi f_o t + \hat{n}(t)\sin 2pi9ff_o t = x(t)[\cos^2 2\pi f_o t + \sin^2 2\pi f_o t] \tag{C.77}$$

$$= x(t)$$

Similarly, we can reverse the multiplications to find

$$y(t) = \hat{n}(t)\cos 2\pi f_o t - n(t)\sin 2\pi f_o t \tag{C.78}$$

The autocorrelation of $x(t)$ and $y(t)$ can now be derived from Eqs. (C.77) and (C.78).

$$R_x(\tau) = R_y(\tau) = R_n(\tau)\cos 2\pi f_o \tau + \left(R_n(\tau) * \frac{1}{\pi\tau}\right)\sin 2\pi f_o \tau \tag{C.79}$$

Finally, we apply the modulation theorem to Eq. (C.79) to get

$$G_x(f) = G_y(f) = G_n(f - f_o) + G_n(f + f_o)$$

$$for \ f_o - f_m < |f| < f_o + f_m$$

(C.80)

Equation (C.80) is the key result which will enable us to calculate the effects of noise on AM and FM communication systems.

Example
Express the three narrowband noise processes of Fig. C.15 in quadrature form using f_o as the center frequency.

Solution:
We use Eq. (C.80) to immediately sketch the power spectral densities of $x(t)$ and $y(t)$. These are shown in Fig. C.16. The noise is then expressed as

$$n(t) = x(t)\cos 2\pi f_o t - y(t)\sin 2\pi f_o t$$

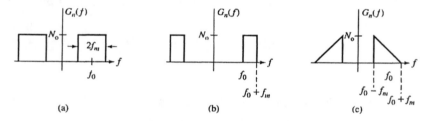

Figure C.15 - Noise Processes for Example

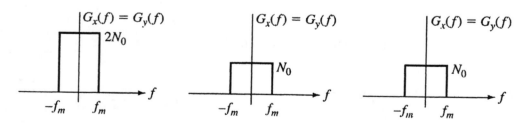

Figure C.16 - Power Spectral Density of Quadrature Components

C.6 Matched Filter

There are a variety of operations we may wish to perform on a received signal. In some cases, we wish to filter the signal in order to remove as much noise as possible, and therefore be left with a time function that resembles the desired signal as closely as possible. In other situations, we may wish to maximize the output signal to noise (power) ratio without regard to preserving the shape of the signal waveform. That is, we may use a filter which significantly alters the shape of both the signal and of the noise, and does so in a way that increases the signal to noise ratio. Such a filter distorts the signal.

Analog receivers typically try to reconstruct the waveform as closely as possible, while digital receivers attempt to "pull" the signal out of background noise without regard to distortion.

In order to motivate this study, let's get way ahead of the game. Suppose you wish to send a list of ones and zeros, and do so by speaking the words "one" and "none" into a microphone. To send 1010, you would speak "one-none-one-none". Noise. adds to the transmitted signal, and let's assume that the receiver has difficulty in distinguishing between the two words. Now suppose that you filter the received signal plus noise in a manner that blocks a good portion of the noise. But in the process, you change the transmitted "one" signal into a waveform that, when placed into a speaker, generates the word "start". The same filter changes "none" to "halt". Therefore, instead of hearing a highly noise-corrupted sequence of the words "one" and "none", you hear a relatively uncorrupted signal consisting of "start-halt-start-halt". You could still recover the original binary sequence in spite of what would be considered severe distortion of the signal waveform.

The *matched filter* is a linear system that maximizes the output signal to noise ratio. We designate the input to the filter as $s(t) + n(t)$, and the resulting output is $s_o(t) + n_o(t)$. This is shown in Fig. C.17.

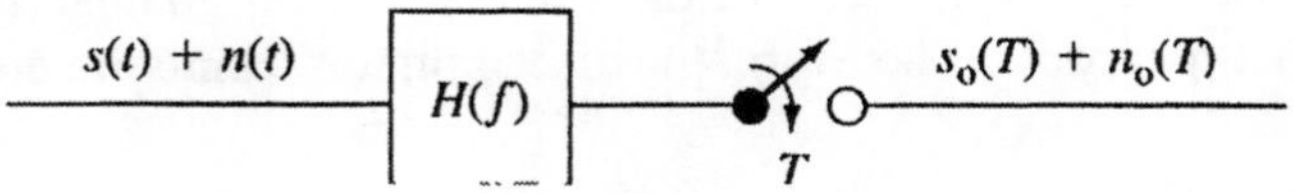

Figure C.17 - The Matched Filter

Since the system is assumed to be linear, $s_o(t)$ is the output due to an input of $s(t)$ and $n_o(t)$ is the output due to $n(t)$. The filter is designed to maximize the ratio, $s^2_o(T)/n^2_o(T)$. Since the denominator of this expression is random, we use the average value. The output signal to noise ratio, ρ, is then given by

$$\rho = \frac{s_o^2(T)}{n_o^2(T)} = \frac{\left| \int_{-\infty}^{\infty} S(f)H(f)e^{j2\pi fT} \, df \right|^2}{\int_{-\infty}^{\infty} |H(f)|^2 G_n(f) \, df} \tag{C.81}$$

The numerator of Eq. (C.81) is the square of the inverse Fourier transform of the product of the input transform with the system function. Thus, it is the square of the deterministic time sample, $s_o(T)$. The $G_n(f)$ in the denominator is the power spectral density of the input noise. Thus the denominator integrand is the power spectral density of the output noise. Note that we are integrating from $-\infty$ to $+\infty$ instead of doubling the integral from zero to ∞. We do this for reasons that will soon become clear.

We wish to choose $H(f)$ to maximize the expression of Eq. (C.81). This is a difficult maximization problem (you cannot simply take the derivative and set it to zero since we are trying to find a function rather than a value). The choice is simplified if we apply *Schwartz's inequality* to the numerator. In doing so, we will be able to solve for $H(f)$ almost by inspection.

Schwartz's inequality states that for all functions, $f(x)$ and $g(x)$,

$$\left| \int f(x)g(x)\ dx \right|^2 \leq \int |f^2(x)|\ dx \int |g^2(x)|\ dx \qquad (C.82)$$

We will derive Schwartz's inequality for the special case of real functions by starting with the observation that

$$\int [f(x) - Tg(x)]^2\ dx \geq 0$$

for all real $f(x)$, $g(x)$ and T (i.e., we are integrating a non-negative function). Expanding this, we find

$$T^2 \int g^2(x)dx - 2T \int f(x)g(x)\ dx + \int f^2(x)\ dx \geq 0 \qquad (C.83)$$

The left side of Eq. (C.83) is a quadratic in T. Since the value can never go negative, the quadratic cannot have distinct real roots. Therefore, the discriminator cannot be positive. Thus,

$$4\left[\int f(x)g(x)\ dx \right]^2 - 4 \int f^2(x)\ dx \int g^2(x)\ dx \leq 0 \qquad (C.84)$$

and the inequality is established. Proving the inequality for complex functions is more difficult, so we ask you to accept the result in Eq. (C.82).

We now apply Schwartz's inequality to Eq. (C.81). Hindsight is a wonderful thing. Had you already solved this, you would know that we wish to cancel terms from the numerator and denominator. To begin this process, we rewrite the numerator of Eq. (C.81) as follows:

$$\left| \int_{-\infty}^{\infty} S(f)H(f)e^{j2\pi fT}df \right|^2 = \left| \int_{-\infty}^{\infty} \frac{S(f)}{\sqrt{G_n(f)}} H(f)\sqrt{G_n(f)}e^{j2\pi fT}df \right|^2 \qquad \text{(C.85)}$$

The square root operation is unambiguous since $G_n(f)$ can never be negative. Now applying Schwartz's inequality, we find

$$\left| \int_{-\infty}^{\infty} S(f)H(f)e^{j2\pi fT}df \right|^2 \leq \int_{-\infty}^{\infty} |H(f)|^2 G_n(f)df \int_{-\infty}^{\infty} \frac{|S(f)|^2}{G_n(f)}df \qquad \text{(C.86)}$$

Combining this with Eq. (C.81) we have,

$$\rho <+= \frac{\int_{-\infty}^{\infty} |H(f)|^2 G_n(f)df \int_{-\infty}^{\infty} |S(f)|^2 / G_n(f)df}{\int_{-\infty}^{\infty} |H(f)|^2 G_n(f)df} \qquad \text{(C.87)}$$

$$= \int_{-\infty}^{\infty} \frac{|S(f)|^2}{G_n(f)} df$$

Equation (C.87) fixes an upper bound on the signal to noise ratio at the output of the filter. If we can somehow guess at an $H(f)$ which yields this maximum, we need look no further.

Before attempting the guess, let's recap the approach we are taking. We wish to choose $H(f)$ to maximize the expression of Eq. (C.81). This is a difficult mathematical problem to solve. Instead of direct solution, we have placed an upper bound upon the expression in Eq. (C.81). If the signal to noise ratio cannot exceed that bound, and we somehow find an $H(f)$ that achieves that bound, we have solved the original problem (in general, there is no guarantee that we can even achieve a bound of this type–however, in this case, we can).

We have reduced the problem to one of finding an $H(f)$ that reduces Eq. (C.81) to the expression in Eq. (C.87). We are asking you to be creative, and there is no road map for doing so. You need to stare at the two expressions hoping for an inspiration [indeed, except by hindsight we have no assurance that an $H(f)$ exists that will achieve the bound of Eq. (C.87)].

The answer is (if you figured this out, your insight is excellent)

$$H(f) = e^{-j2\pi fT} \frac{S^*(f)}{G_n(f)} \tag{C.88}$$

$S^*(f)$ is the complex conjugate of $S(f)$. If you were not able to see this answer, you might wish to go back and assume that the noise is white (as we shall in a moment). That is, assume $G_n(f)$ is a constant. This makes the creative inspiration easier to achieve.

Since $H(f)$ appears as a square in both the numerator and denominator of Eq. (C.81), any scaling factor can be applied to $H(f)$ without affecting the signal to noise ratio. That is, the $H(f)$ of Eq. (C.88) can be multiplied by any constant. We shall therefore rewrite this equation inserting a "C" for an arbitrary constant.

$$H(f) = Ce^{-j2\pi fT} \frac{S^*(f)}{G_n(f)} \tag{C.89}$$

Alas, simple amplification does not improve the signal to noise ratio since both the signal and noise are multiplied by the same amount.

Now let's assume that the input noise is white so $G_n(f) = N_o/2$. The matched filter of Eq. (C.89) then becomes

$$H(f) = \frac{2C}{N_o} e^{-j2\pi fT} S^*(f) \; "=" \; Ce^{-j2\pi fT} S^*(f) \tag{C.90}$$

Note that since C is an arbitrary constant, it would create unnecessary bookkeeping to write $2C/N_o$ in Eq. (C.90). That's why we have replaced $2C/N_o$ with C (we put the equality sine in quotes so you don't draw the conclusion that C must be zero).

We can find the signal to noise ratio at the output of the matched filter (with white noise at the input) directly from Eq. (C.87).

$$\rho = \frac{2}{N_o} \int_{-\infty}^{\infty} |S(f)|^2 \, df = \frac{2}{N_o} \int_{-\infty}^{\infty} s^2(t) \, dt \tag{C.91}$$

The final equality in Eq. (C.91) results from Parseval's theorem.

The inverse Fourier transform of the Eq. (C.90) yields the impulse response of the matched filter.

$$h(t) = Cs(T - t) \tag{C.92}$$

This is found by noting that the inverse transform of $S*(f)$ is $s(-t)$, and the exponential leads to a time shift. At this point, there is no assurance that this filter is physically realizable (i.e., causal).

Example

Find the impulse response of the matched filter for the two time functions shown in Fig. C.18.

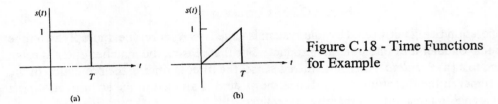

Figure C.18 - Time Functions for Example

Solution:

$h(t)$ is derived directly from Eq. (C.92). The result is shown in Fig. C.19.

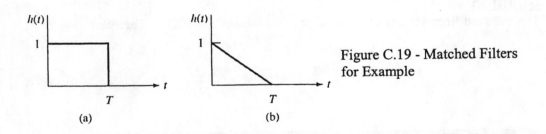

Figure C.19 - Matched Filters for Example

The actual time function at the output of the matched filter can be found by convolving the input time function with the impulse response. Therefore,

$$s_o(t) + n_o(t) = [s(t) + n(t)] * h(t)$$

and at time $t=T$, we have

$$s_o(T) + n_o(T) = \int_0^T [s(\tau) + n(\tau)]s(\tau) \, d\tau$$

Thus, the matched filter is equivalent to the system of Fig. C.20.

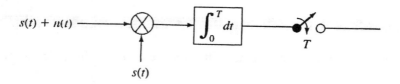

Figure C.20 - Correlator

The operation being performed by this system is called *correlation* (i.e. multiply two time functions together and integrate the product). For that reason, the matched filter is often referred to as a *correlator*. In a generalized sense, the filter is finding the projection of the input signal in the direction of *s(t)*. Since the system is aligned in the s(t) direction, the output signal to noise ratio is thereby maximized.

Example

Find the output signal to noise ratio of a matched filter where the signal is

$s(t) = A$,   for $0 < t < T$

The noise is white with power spectral density $N_o/2$.

Solution:

The matched filter achieves the signal to noise ratio of Eq. (C.91). Therefore,

$$SNR = \frac{2}{N_o} \int_{-\infty}^{\infty} s^2(t) \; dt = \frac{2A^2T}{N_o}$$

APPENDIX D

Table of Fourier Transforms

GENERAL PROPERTIES

$$S(f) = \int_{-\infty}^{\infty} s(t)e^{-j2\pi ft}dt$$

$$s(t) = \int_{-\infty}^{\infty} S(f)e^{+j2\pi ft}df$$

Function	Fourier Transform
$s(t - t_0)$	$e^{-j2\pi ft_0}S(f)$
$e^{j2\pi f_0 t}s(t)$	$S(f - f_0)$
$s(t)\cos 2\pi f_0 t$	$\frac{1}{2}[S(f - f_0) + S(f + f_0)]$
$\dfrac{ds}{dt}$	$j2\pi f\, S(f)$
$\displaystyle\int_{-\infty}^{t} s(\tau)d\tau$	$\dfrac{S(f)}{j2\pi f}$
$r(t)*s(t)$	$R(f)S(f)$
$r(t)s(t)$	$R(f)*S(f)$
$s(at)$	$\dfrac{1}{a}S\!\left(\dfrac{f}{a}\right)$
$\dfrac{1}{a}s\!\left(\dfrac{t}{a}\right)$	$S(af)$

Function	_Fourier Transform_
$e^{-at}U(t)$	$\dfrac{1}{a + j2\pi f}$ $a > 0$
$te^{-at}U(t)$	$\dfrac{1}{(a + j2\pi f)^2}$ $a > 0$
e^{-at^2}	$\sqrt{\dfrac{\pi}{a}}\ \exp\left(-\dfrac{\pi^2 f^2}{a}\right)$ $a > 0$
$\dfrac{\sin at}{\pi t}$	$\begin{cases} 1, & \lvert f \rvert < a/2\pi \\ 0, & \text{otherwise} \end{cases}$
$\begin{cases} \frac{1}{2}, & \lvert t \rvert < T \\ 0, & \text{otherwise} \end{cases}$	$\dfrac{\sin 2\pi f T}{2\pi f}$
$\begin{cases} 1 - \dfrac{\lvert t \rvert}{T}, & \lvert t \rvert < T \\ 0, & \text{otherwise} \end{cases}$	$\dfrac{\sin^2 \pi f T}{T\pi^2 f^2}$
$e^{-a\lvert t \rvert}$	$\dfrac{2a}{a^2 + 4\pi^2 f^2}$
$\operatorname{sgn}(t)$	$\dfrac{1}{j\pi f}$
$\delta(t)$	1
1	$\delta(f)$
$e^{j2\pi f_0 t}$	$\delta(f - f_0)$
$\cos 2\pi f_0 t$	$\frac{1}{2}[\delta(f - f_0) + \delta(f + f_0)]$
$\sin 2\pi f_0 t$	$\dfrac{j}{2}[\delta(f + f_0) - \delta(f - f_0)]$
$U(t)$	$\frac{1}{2}\delta(f) + \dfrac{1}{j2\pi f}$

APPENDIX E

Q Function and Error Function

The following two tables plot Q functions and error functions where

$$Q(x) = \frac{1}{\sqrt{2\pi}} \int_x^\infty e^{-\frac{u^2}{2}} dx \qquad erf(x) = \frac{2}{\sqrt{\pi}} \int_0^x e^{-u^2} du$$

$$erfc(x) = \frac{2}{\sqrt{\pi}} \int_x^\infty e^{-u^2} du$$

Q FUNCTION

x	Q(x)	x	Q(x)
0	0.500	0.05	0.480
0.10	0.460	0.15	0.440
0.20	0.421	0.25	0.401
0.30	0.382	0.35	0.363
0.40	0.345	0.45	0.326
0.50	0.309	0.55	0.291
0.60	0.274	0.65	0.258
0.70	0.242	0.75	0.227
0.80	0.212	0.85	0.198
0.90	0.184	0.95	0.171
1.00	0.159	1.05	0.147
1.10	0.136	1.15	0.125
1.20	0.115	1.25	0.106
1.30	0.097	1.35	0.089
1.40	0.081	1.45	0.074

X	Q(x)	x	Q(x)
1.50	0.067	1.55	0.061
1.60	0.055	1.65	0.049
1.70	0.045	1.75	0.04
1.80	0.036	1.85	0.032
1.90	0.029	1.95	0.026
2.00	0.023	2.05	0.020
2.10	0.018	2.15	0.016
2.20	0.014	2.25	0.012
2.30	0.011	2.35	0.009
2.40	0.008	2.45	0.007
2.50	0.006	2.55	0.005
2.60	0.005	2.65	0.004
2.70	0.003	2.75	0.003
2.80	0.003	2.85	0.002
2.90	0.002	2.95	0.002
3.00	0.001	3.05	0.001
3.10	9×10^{-4}	3.15	8×10^{-4}
3.20	7×10^{-4}	3.25	6×10^{-4}
3.30	6×10^{-4}	3.35	5×10^{-4}
3.40	4×10^{-4}	3.45	3×10^{-4}
3.50	3×10^{-4}	3.55	3×10^{-4}

erf and erfc FUNCTION

x	erf(x)	erfc(x)	x	erf(x)	erfc(x)
0	0.0000	1.0000	0.05	0.0564	0.9436
0.10	0.1125	0.8875	0.15	0.1680	0.8320
0.20	0.2227	0.7773	0.25	0.2763	0.7237
0.30	0.3286	0.6714	0.35	0.3794	0.6206
0.40	0.4284	0.5716	0.45	0.4755	0.5245
0.50	0.5205	0.4795	0.55	0.5633	0.4367
0.60	0.6039	0.3961	0.65	0.6420	0.3580
0.70	0.6778	0.3222	0.75	0.7112	0.2888
0.80	0.7421	0.2579	0.85	0.7707	0.2293
0.90	0.7969	0.2031	0.95	0.8209	0.1791
1.00	0.8427	0.1573	1.05	0.8624	0.1376
1.10	0.8802	0.1198	1.15	0.8961	0.1039
1.20	0.9103	0.0897	1.25	0.9229	0.0771
1.30	0.9340	0.0660	1.35	0.9438	0.0562
1.40	0.9523	0.0477	1.45	0.9597	0.0403
1.50	0.9661	0.0339	1.55	0.9716	0.0284
1.60	0.9763	0.0237	1.65	0.9804	0.0196
1.70	0.9838	0.0162	1.75	0.9867	0.0133
1.80	0.9891	0.0109	1.85	0.9911	0.0089
1.90	0.9928	0.0072	1.95	0.9942	0.0058
2.00	0.9953	0.0047	2.05	0.9963	0.0037
2.10	0.9970	0.0030	2.15	0.9976	0.0024
2.20	0.9981	0.0019	2.25	0.9985	0.0015
2.30	0.9989	0.0011	2.35	0.9991	0.0009
2.40	0.9993	0.0007	2.45	0.9995	0.0005
2.50	0.9996	0.0004	2.55	0.9997	0.0003
2.60	0.9998	0.0002	2.65	0.9999	0.0001
2.70	0.9999	0.0001	2.75	0.9999	0.0001

APPENDIX F

Simulation Software

The CD packaged with this textbook contains both MATLAB programs and Tina simulation software.

MATLAB Examples

MATLAB examples are included throughout this text. Some of the examples present the code to plot performance curves that appear in the chapters. Other examples allow you to input parameters and view a variety of results.

In all cases, the MATLAB instructions are given directly in the text. This is done so you can study the instructions to get practice in using MATLAB. But to save you the trouble of typing in all of the instructions, the CD contains M-files of all of the programs.

TINA Simulation Software

The CD also contains a trial version of TINA PRO from DesignSoft. When you first insert the CD, you should automatically go to the setup program. If you want to do this manually, just run **setup.exe** on the CD.

When you run the Tina simulation program, you will get an opening schematic editor screen that looks like that shown on the following page.

Use the file pull down menu, or the open file icon on the tool bar directly below that menu to open the various files we have created to match examples in this book. All of the simulation examples are included in the *Communications* folder. For the examples we give in this text, we are only using a small portion of the powerful capability of this software. The help screens will give you additional information. You can also find product information at *www.designsoftware.com*

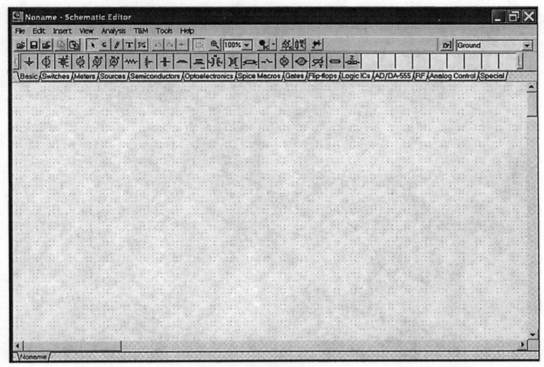

Opening Schematic Editor Screen

APPENDIX G

References

This appendix lists some of the general textbooks that overlap significantly with this book. It is always useful to get additional approaches to the subject matter. For application-oriented material, it's hard to beat the web with a good search engine. To list any particular web sites in this appendix would be of limited value since the field is advancing so rapidly.

With the exception of Jack Kurzweil, each of the authors on this list have written several texts (in addition to the books we have listed below) related to the subject matter.

Carlson, A. Bruce, Crilly, P.B., Ratledge, J.C., *Communication Systems Fourth Edition*, McGraw Hill Higher Education, 2001

Couch, L.W., *Digital and Analog Communication Systems, Sixth Edition*, Prentice Hall, 2000

Haykin, S., *Communication Systems, Fourth Edition*, John Wiley and Sons, 2000

Kurzweil, J., *An Introduction to Digital Communications*, John Wiley and Sons, 1999

Lathi, B. P. *Modern Digital and Analog Communication Systems, Third Edition,* Oxford University Press, 1998

Proakis, J. G., Salehi, M., *Communication Systems Engineering, Second Edtion*, Prentice Hall, 2001

Index